Y0-AEZ-225

Wolfgang König

Peptide and Protein Hormones

Distribution:
VCH, P. O. Box 101161, D-6940 Weinheim (Federal Republic of Germany)
Switzerland: VCH, P. O. Box, CH-4020 Basel (Switzerland)
United Kingdom and Ireland: VCH (UK) Ltd., 8 Wellington Court, Cambridge CB1 1HZ (England)
USA and Canada: VCH, 220 East 23rd Street, New York, NY 10010–4606 (USA)

ISBN 3-527-28417-6 (VCH, Weinheim) ISBN 1-56081-211-7 (VCH, New York)

Wolfgang König

Peptide and Protein Hormones

Structure, Regulation, Activity

A Reference Manual

Weinheim · New York · Basel · Cambridge

Dr. Wolfgang König
Steinernkreuz 2
D-8441 Stallwang
Federal Republic of Germany

Published jointly by
VCH Verlagsgesellschaft, Weinheim (Federal Republic of Germany)
VCH Publishers, New York, NY (USA)

Editorial Director: Dr. Hans-Joachim Kraus
Production Manager: Dipl.-Wirt.-Ing. (FH) Hans-Jochen Schmitt

Library of Congress Card No.: 92-24855

British Library Cataloguing-in-Publication Data:
A catalogue record for this book is available from the British Library

Die Deutsche Bibliothek – CIP-Einheitsaufnahme

König, Wolfgang:
Peptide and protein hormones : structure, regulation, activity ; a reference manual / Wolfgang König. – Weinheim ; New York ; Basel ; Cambridge : VCH, 1993
ISBN 3-527-28417-6

Printed on acid-free and low-chlorine paper.

Composition: Allied Publishers, 110002 New Delhi, India
Printing: betz-druck GmbH, D-6100 Darmstadt
Bookbinding: Großbuchbinderei J. Schäffer, D-6718 Grünstadt
Printed in the Federal Republic of Germany

Preface

Since the publication of The Peptides, Volume II by E. Schröder and K. Lübke in 1966, which described the synthesis, occurrence and action of biologically active peptides, no comparable book has been written. This edition tries to fill up the large gap of about 25 years of intensive research on peptide and protein hormones.

Like the chinese Yin and Yang there are no bad or good hormones. A balance of hormones is important for a healthy life. Only a unbalance leads to diseases. With reference to Yin and Yang, the important sex hormone cycle is associated with Yin, the protecting, conceiving and cooperating system, whereas the stress hormone cascade belongs to the expansive, demanding and agressive Yang. As a consequence of western culture, which favors Yang more than Yin, also most of our diseases, e.g. frequent cardiovascular diseases, depend on stress factors and an increased secretion of stress hormones. Restoring the balance of hormones may be considered as the ultimate medical skill. Thereby peptide and protein hormones as well as their antagonists and secretion regulators may be of great value as therapeutics.

Stallwang,
January 1993

W. König

Contents

1. Introduction

In the past, peptide hormones were mainly classified according to the site of synthesis or release, e.g., peptide hormones of the hypothalamus, hypophysis, gastrointestinal tract or of the thyroid gland. This classification is no longer expedient because peptide hormones are not produced at only one site in the body. Classification according to the indication is also unsuitable because these hormones can develop different activities depending on the site of action and on the dose. Peptides that are formed in endocrine cells can have an endocrine or a paracrine function. An endocrine peptide is secreted directly into the bloodstream and produces an effect at a distant receptor. A peptide exerts a paracrine effect if after being released, it affects the metabolism of either its own cell or a neighbouring cell. Peptides that exert a paracrine effect are also referred to as peptide regulatory factors (PRFs) [1.1].

An attempt will be made to classify structurally similar peptides into families and, at the same time, to integrate the activity cascades [1.2].

1.1. Synthesis of Peptide and Protein Hormones [1.3]

Peptide and protein hormones are synthesized in the cell in the form of large prepropeptides which are then cleared and modified by enzymes to give active products [1.4]. Many biologically active peptides end with an amide group at the C-terminus. This amide group is formed by reaction with the C-terminus of glycine using peptidylglycine-α-amidating monooxygenase.

Naturally occurring peptides can be isolated by conventional separation techniques (precipitation, chromatography etc.). However, the demand for increased purity and economic considerations have resulted in the development of methods for synthesizing peptides and proteins. R. Geiger calculated that peptides up to a chain length of about 20 amino acids could be produced economically using peptide chemistry [1.3]. For peptides and proteins with a chain length bigger than 20 amino acids, genetic engineering methods are preferred.

Peptides are usually synthesized from the C-terminus. Two synthesis methods are used – the classical peptide synthesis in solution and peptide synthesis in the solid phase whereby the C-terminus of the amino acid is bound to a polymeric carrier. The latter method can be automated. The synthesis used depends on the protecting groups of the side-chain groups. The hydroxyl groups of tyrosine, serine, and threonine are mostly protected by the formation of ethers. Amino groups are blocked by urethane protecting groups and carboxyl groups are esterified with alcohols. Two strategies are usually followed in peptide synthesis.

A. If acid-labile tert-butyl typ protecting groups are used in side chains either the benzyloxycarbonyl (*Z*) group or the 9-fluorenylmethyloxycarbonyl (Fmoc) group are used as α-amino protecting groups. The Z group can be eliminated by catalytic hydrogenation and the Fmoc group by secondary amines.

B. If, however, the groups in the side chains are protected by benzyl-type groups, which are eliminated under strongly acidic conditions or by catalytic hydrogenation, then tertiary butoxycarbonyl (Boc) groups are usually used as α-amino protecting groups which can be selectively eliminated under mild acidic conditions.

Dicyclohexylcarbodiimide and diisopropylcarbodiimide are still important condensation reagents for the formation of the peptide bond in conventional peptide synthesis and in the solid-phase synthesis reaction, respectively. Addition of *N*-hydroxysuccinimide (HONSu), 1-hydroxybenzotriazole (HOBt), or 3-hydroxy-4-oxo-3,4-dihydro-1,2,3-benzotriazine (HOObt) prevent most side reactions, e.g., formation of *N*-acyl urea and racemization. n-Propylphosphonic acid anhydride is suitable for the synthesis of peptides and esters with little racemization. Frequently used reagents for peptide synthesis include benzotriazol-1-yl-oxy-tris(dimethylamino)phosphonium hexafluorophosphate (BOP reagent), *O*-benzotriazoyl-1,1,3,3-tetramethyluronium hexafluorophosphate (HBTU), [1.5], and *O*-[(cyano(ethoxycarbonyl)methylidene)-amino]-1,1,3,3-tetramethyluronium tetrafluoroborate (TOTU) [1.6].

Peptide synthesis using proteases is becoming increasingly important alongside the purely chemical methods. Synthesis of h-insulin from p-insulin was carried out successfully using proteases.

Peptides and proteins can also be produced in genetically altered bacteria and yeasts [1.7], [1.8]. Plasmids or cosmids are used to introduce synthetic [1.9], semisynthetic, or natural genes into microorganisms such as *E. coli* (transformation). Plasmids are small double-stranded DNA rings and cosmids are genetic hybrids composed of a bacterial plasmid and a lambda bacteriophage. The plasmids are split using restriction enzymes and a part of them is replaced by foreign DNA sequences in an enzymatic linkage reaction (ligation). Development of the gene to form the required protein (expression) takes place in two steps. In the first step a RNA transcription is made from the DNA sequence. The initiation and termination of the transcript is determined by signals (specific DNA sequences) which are known as promoters and terminators, respectively. The messenger RNA (mRNA) so formed contains a sequence which can bind to ribosomes. After binding, the ribosomes then allow the programmed synthesis of the required protein (translation). The protein is synthesized from the translation initiator codon (usually AUG for formylmethionine, fMet) to one of the three codons for termination of the chain (UAG, UGA, or UAA).

DNA recombination using genetical engineering methods is also often used for cloning DNA sequences whose structure is to be elucidated. The structure of various proteins and peptides has been determined by knowledge of their DNA.

1.2. Mechanism of Action of Peptide and Protein Hormones

Peptide and protein hormones bind to specific receptors where they stimulate the synthesis and activation of enzymes via second messengers such as intracellular G-proteins, calcium, cyclic 3:5'-adenosine monophosphate (cAMP), cyclic 3:5'-guanosine monophosphate (cGMP), inositol triphosphate, diacylglycerol, or lysophosphatidylinositol. The hormones also stimulate the release of prostaglandins, steroid hormones, thyroid hormones, peptides, and glycoproteins. Occasionally, internalization of the receptor may occur. Here, excess peptide hormones lead to disappearance of the receptors (receptor down regulation) which results in temporary stimulation followed by inhibition of the target cells. Peptide agonists are analogs of these peptide hormones which, like them, stimulate the cell. Competitive peptide inhibitors or antagonists are, in contrast, analogs which occupy the peptide receptors and thus try to displace the agonists from their receptors, but do not stimulate the cell to further synthesis. Competitive antagonists thus inhibit the activation of the cell stimulated by agonists without a receptor down regulation.

1.3. Nomenclature and Abbreviations

According to IUPAC, amino acids can be abbreviated with three-letter [1.10] or one-letter symbols [1.11] (Table 1). The one-letter nomenclature is becoming increasingly popular and will be used in this article.

Table 1. Abbreviations for L-amino acids*

One-letter symbol	Three-letter symbol	Amino acid
	Abu	α-Aminobutyric acid
	Acc	1-aminocyclopropan-1-carboxylic acid
	Ada	adamantylalanine
	Aib	α-aminoisobutyric acid
A	Ala	alanine
	Aoc	(1S,3S,5S)-2-azabicyclo[3.3.0]octan-3-carboxylic acid
B	Asx	aspartic acid or asparagine**
	Car	carboranylalanine
	Cit	citrulline
C	Cys	cysteine
D	Asp	aspartic acid
E	Glu	glutamic acid
F	Phe	phenylalanine
	Phe(Cl_2)	3.4-dichlorophenylalanine
	p-Cl-Phe	p-chlorophyenylalanine
	p-F-Phe	p-fluorophenylalanine
	Bip	biphenylalanine
	Dip	$\beta.\beta$-diphenylalanine
	Nal	3-(2-naphthyl-)alanine
	Pal	3-(3-pyridyl-)alanine
G	Gly	glycine
	Chg	cyclohexylglycine

One-letter symbol	Three-letter symbol	Amino acid
	Pgl	phenylglycine
	Har	homoarginine
H	His	histidine
I	Ile	isoleucine
	ILys	N^{ε}-Isopropyl-lysin
K	Lys	lysine
L	Leu	leucine
	Nle	norleucine
M	Met	methionine
	Met(O)	methionine sulfoxide
N	Asn	asparagine
Ng		asparagine with a N-glycosidic carbohydrate residue
	Oic	octahydroindole carboxylic acid
	Orn	ornithine
	Pen	penicillamine
P	Pro	proline
Ph	Hyp	4-hydroxyproline
Q	Gln	glutamine
*Q	pGlu	pyroglutamic acid
R	Arg	arginine
	Sar	sarcosine
S	Ser	serine
Sg		serine with a O-glycosidic carbohydrate residue
	Sta	statine
	Thi	thienylalanine
T	Thr	threonine
	Tic	tetrahydroisoquinoline carboxylic acid
V	Val	valine
W	Trp	tryptophan
X	Ams	unknown amino acid
Xa	Ams-NH_2	amino acid amide
X-ol	Ams-ol	amino acid with carboxyl group reduced to alcohol function
Y	Tyr	tyrosine
Ys	Tyr (So_3H)	tyrosine, O-sulfated
Z	Glx	glutamic acid or glutamine**
/	///	amino acid deletion

*D-amino acids are described by the corresponding small letters.

**Uncertainty in the structural elucidation.

Aza amino acids are amino acids in which the α-CH-group is substituted by N.

The regulatory peptides [1.12] and their analogs [1.13] will also be designated in accordance with IUPAC, if possible. The abbreviations for the peptides will be mentioned at the beginning of each chapter. Small letters in front of the name of the peptide refer to the origin of the peptide and provide information on the species. The first letters of the latin name are used for this purpose (Table 2).

The position of an amino acid in a peptide chain is denoted by a superscript, after the three-letter symbol (e.g., His^5). Regions within peptide chains are indicated in parentheses, e.g., β-EP(1-27) denotes amino acids in positions 1 to 27 of β-endorphin.

Table 2. Prefixes - Abbreviations

Abbreviation	Species (latin name)
al-	alligator (*Alligator*)
an-	duck (*Anas*)
ans-	goos (*Anser*)
anq-	eel (*Anquilla)*
ant-	frogfish (*Antennarioidei*)
b-	cow (*Bos*)
ba_1-	fin whale (*Balaenoptera physalus*)
ba_2-	Sei whale (*Balaenoptera borealis*)
ba_3-	*Balaenoptera edeni*
ba_4-	*Balaenoptera acutorostrata*)
bl-	German cockroach (*Blatella*)
br-	amphioxus (*Branchiostome californiensis*)
ca-	dog (*Canis*)
cam-	camel (*Camelus*)
cap-	goat (*Capra*)
cat-	white sucker (*Catostomus*)
cav-	guinea pig (*Caviidae*)
ch-	chinchilla (*Chinchilla*)
co-	daddy sculpin (*Cottus scorpius*)
cr-	hamster (*Cricetus*)
crb-	chain recombinant bacterial
cro-	rattle snake (*Crotalus*)
cu-	rabbit (*Cuniculus*)
cy-	carp (*Cyprinus*)
di-	opossum (*Didelphis*)
e-	horse (*Equus*)
el-	elephant (*Elephantidae*)
emp-	enzymatic modified porcine
f-	cat (*Feles*)
g-	chicken (*Gallina)*
ga-	common cod (*Gadus*)
go-	gudgeon (*Gobius*)
h-	man (*Homo*)
he-	*He*liothis
hy-	porcupine (*Hystrix*)
ic-	channel catfish (*Ictalurus*)
lo-	angler (*Lophius*)
loc-	*Locusta*
m-	mouse (*Mus*)
ma-	*Manduca*
mac-	rhesus monkey (*Macaca mulatta*)
me-	turkey (*Melagris*)
mu-	mink (*Mustela vision*)
my-	hagfish (*Myxine glutinosa*)
myo-	coypu (*Myocastor coypus*)

Abbreviation	Species (latin name)
o-	sheep (*Ovins*)
oc-	degu (*Octodon degus*)
on_1-	coho salmon (*Oncorhynchus kisutch*)
on_2-	chum salmon (*Oncorhynchus keta*)
p-	pig (*Porcus*)
pe-	American cockroach (*Periplaneta*)
pet-	lamprey (*Petromyzon marinus*)
ph-	sperm whale (*Physter catodon)*
pl-	flounder (*Platichthys flesus*)
pr-	shark (*Pristis*)
pro-	casiragua (*Proechimys guairae*)
ps-	turtle (*Pseudemys*)
r-	rat (*Rattus*)
ra-	frog (*Rana*)
ra_1-	bullfrog (*Rana catesbeiana*)
ra_2-	European common frog (*Rana temporaria*)
sa-	salmon (*Salmo*)
sai-	squirrel monkey (*Saimiri sciurea*)
sc-	common dogfish (*Scyliorhinus*)
sch-	*Schistocera*
sco-	tuna (*Scombridae*)
se-	yellowtail (*Seriola dumerili*)
si-	monkey (*Simia*)
sil-	real sheatfish (*Silurus*)
sq-	spiny dogfish (*Squalus*)
st-	ostrich (*Struthiocamelus*)
th-	Thunnies (*Thunnus*)
to-	electric ray (*Torpedo*)
um-	mud fish (*Umbra*)
xe-	*Xenopus laevis*

In addition to the abbreviations of amino acids and species further abbreviations are used for peptides, protecting groups and other items (Table 3).

Table 3. Abbreviations for Peptides, Protecting groups and other Items

Abbreviation	Peptide
A4	amyloid A4 protein = amyloid β-protein
AAP	antiarrhytmic peptide
Ac	acetyl
ACE	angiotensin converting enzyme
Acm	acetamidomethyl
ACTH	corticotropin (adenocorticotropic hormone)
AKH	adipokinetic hormone
ALS	amyotrophic lateral sclerosis
ANF	atrial natriuretic factor
ANP	atrial natriuretic peptide
AP	anthral peptide

Abbreviation	Peptide
APP	A4 precursor protein
AT	angiotensin
ATG	angiotensinogen
AXP	anorexigenic peptide
BB	bombesin
BK	bradykinin
BMP	bone morphogenetic protein
BNP	brain natriuretic peptide
Boc	tert.-butyloxycarbonyl
BPP	bothrops jararaca's pentapeptide
Bt_2cGMP	Dibutyryl-cGMP
Bzl	benzyl
cAMP	cyclic adenosine monophosphate
CCK	cholecystokinin
CCK-RP	cholecystokinin releasing peptide
CG	chorionic gonadotropin
cGMP	cyclic guanosine monophosphate
CGRP	calcitonin gen related peptide
CHH	crustacean hyperglycemic hormone
CLIP	corticotropin like intermediate lobe peptide
CM	casomorphine
CNS	central nervous system
CNP	C-type natriuretic peptide
CPON	C-flanking peptide of neuropeptide Y
CRF	corticoliberin (corticotropin releasing factor)
CRH	corticoliberin (corticotropin releasing hormone)
CRL	caerulein
CSF	cerebrospinal fluid
CT	calcitonin
CTP	cell-tropin
DAG	diacylglycerol
DAMME	[2-D-alanin, 4-N-methyl-phenylalanin, 5-methionin-sulfoxid]-enkephalin
DM	dermorphin
DP	dynorphin
DRPH	light adapting distal retinal pigment hormone
DSIP	delta-sleep inducing peptide
DT	deltorphin (= dermenkephalin)
DTF	dental fluid transport stimulating peptide
EDRF	epithelial derived relaxing factor
EGF	epithermal growth factor
EK	enkephalin
ELE	eledoisin
EP	endorphin
ET	endothelin
Et	ethyl
FGF	fibroblast growth factor
FGn-RH	gonadoliberin from follicle fluids
FMRF-amid	femarfarm
FSH	follitropin (follicle stimulating hormone)
FTS	Bach factor (facteur thymique serique)
GABA	gamma amino butyric acid

Abbreviation	Peptide
GAP	gonadotropin releasing hormone associated peptide
GG	glucagon
GH	growth hormone (= somatotropin, STH)
GIP	gastric inhibitory peptide = glucose dependent insulinotropic peptide
GLI	glicetin
GLP	glucagon like peptide
GnRH	gonadoliberin (= gonadotropin releasing hormone)
GnTH	gonadotropin (= gonadotropic hormone)
GPI	guinea pig ileum
GRP	gastrin releasing peptide
GT	gastrin
HAH	head activator of hydra
HD	helodermin
HK	high molecular weight kininogen
HS	helospectin
5-HT	serotonine (= 5-hydroxy-tryptamine)
HTH	hypertrehalosemic hormone
IAPP	islet amyloid polypeptide (= amylin)
IFN	interferon
Ig	immunoglobulin
IGF	insulin-like growth factor
IGF-BP	insulin-like growth factor binding protein
IGF-R	insulin-like growth factor receptor
IHB	inhibin
IL	interleukin
INS	insulin
INSR	insulin receptor
IP_3	inositol triphosphate
KAS	kassinin
KG	kininogen
KP	kyotorphin
LANT 6	like avian neurotensin hexapeptide
LH	lutropin (= luteinizing hormone)
LK	low molecular weight kininogen
LPH	lipotropin (= lipolytic hormone)
MCD peptide	mast cell degranulating peptide
MCH	melanophores concentrating peptide
Me	methyl
MG	menopausal gonadotropin (= urogonadotropin)
MIF	melanostatin (= melanotropin inhibiting factor)
MIS	Muellerian inhibiting substance
MMC	migratory motorial complex
MND	motor neuron disease
MPF	melanotropin potentiating factor
MRCH	melanization and reddish coloration hormone
MRF	melanoliberin (= melanotropin releasing factor)
MSH	melanotropin (= melanocytes stimulating hormone)
MT	motilin
MTO	myotropin
NGF	nerve growth factor

Abbreviation	Peptide
neo-KP	neo-kyotorphin
Nic	nicotinoyl
NKA	neurokinin A
NKB	neurokinin B
NPK	neuropeptide K
NPY	neuropeptide Y
NRP	neurotensin related peptide
NRS	neurotensin related substance
NT	neurotensin
OP	osteogenic protein
OT	oxytocin
PAF	platelet activating factor
PBAN	pheromon biosynthesis activating neuropeptide
PDGF	platelet derived growth factor
PG	prostaglandin
PGL	peptide with N-terminal glycin and C-terminal leucinamide
PH	parotid hormone
PHI	peptide with N-terminal histidin and C-terminal isoleucinamide
PHM	peptide with N-terminal histidin and C-terminal methioninamide
PHY	physalaemin
Pic	picolyl
PIF	prolactin inhibiting factor
PL	placenta lactogen
PLA_2	phospholipase A_2
PLC	phospholipase C
PLF	proliferin
PLP	prolactin like proteins
PMSG	pregnant mare serum gonadotropin
POMC	proopiomelanocortin
PP	pancreatic peptide
Rha	α-L-rhamnopyranosyl
PRC	prolactin related cDNA
PRL	prolactin
PRLR	prolactin receptor
PRP	proliferin related protein
PSP	pancreatic spasmolytic peptide
PTH	parathyroid hormone
PTHrP	parathyroid hormone related protein
PYL	peptide with N-terminal tyrosin and C-terminal leucinamide
PYY	peptide with N-terminal tyrosin and C-terminal tyrosinamide
RLX	relaxin
RPA	rabbit pulmonary artery
RPCH	red pigment concentrating hormone
RPV	rat portal vein
SEC	secretin
SL	somatolactin
SM	somatomedin

Abbreviation	Peptide
SP	substance P
SRF	somatoliberin (= somatotropin releasing factor)
SRIF	somatostatin (= somatotropin release inhibiting factor)
STH	somatotropin (= somatotropic hormone, growth hormone)
STHR	somatotropin receptor
SV	sauvagin
T_3	triiodothyronine
T_4	thyroxine
TBU	tert.-butyl
TGF	transforming growth factor
TNF	tumor necrosis factor
TP	thymopoietin
TP-5	thymopentin
TPPV	transforming protein of polyoma virus
TRH	thyroliberin (= thyrotropin releasing hormone)
TS	thymosin
TSH	thyrotropin (= thyroid stimulating hormone)
UT	urotensin
VIP	vasoactive intestinal peptide
VP	vasopressin
VT	vasotocin
Z	benzyloxycarbonyl

2. Gonadoliberin, Thyroliberin, Gonadotropins, Thyrotropin

Thyroliberin (TRH) as well as gonadoliberin (GnRH) are formed in the hypothalamus and release thyrotropin (TSH) and the gonadotropins follitropin (follicle stimulating hormone, FSH) or lutropin (luteinizing hormone, LH)) respectively from the anterior pituitary gland. TSH stimulates the formation of the thyroid hormones, triiodothyronine (T_3) and thyroxine (T_4), which, in turn, inhibit the release of TSH but stimulate TRH. LH and FSH are responsible for the synthesis of the steroid sex hormones in the gonads. These hormones, in turn, have an inhibitory effect on the release of gonadotropins, but stimulate the release of GnRH. In the gonads, FSH stimulates the release of the inhibins (IHB-A and IHB-B), which exert a negative feedback to inhibit further secretion of FSH from the hypophysis. The hormone follistatin formed in the gonads also inhibits the release of FSH. On the other hand, the activins (homo- and heterodimers of the IHB β-chains), like the transforming growth factor β (TGF-β), stimulate the formation of FSH, see Figure 1.

The tripeptide TRH and the decapeptide GnRH have structural similarities:

GnRH	*Q	H	W	S	Y	G	L	R	P	Ga
TRH	*Q	H	/	/	/	/	/	/	Pa	

The gonadotropins (LH and FSH) and TSH also belong to one family. They consist of two glycoproteins in a non-covalent association, an α- and a β-chain. TSH and the gonadotropins have identical α-chains.

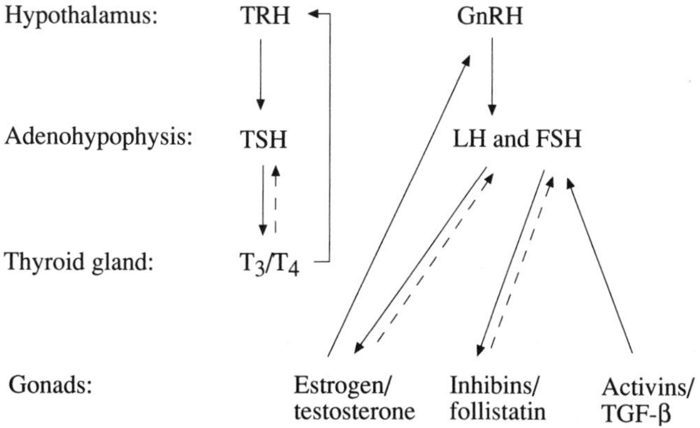

Figure 1. Gonadoliberin/thyroliberin gonadotropins/thyrotropin cascade

TGF-β, the inhibins, the muellerian inhibiting substance and the bone morphogenetic proteins are all structurally related. The inhibins consist of two glycoprotein chains that are linked together by disulfide chains. The α-chains of both inhibins are identical. The homo- and heterodimers of the IHB β-chains are the activins. TGF-β is a homodimeric compound.

Follistatin is a single-chain glycoprotein which has no structural similarity to the TGF-β/inhibin family.

Another protein (7B2) that is released by GnRH should be mentioned here.

2.1. Gonadoliberin [2.1]

2.1.1. Occurrence

The decapeptide gonadoliberin (GnRH) is formed as a precursor peptide consisting of 92 amino acids (prepro-GnRH) [2.2]. The C-terminal peptide of prepro-GnRH (gonadotropin-releasing hormone associated peptide, GAP) consists of 56 amino acids and exhibits a prolactin inhibiting effect [2.3].

GnRH has not only been found in the hypothalamus, but also in the brain, heart, liver, pancreas, kidneys and adrenals, small intestine, gonads and in the milk of lactating rats [2.4].

Until now, no differences have been observed in the GnRHs of different mammalian species. Two GnRHs (g-GnRH I and II) have been found in chickens, fish [2.5] and amphibians [2.6]. The sa-GnRH found in salmon has also been observed in other fish and reptiles [2.7]. The pet-GnRH of the lamprey (*Petromyzones*) has the least similarity to the GnRH of mammals [2.8]. A similar compound has been discovered in the hypothalami of humans [2.9]. Factors that are similar to GnRH (α-factors) have been isolated from microorganisms, e.g., brewer's yeast. These factors cause the aggregation of cells and the exchange of chromosomes.

h-, p-, o-GnRH	*Q	H	W	S	Y	G	L	R	P	Ga				
g-GnRH-I	*Q	H	W	S	Y	G	L	Q	P	Ga				
g-GnRH-II	*Q	H	W	S	H	G	W	Y	P	Ga				
sa-GnRH	*Q	H	W	S	Y	G	W	L	P	Ga				
pet-GnRH	*Q	H	Y	S	L	E	W	K	P	Ga				
α-factor α1	W	H	W	/	L	Q	L	K	P	G	Q	P	M	Y
α-se-Pheromon	W	H	W	/	L	R	L	S	Y	G	Q	P	I	Y

2.1.2. Release [2.10]

The synthesis of GnRH from its precursor requires a peptidylglycine α-amidating monooxygenase that is dependent on Cu^{2+} and ascorbic acid for activity [2.11]. GnRH is secreted from the hypothalamus by pulses of short duration at intervals of 60 to 120 minutes and reaches the anterior pituitary via the hypophyseal portal system.

This results in the release of the gonadotropins LH and FSH.

Gonadotropins, which should actually inhibit GnRH by a negative feedback mechanism, showed a stimulating effect on the release of GnRH from GnRH-containing neurons of normal male rats. The action of LH on this system is more pronounced than that of FSH [2.12].

The steroid hormones of the gonads (testosterone, estrogen and progesterone) stimulate the release of GnRH [2.13]-[2.15]. The post-mortem hypothalamic level of GnRH in women after menopause is lowered [2.16]. VIP [2.17], [2.18], NPY [2.19], thymosin-β_4 and compounds of the arachidonic acid cascade (PGE_2, 5- or 12-hydroxyeicosatetraenoic acid [2.20]) also stimulate the release of GnRH.

Prolactin [2.21], [2.22], stress and the stress hormones (CRH, ACTH, β-EP, oxytocin and glucocorticoids) inhibit the formation of GnRH [2.10], [2.23]-[2.26].

The effect of the endogenous opioids is, however, dependent on the steroid hormones of the gonads. Indeed, testosterone stimulates the release of GnRH when the endogenous opioides are blocked and inhibits GnRH release when the endogenous opiates are stimulated [2.27].

2.1.3. Receptors [2.28]

Receptors for GnRH are found not only in the hypophysis, but also in the gonads, placenta [2.29], corpus luteum [2.30], [2.31], and on the oocytes [2.32]. In human breast cancer tissue, two classes of [D-Trp^6] GnRH receptors were detected in 52% of the cases: one of high-affinity with a low capacity and the other of low-affinity with a high capacity. These receptors positively correlate with estrogen and progesterone binding in these tissues [2.33].

In female rats, physiological doses of GnRH stimulate the formation of GnRH receptors in the hypophysis [2.34], while high doses have the opposite effect [2.35]. Highly active GnRH agonists reduce the number of GnRH receptors, while the GnRH antagonists have no effect [2.35]. GnRH agonists are rapidly taken up into the cell by endocytosis, while the antagonists remain on the membrane for a long time [2.36].

2.1.4. Biological Effects

Under physiological conditions (pulsatile application every 60-120 minutes, GnRH stimulates the hypophyseal synthesis and release [2.37] of lutropin (LH) and follitropin (FSH) in a calcium dependent process. On the other hand, the secretion of LH and FSH is suppressed in rats and human beings by a down regulation of GnRH receptors by high non-physiological doses in a process that is independent of calcium [2.35].

Estrogen stimulates the GnRH induced secretion of LH when low doses of GnRH are infused, but inhibits the secretion at high GnRH doses [2.38]-[2.40]. Estradiol, insulin [2.35], adrenalin, nonadrenalin [2.41] and NPY [2.42] stimulate the LH-releasing effect of GnRH. On the other hand progesterone or dihydrotestosterone and, in female rats, stress and glucocorticoids [2.43]-[2.45] inhibit the GnRH-induced release of LH [2.35]. GnRH not only affects LH and FSH, but also exerts a direct effect on steroid synthesis in the gonads of experimental animals. Small doses or a short treatment have a stimulating effect [2.46]-[2.50] and high doses or long-term treatment exert an inhibitory effect [2.51]-[2.53] on steroidogenesis in the gonads.

In chondrasarcoma and other tumors [2.54]-[2.57], high doses of GnRH agonists inhibit the growth of the tumor.

Similar to castration, treatment with high doses of GnRH agonists (chemical castration) regenerates the thymus in old male rats [2.58].

2.1.5. Structure-Activity Relationships [2.59], [2.60]

GnRH Agonists. The GnRHs that occur in other species are less active in mammals. Substitution of His^5 for Tyr^5 and Trp^7 for Leu^7 give rise to peptides with full GnRH activity. In fact, [His^5] GnRH is twice as effective as GnRH in releasing FSH [2.61].

The agonistic effect of GnRH is increased in particular by substituting an ethylamine residue for $Gly\text{-}NH_2^{10}$ and by replacing Gly^6 by lipophilic D-amino acids. The resultant compounds are about 100 to 200 times as active as GnRH. It is important for the substitution at position 6 that the D-amino acid contains an α-CH and a β-CH_2 group. Still higher activites can be achieved by the use of synthetic aromatic D-amino acids, e.g., D-3-(2,4,6-trimethylphenyl)alanine (2.62). A combination of the hydrophilic Azagly-NH_2 in position 10 with these hydrophobic amino acids led to [D-3-(2-naphthyl)-Ala^6, $Azagly^{10}$] GnRH, one of the most active compounds available until now (230 $\times$ GnRH) [2.63]. Hydrophilic D-serine-O-glycosides [2.64] or β-Asp-α-esters [2.65] in position 6 also give rise to very active compounds.

GnRH Antagonists. In addition to the modification of position 6, changes at positions 1, 2, 3, and 10 lead to GnRH antagonists. Highly active compounds are:

1) Ac-[D-Nal^1, D-p-Cl-Phe^2, D-Trp^3, D-$Har(Et_2)^6$, D-Ala^{10} GnRH (*Detirelix* [2.66], [2.67]
2) Ac-[D-Nal^1, D-pCl-Phe^2, D-Pal^3, D-$Har(Et_2)^6$, $Har(Et_2)^8$, D-Ala^{10}] GnRH (RS-26306) [2.60]
3) Ac-[D-Nal^1, D-p-Cl-Phe^2, D-Trp^3, D-$Ser(Rha)^6$, $AzaGly^{10}$] GnRH (HOE 013, Ramorelix) [2.68]
4) Ac-[D-Nal^1, D-p-Cl-Phe^2, D-Pal^3, Arg^5, D-4-p-methoxybenzoyl-2-aminobutyric $acid^6$, D-Ala^{10}] GnRH (ORF 21243, "Nal-Glu") [2.69], [2.70]
5) Ac-[D-Nal^1, D-pClPhe^2, D-Pal^3, $Lys(Pic)^5$, D-$Lys(Pic)^6$, $ILys^8$, D-Ala^{10}] GnRH (ORF 23541, "Nal-LYS" Antide) [2.71]
6) Ac-[D-Nal^1, D-p-Cl-Phe^2, D-Pal^3, D-Cit^6, D-Ala^{10}] GnRH (SB-75, D 20 453, Cetrorelix) [2.72]

A problem associated with GnRH antagonists is the anaphylactic, mediator triggering effect, which is caused, in particular, by the additional basic charges at position

6 and 5 and by the strongly lipophilic N-terminal amino acids [2.73]. This mediator releasing effect is reduced if the Arg in position 6 is moved to position 5 (e.g., Nal-Glu), D-Trp^3 is replaced by D-Pal^3 (e.g., Nal-Glu, Nal-Lys, SB-75 and RS-26306) and the arginine in position 6 or 8 is modified (e.g., RS-26306) or suitably substituted (e.g., Nal-Lys, HOE 013, SB-75) [2.72], [2.74]-[2.77].

These antagonists (e.g., N-Ac-[D-p-ClPhe^{1,2}, D-Trp^3, D-Arg^6, D-Ala^{10}] GnRH) protect germ cells from X-radiation by reversibly blocking the GnRH receptors [2.78].

2.1.6. Applications [2.60]

Cryptorchism is treated by the intranasal application of GnRH (gonadorelin, Kryptocur) [2.79]-[2.80].

Delayed puberty, hypogonadism, anorexia nervosa, Kallmann syndrome and secondary amenorrhea can be treated with a pulsatile gonadorelin infusion (Fertiral, Cyclomat).

Gonadorelin is also used in the differential diagnosis of fertility disturbances.

In the case of anorexia nervosa, the parenteral application of 5μg of triptorelin every second day produced a favorable effect on the release of LH and on the menstrual cycle [2.81]. A case of acute intermittent porphyria could be controlled by the subcutaneous administration of 5 μg/day of histrelin [2.82].

Analogs that have a longer and stronger effect are used predominantly in high doses for the suppression of the gonads. The following GnRH analogs are either on the market or being clinically tested.

1) [D-Ser(TBU)^6]GnRH-(1-9)-nonapeptide-ethylamide (buserelin)
2) [D-Ser (TBU)^6, AzaGly^{10}] GnRH (goserelin) [2.83]
3) [D-Trp^6] GnRH (triptorelin) [2.84]
4) [D-Trp^6] GnRH-(1-9)-nonapeptide-ethylamide (deslorelin)
5) [D-Leu^6] GnRH-(1-9)-nonapeptide-ethylamide (leuprolid acetate, leuprorelin) [2.85]
6) [D-Trp^6, N-MeLeu^7] GnRH-(1-9)-nonapeptide-ethylamide (lutrelin) [2.86]
7) [D-Nal^6]GnRH (nafarelin) [2.87]
8) [D-Nal^6, AzaGly^{10}] GnRH·(Syntex RS 49 947 [2.66])
9) [D-His(imBzl)^6] GnRH-(1-9)-nonapeptide-ethylamide (histrelin) [2.88], [2.89]

GnRH agonists can be applied parenterally, nasally or as a biodegradable implant. They are being clinically tested or applied for the indications: inhibition of testosterone and estrogen dependent tumors (carcinoma of the prostate and the breast), in endometriosis, as contraceptives [2.90], [2.91], in pubertas precox [2.92] and for *in vitro* fertilization [2.93]. Other indications are: regression of benignant uterus myoma and leiomyoma [2.94]-[2.97]), hirsutism [2.98], premenstrual intermittend porphyria (disturbance of heme synthesis [2.99]) and normal prostate hyperplasia [2.100].

In women, this suppressive therapy with GnRH agonists lowers the mineral content of the bones [2.101], [2.102].

2.2. Gonadoliberin from Follicle Fluids

A peptide containing 14 amino acids was isolated from human follicle fluid and was found to be identical to h-α_1-antitrypsin-(11-24) [2.103]

		11												24	
h-α_1-Antitrypsin	..K	T	D	T	S	H	H	D	Q	D	H	P	T	F	N	K
h-FGnRH		T	D	T	S	H	H	D	Q	D	H	P	T	F	N	

In comparison with GnRH, h-gonadoliberin from follicle fluids (h-FGnRH) is substantially less effective in releasing gonadotropins from the mouse hypophysis.

2. 3. Thyroliberin [2.104]

2. 3.1. Occurrence

Thyroliberin (TRH) is found primarily in the nervous system of the brain, gastrointestinal tract, pancreas, and in the prostate.

TRH is formed as prepro-TRH. Until now, the structures of r-prepro-TRH and an amino terminal fragment of xe-prepro-TRH have been elucidated via the cDNA. It has been found that r-prepro-TRH consists of 225 amino acids and contains 5 copies of the TRH sequence.

r-prepro-TRH-(75-82)	. . K R	Q	H	P	G	K	R . .
r-prepro-TRH-(200-207)	. . K R	Q	H	P	G	K	R . .
r-prepro-TRH-(107-114)	. . K R	Q	H	P	G	R	R . .
r-prepro-TRH-(152-159)	. . K R	Q	H	P	G	R	R . .
r-prepro-TRH-(170-177)	. . K R	Q	H	P	G	R	R . .
TRH		*Q	H	Pa			

This precursor molecule is cleaved between the basic amino acids and the TRH sequence by endopeptidases. The N-terminal glutamine is cyclized to pyroglutamic acid either spontaneously or by enzymatic catalysis and the C-terminal glycine is converted to the amide function with the help of a peptidylglycine-α-amidating monooxygenase. The peptides located between these sequences are called the prepro-TRH connecting peptides. Two of these connecting peptides (prepro-TRH-(160-169) and prepro-TRH-(178-199)) together with TRH are the main storage forms of this hormone in the hypothalamus, spinal cord [2.105], and in the pancreas [2.106]

r-prepro-TRH-(160-169) S F P W M E S D V T

r-prepro-TRH-(178-199) F I D P E L Q R S W E E K E G E G V L M P E.

The anorexigenic peptide (pGlu-His-Gly-OH), which was isolated from the urine of women suffering from anorexia nervosa, and a metabolite of TRH, cyclo-(His-Pro), have a central appetite-inhibiting effect [2.107]. There is a temporary dramatic increase in the plasma level of cyclo-His-Pro after a glucose meal [2.108] which correlates positively with the weight loss in bulimia and anorexia nervosa [2.109]. Another structurally similar peptide, pGlu-Glu-Pro-amide, was found in rabbit prostate and in human sperma [2.110].

2. 3. 2. Release

TRH is released centrally by electric shocks [2.111], 6-hydroxy-dopamine [2.112], dopamine (D_2-receptors) [2.113], cold [2.114], and by the hormones of the thyroid gland [2.115]. It is inhibited by opioid peptides [2.116]. In the periphery, TRH is raised by histamine via H_2-receptors [2.117], testosterone [2.118], and serotonin [2.119] and decreased by T_3, [2.120], carbachol [2.119], streptozotocin [2.106], and dopamine [2.121].

An enzyme that is apparently specific for TRH (thyro-liberinase, pyroglutamyl-peptidase II, E.C. 3.4.19) and cleaves the pyroglutamyl-histidine bond is found in brain homogenates (brain synaptosomal membrane) [2.122]. An inhibitor of pyroglutmylpeptidase II (N-1-carboxy-2-phenethyl-(N^{im}-benzyl)-histidyl-β-naphthylamide) increases *in vitro* the TRH content of sliced hypothalami [2.123].

2. 3. 3. Receptors

The receptors that are responsible for TSH secretion are down regulated by chronic pulsatile TRH application, whereas the receptors for PRL secretion are not as sensitive [2.124]-[2.126].

Both thyroidectomy [2.127] as well as thyroxine treatment [2.128] lead to an increase in the TRH binding sites in the hypophysis of the rat. It has also been found that κ- and δ-opioids inhibit the binding of [^{3}H] MeTRH on brain mebranes, but μ-opioids are inactive [2.129]. The blood pressure of spontaneously hypertensive rats correlates positively with the TRH receptors in the brain. Indeed, these TRH receptors could be jointly responsible for the pathophysiology of high blood pressure [2.130].

Chlorodiazepoxide is a competitive antagonist of the TRH receptor (K_i = 1.5 $\times$ 10^{-5}M) in GH3 cells [2.131] and in the [^{3}H] Me-TRH receptor binding assay in rat amygdala (IC_{50} = 3.6 $\times$ 10^{-7}M) [2.132]. Other TRH receptor antagonists are diazepam and midazolam. The 4-methyl-midazolam-(−) isomer appears to be a specific TRH antagonist [2.133].

2. 3.4. Biological Effects [2.134]

Biochemical/metabolic functions

These functions comprise the stimulation of 1,2-diacylglycerol and inositol-1,4,5-triphosphate, the increase of intracellular calcium and the decrease of serum calcium.

Endocrine functions

Pituitary-hypothalamus: release of thyrotropin (TSH) and prolactin (PRL). The TRH-stimulated TSH-secretion can be inhibited by somatostatin (SRIF) [2.135], dopamine [2.135] and calcitonin [2.136]. The TRH-stimulated release of PRL is stimulated by estrogen and inhibited by testosterone [2.137]. Modulation of somatotropin release (e.g., release in patients suffering from acromegaly [2.138] and diabetes [2.139]) and inhibition of β-endorphin [2.140], [2.141] are observed.

Thyroid gland: increase in T_4, T_3 and throglobulins via TSH [2.142] or directly [2.143].

Pancreas: inhibition of glucagon.

Digestive system: The effects on the digestive system include the relaxation of the lower esophageal sphincter [2.144], increase in stomach acid secretion [2.145], a spasmolytic effect in the small intestine of the dog caused by very small doses (30 ng/kg i.v.) [2.146], and the emptying of the gallbladder [2.147].

Urogenital tract: Contraction of the urinary bladder [2.144].

Cardiovascular system: A species-specific vascular effect is observed [2.148]. Vasodilatation is produced via activation of the vagus and vascular contraction occurs via activation of the sympathetic nervous system [2.149]. Blood pressure is increased [2.150].

Central nervous system [2.134], [2.151]: The effects in the central nervous system include the potentiation of K^+-stimulated acetylcholine release [2.152], reversal of narcotic/alcohol induced sedation, reversal of natural sedation (hibernation, sleep), wet-dog shaking activity, locomotor activity, forepaw tremor (rats), muscle tone improvement (antagonism of induced relaxation), increased blood pressure and heart rate, increased cerebral blood flow, renal renin release [2.153], increased catecholamine release from the adrenal medulla. Thermoregulation comprises induction of hypo- or hyper-thermia (depending on species) and reversal of induced hypothermia. Increased respiratory rate and minute volume are observed. Anticonvulsant activity. Antinociceptive action (release of opioids [2.140], and inhibition of cortical neuron excitability. Central gastrointestinal effects comprise a decreased food and water intake, increased gastric secretion and gastric emptying, increased or decreased motility (depending on species), vagus mediated contraction of the stomach [2.154], and increased pancreatic exocrine secretion. Hyperglycemia in rats and hypoglycemia in mice.

2.3.5. Structure-Activity Relationships

Of the numerous analogs of TRH that have been synthesized, only four compounds are more effective than TRH in releasing TSH. The analogs [2-(L-β-(pyrazolyl-1)Ala] TRH and [1-(3-carboxy-5-oxo-thiamorpholine)]TRH are 1.5 times as active as TRH. The compounds [2-(N-MeHis)]TRH and the corresponding N-amylamide [2.155] have eight times and approximately ten times the potency of TRH respectively.

A large number of TRH analogs exhibit, above all, the central activities of TRH, without releasing TSH to the same extent:

1. L-Pyro-2-aminoadipyl-His-Thiaproline-amide (MK 771),
2. (1.2.3.6-Tetrahydro-2.6-dioxo-4-pyrimidinyl)-carbonyl-His-Pro-amide, Orotyl-His-Pro-amide (CG-3509, CAS-62305-86-6) [2.156].
3. 6-Methyl-5-oxo-thiamorpholinyl-3-carbonyl-His-Pro-amide (Montirelin, CG 3703) [2.157],
4. (3S)-1-Oxo-1.2.3.4-tetrahydroisochinolin-3-carbonyl-His-Pro-amide [2.158]
5. ((S)-4-Oxo-2-azetidinyl)-carbonyl-His-Pro-amide (YM-14673, azetirelin dihydrate, CAS-95729-65-0) [2.159].
6. γ-Butyrolacton-γ-carbonyl-His-Pro-amide (DN-1417) [2.160].
7. 1-Benzyl-2-oxo-imidazolidin-4-carbonyl-His-Pro-amide [2.161],
8. pGlu-His-(3.3'-dimethyl)Pro-amide (RX 77368) [2.162],
9. L-Pyro-2-aminoadipyl-Leu-Pro-amide (Posatirelin, RGH-2202, CAS-78664,73-0) [2.163], [2.164].

The analogs, MK-771 and CG-3703 are stable to the TRH-specific pyroglutamylpeptidase II. On the other hand, DN-1417 is degraded faster than TRH. The compounds, CG-3703, DN-1417, and MK-771 act as competitive inhibitors of pyroglutamylpeptidase II [2.165].

2. 3.6. Applications

TRH (generic name: Protirelin) is used as a diagnosticum for thyroid and hypophyseal functions (Relefact TRH, Antepan, Thyroliberin/TRF Merck, TRH). TRH preparations for parenteral (0.2-0.4 mg), oral (40 mg), and intranasal (1 mg, Relefact TRH-nasal) application are available.

The central effect of TRH is tested preferably with TRH analogs that exhibit central activity. Possible applications are listed below [2.134]:

1. motor neuron disease (MND) [2.166]/amyotrophic lateral sclerosis (ALS) [2.167] [2.168]
2. spinal injury
3. cerebral ischaemia (stroke) [2.169]
4. circulatory shock [2.169]
5. CNS trauma [2.170], [2.171]
6. Alzheimer's disease/senile dementia [2.172]
7. Down's syndrome
8. schizophrenia
9. depression
10. antinociception/analgesia (with morphine)
11. respiration [2.173]
12. reversal of anesthesia and narcotic overdose
13. spinocerebellar degeneration
14. epilepsy

In the case of MND, clinical studies have shown that RX 77368 produced a marked improvement [2.134].

2.4. Gonadotropins [2.174], [2.175]

2.4.1. Occurrence

The gonadotropins (gonadotropic hormones, GnTHs) include the glycoproteins, lutropin (luteinizing hormone, LH), follitropin (follicle stimulating hormone, FSH) and chorionic gonadotropin (CG). The hormones LH and FSH are formed in the adenohypophysis and CG is produced in the chorion. They consist of two protein chains, each chain contains disulfide bridges. Disulfide bridges between the two chains do not exist. The α-chains of these glyco-proteins are identical [2.176]. Only the β-chains, which are responsible for activity, are different [2.177]. These chains are synthesized separately as procompounds.

The structure of GnTHs from man, horse, sheep, cow, and pig as well as LH sequences from rat, rabbit, salmon, and pike eel (*Muraenesox cinereus*) (mur-GnTH) have been elucidated [2.178]: see Table 4 and 5.

Thyrotropin (TSH) is structurally related to these glycoproteins. The α-chain of TSH is identical to that of the GnTHs.

2.4.2. Release

The action of pulsatile GnRH on the gonadotropic cells of the hypophysis controls the release of the gonadotropins, LH and FSH. More FSH is released at a decreased pulse frequency and LH is preferentially released at an increased GnRH frequency. On the other hand, centrally applied GnRH reduces the secretion of LH [2.179]. Endothelin from anterior pituitary cells is as effective as GnRH in releasing GnTHs in a process that is independent of GnRH receptors [2.180].

Low doses of estrogen and testosterone stimulate the release of GnTH via the activation of GnRH formation [2.181]-[2.183]. However, the long-term application of higher doses results in the direct inhibition (negative feedback) of the release of GnTH in the hypophysis [2.182], [2.184]. Calcitriol (2.25$(OH)_2D_3$) also potentiates the GnRH induced release of GnTH [2.185]. Phospholipase A_2 [2.186] PGE_2 [2.187] and diacylglycerol [2.188] stimulate the hypophyseal release of the GnTHs, while inhibitors of cyclooxygenase inhibit this release [2.189].

The frequency of LH secretion in women increases from the early follicular phase to the mid follicular phase and the late follicular phase. The LH amplitude decreases in the midfollicular phase (negative feedback by estrogen). The LH pulse frequency and the amplitude decrease during the luteal phase (effect of progesterone). FSH correlates here with LH [2.190]. The inhibitory influence of the steroid hormones is lacking in menopause and the level of LH and FSH rises. In this case, treatment with progesterone [2.191] or ethinylestradiol [2.192] lowers the raised plasma GnTH levels. Testosterone inhibits LH and FSH release in men [2.193]. The release of FSH is inhibited by the inhibins and stimulated by the activins and by TGF-β.

Table 4. β-Chains of gonadotropins and thyrotropin [2.177]

β-chain	Amino acid sequence
	1 … 25
h-CG-β	S K E P L R P R C R P I Ng A T L A V E K E G C P V-
e-CG-β	S R G P L R P L C R P I Ng A T L A A E K E A C P I-
h-LH-β	S R E P L R P W C H P I Ng A I L A V E K E G C P V-
e-LH-β	S R G P L R P L C R P I Ng A T L A A E K E A C P I-
o-LH-β	S R G P L R P L C Q P I Ng A T L A A E K E A C P V-
b-LH-β	S R G P L R P L C Q P I Ng A T L A A E K E A C P V-
p-LH-β	S R G P L R P L C R P I Ng A T L A A E D E A C P V-
r-LH-β	S R G P L R P L C R P V Ng A T L A A E N E F C P V-
cu-LH-β	*Q P A R G P L R P L C R P V Ng A T L A A E N E A C P V-
sa-GnTH-β	S L M Q / / P / C Q P I Ng Q T V S L E K E G C P T-
mur-GnTH-β	S V L Q / / P / C Q P I N E T I S V E K D G C P K-
	1 … 19
h-FSH-β	N S C E L T Ng I T I A I E K E E C R F-
e-FSH-β	N S C E L T Ng I T I A V E K E G C R F-
o-FSH-β	S C E L T Ng I T I A V E K E E C S F-
b-FSH-β	S C E L T Ng I T I T V E K E E C G F-
p-FSH-β	C E L T Ng I T I T V E K E E C T F-
h-TSH-β	F C I P T E Y M T H I E R R E C A Y-
b-TSH-β	F C I P T E Y M M H V E R K E C A Y-
p-TSH-β	F C I P T E Y M M H V E R K E C A Y-
m-TSH-β	F C I P T E Y T M Y V D R R E C A Y-
r-TSH-β	F C I P T E Y M M Y V D'R R E C A Y-
	26 … 50
h-CG-β	C I T V Ng T T I C A G Y C P T M T R V L Q G V L P-
e-CG-β, LH-β	C I T F T T S I C A G Y C P S M V R V M P A A L P-
h-LH-β	C I T V Ng T T I C A G Y C P T M M R V L Q A V L P-
o-LH-β	C I T F T T S I C A G Y C P S M K R V L P V I L P-
b-LH-β	C I T F T T S I C A G Y C P S M K R V L P V I L P-
p-LH-β	C I T F T T S I C A G Y C P S M R R V L P A A L P-
r-LH-β	C I T F T T S I C A G Y C P S M V R V L P A A L P-
cu-LH-β	C I T F T T S I C A G Y C P S M V R V L P A A L P-
sa-GnTH-β	C L V I R A P I C S G H C V T K E P V F K S P F S-
mur-GnTH-β	C L V F Q T S I C S G H C I T K D P S Y K S P L S-
	20 … 44
h-FSH-β	C I S I Ng T T W C A G Y C Y T R D L V Y K N P A R-
e-FSH-β	C I T I Ng T T W C A G Y C Y T R D L V Y K D P A R-
o-FSH-β	C I S I Ng T T W C A G Y C Y T R D L V Y K B P A R-
b-FSH-β	C I S I Ng T T W C A G Y C Y T R D L V Y K D P A R-
p-FSH-β	C I S I Ng T T W C A G Y C Y T R D L V Y K B P A R-
h-TSH-β	C L T I Ng T T I C A G Y C M T R D I N G K L F L P-
b-TSH-β	C L T I Ng T T V C A G Y C M T R D V N G K L F L P-
p-TSH-β	C L T V Ng S T I C A G Y C M T R D F D G K L F L P-
m-, r-TSH-β	C L T I Ng T T I C A G Y C M T R D I N G K L F L P-

β-chain	Amino acid sequence
	51 75
h-CG-β	A / L P / Q V V C N Y R D V R F E S I R L P G C P R G-
e-CG-β, LH-β	A / I P / Q P V C T Y R E L R F A S I R L P G C P P G-
h-LH-β	P / L P / Q V V C T Y R D V R F E S I R L P G C P R G-
o-LH-β	P / M P / Q R V C T Y H E L R F A S V R L P G C P P G-
b-LH-β	P / M P / Q R V C T Y H E L R F A S V R L P G C P P G-
p-LH-β	P / V P / Q P V C T Y R E L I F A S S R L P G C P P G-
r-LH-β	P / V P / Q P V C T Y R E L R F A S V R L P G C P P G-
cu-LH-β	P / V P / Q P V C T Y R E L R F A S I R L P G C P P G-
sa-GnTH-β	T / V T / Q H V C T Y R D V R Y E M I R L P D C R P W-
mur-GnTH-β	T / V Y / Q R V C T Y R D V R Y E T V R L P D C R P G-
	45 69
h-FSH-β	P / K I / Q K T C T F K E L V Y E T V R V P G C A H H-
e-FSH-β	P / K I / Q K T C T F K E L V Y E T V K V P G C A H H-
o-FSH-β	P / B I / Q K T C T F K E L V Y E T V K V P G C A H H-
b-FSH-β	P / N I / Q K T C T F K E L V Y E T V K V P G C A H H-
p-FSH-β	P / B I / Z K T C T Y R Z L V Y Z T V K V P G C A H H-
h-TSH-β	K Y A L S Q D V C T Y R D F I Y R T V E I P G C P L H-
b-TSH-β	K Y A L S Q D V C T Y R D F M Y K T A E I P G C P R H-
p-TSH-β	K Y A L S Q D V C T Y R D F M Y K T V E I P G C P H H-
m-TSH-β	K Y A L S Q D V C T Y R D F I Y R T V E I P G C P H H-
r-TSH-β	K Y A L S Q D V C T Y R D F T Y R T V E I P G C P H H-
	76 100
h-CG-β	V N P V V S Y A V A L S C Q C A L C R R S T T D C-
e-CG-β, LH-β	V D P M V S F P V A L S C H C G P C Q I K T T D C-
h-LH-β	V D P V V S F P V A L S C R C G P C R R S T S D C-
o-LH-β	V D P M V S F P V A L S C H C G P C R L S S T D C-
b-LH-β	V D P M V S F P V A L S C H C G P C R L S S T D C-
p-LH-β	V D P T V S F P V A L S C H C G P C R L S S S D C-
r-LH-β	V D P I V S F P V A L S C R C G P C R L S S S D C-
cu-LH-β	V D P E V S F P V A L S C R C G P C R L S S S D C-
sa-GnTH-β	S E P H V T Y P V A L S C D C S L C N M D T S D C-
mur-GnTH-β	V D P H V T F P V A L S C D C N L C T M D T S D C-
	70 94
h-FSH-β	A D S L Y T Y P V A T Q C H C G K C D S D S T D C-
e-FSH-β	A D S L Y T Y P V A T Z C H C G K C B S D S T B C-
o-FSH-β	A D S L Y T Y P V A T E C H C G K C D S D S T D C-
b-FSH-β	A D S L Y T Y P V A T E C H C S K C K S D S T D C-
p-FSH-β	A B S L Y T Y P V A T Z C H C G K C B S B S T B C-
h-TSH-β	V A P Y F S Y P V A L S C K C G K C D T D Y S D C-
b-TSH-β	V T P Y F S Y P V A I S C K C G K C N T D Y S D C-
p-TSH-β	V T P Y F S Y P V A I S C K C G K C D T D Y S D C-
m-TSH-β	V A P Y F S Y P V A V S C K C G K C N T D N S D C-
r-TSH-β	V A P Y F S Y P V A L S C K C G K C N T D Y S D C-

β- chain	Amino acid sequence
	101 … 125
h-CG-β	G G P K D H P L T C D D P R F Q D S S S Sg K A P P-
e-CG-β, LH-β	G V F R D Q P L A C / A P / / Q A S S S S K D P P-
h-LH-β	G G P K D H P L T C D H P Q L S G L L F L.
o-LH-β	G P G R T Z P L A C B H P P L P D I L.
b-LH-β	G G P R T Q P L A C D H P P L P D I L F L.
p-LH-β	G P G R A Q P L A C D R P P L P G L L.
r-LH-β	G G P R A E P L A C D L P H L P G.
cu-LH-β	G G P R A E P L A C D L P H L P G.
sa-GnTH-β	T I E S L Q P D F C I T Q R V L T D G D.
mur-GnTH-β	A I Q S L R P D F C M S Q R A P.
	95 … 111
h-FSH-β	T V R G L G P S Y C S F G E M K E.
e-FSH-β	T V R G L G P S Y C S F G D M K Z Y P V A L S Y.
o-FSH-β	T V R G L G P S Y C S F S D I E R Z.
b-FSH-β	T V R G L G P S Y C S F R E I K E.
p-FSH-β	T V R G L G P S Y C S F G E.
h-TSH-β	I H E A I K T N Y C T K P Q K S Y.
b-TSH-β	I H E A I K T N Y C T K P Q K S Y M V G F S I.
p-TSH-β	I H E A I K T N Y C T K P E K S Y.
m-TSH-β	I H E A V R T N Y C T K P Q S F Y L G G F S V.
r-TSH-β	T H E A V K T N Y C T K P Q T F Y L G G F S G.
	126 … 145
h-CG-β	P Sg L P S P Sg R L P G P Sg D T P I L P Q
e-CG-β, LH-β	S Q P L T S T S T P T P G A S R R S S H P L P I K T S.

Table 5. Common α- chains of the gonadotropins and thyrotropin [2.176]

α- chain	Amino acid sequence
	1 … 25
h-α	V Q D C P E C T L Q E N P F F S Q P-
e-α	F P D G E F T T Q B C P E C K L R E N K Y F F K L-
o-α	F P N G Q F T M Q G C P Q C K L K E N K Y F S K P-
b-α	F P D G E F T M Z G C P Z C K L K E N K Y F S K P-
p-α	T M Q G C P E C K L K E N K Y F S K L-
mur-α	Y P N N E I S R G G C D E C R L K D N K F F S K P-
	25 … 50
h-α	G A P I L Q C M G C C F S R A Y P T P L R S K K T-
e-α	G V P I Y Q C K G C C F S R A Y P T P A R S R K T-
o-α	D A P I Y Q C M G C C F S R A Y P T P A R S K K T-
b-α	B A P I Y Q C M G C C F S R A Y P T P A R S K K T-
p-α	G A P I Y Q C M G C C F S R A Y P T P A R S K K T-
mur-α	S A P I F Q C V G C C F S R A Y P T P L R S K K T-

α- chain	Amino acid sequence
	51 … 75
h-α	M L V Q K Ng V T S E S T C C V A K S Y N R V T V M-
e-α	M L V P K Ng I T S Z S T C C V A K A F I R V T V M-
o-α	M L V P K Ng I T S E A T C C V A K A F T K A T V M-
b-α	M L V P K Ng I T S Z A T C C V A K A F T K A T V M-
p-α	M L V P K Ng I T S E A T C C V A K A F T K A T V M-
mur-α	M L V P K D I T S E A T C C V A R E V T K L D N M-
	76
h-α	G G F K V E Ng H T A C H C S T C Y Y H K S.
e-α	G N I K L Z Ng H T Z C Y C S T C Y H H K I.
o-α	G N V R V E Ng H T E C H C S T C Y Y H K S.
b-α	G N V R V Z Ng H T E C H C S T C Y Y H K S.
p-α	G N A R V E Ng H T E C H C S T C Y Y H K S.
mur-α	K L E / / / N H T D C H C S T C Y Y H K S.

*h-CG-α is extended by three amino acids at the N-terminal (A P D) and contains D instead of N at position 19.

The release of GnTHs is reduced by several factors, usually via the inhibition of GnRH. These factors include deprivation of food [2.194], stress [2.195], [2.196], morphine [2.197] [2.198], endogenous opioids [2.199], marihuana [2.200], glucocorticoids [2.201], [2.202], dopamine agonists [2.203], [2.204], alcohol [2.205] and lithium salts [2.206], [2.207]. Central noradrenergic mechanisms inhibit the release of LH via the activation of β-adrenergic receptors [2.208].

2.4.3. Receptors

The receptors of h-LH/CG [2.209] and r-LH/CG [2.210] have been characterized. The former consists of 696 amino acids with a signal peptide containing 27 amino acids. The seven transmembrane domains are homologous to the G-protein receptors.

FSH stimulates the formation of LH receptors in rat granulosa cells. Estrogen or testosterone are important for the FSH stimulated synthesis of LH receptors [2.211], [2.212]. On the other hand, androgens [2.213] and interleukin-1 (IL-1) [2.214] inhibit the FSH stimulated synthesis of LH receptors.

2.4.4. Biological Effects

In the ovaries, LH stimulates cAMP and inositol triphosphate [2.215] and reduces cGMP by inhibiting guanylate cyclase [2.216]. The activities of LH and FSH complement each other. In the male organism, spermatogenesis can occur only if both hormones are present. Indeed, follitropin is responsible for the development of the sperma, while lutropin stimulates the secretion of testosterone in the interstitial cells, a process necessary for the maturation of the sperms. Although follitropin activates follicular growth in women, ovulation and formation of the corpus luteum do not occur. Indeed, ovulation, corpus luteum formation and estrogen secretion require the simultaneous presence of lutropin.

The renin-angiotensin and the kallikrein-kinin systems and the prostaglandins play a role in the maturation of the gonads. Bradykinin- and angiotensin-II-antagonists inhibit ovulation.

There are structural similarities between FSH and LH and the active sequence of thioredoxin. FSH and LH have an activity that is 60 and 300 times the disulfide isomerase activity of thioredoxin in the reactivation of denatured ribonuclease [2.217].

Thioredoxin	W	C	G	P	C	K
Protein disulfide isomerase	W	C	G	H	C	K
b-LH-β-(89-94)	H	C	G	P	C	R
h-LH-β-(89-94)	Q	C	A	L	C	R
h-FSH-β-(83-88)	H	C	G	K	C	D

2.4.5. Structure-Activity Relationships

The intercystine loops of h-CG-β-(38-57) and LH-β-(38-57) slightly inhibit (IC_{50} = 1.51×10^4 and 2.03×10^5 M resp.) the binding of ^{125}I-h-CG to rat ovarian membranes and stimulate (3.55×10^{-5} M and 2.18×10^{-5}M resp.) the synthesis of testosterone [2.218].

Even the tetrapeptides h-FSH-β-(34-37) and h-FSH-β-(49-52) inhibit the binding of ^{125}I-h-FSH to the receptors. The peptide h-FSH-(33-53) exhibits stronger binding and stimulates the basal synthesis of estradiol [2.219]. The peptides, h-FSH-β-(81-95)-amide and h-FSH-β-(31-45)-amide bind tightly to the FSH receptor, the former being a stimulant of estradiol synthesis in Sertoli cell cultures [2.220].

h-CG-β-(38-57)	C P T M T R V L Q G V L P A L P Q V V C.
h-LH-β-(38-57)	C P T M M R V L Q A V L P P L P Q V V C.
h-FSH-β-(31-45)-NH_2	Y C Y T R D L V Y K N P A R Pa
h-FSH-β-(33-53)-NH_2	Y T R D L V Y K N P A R P K I Q K T C T Fa
h-FSH-β-(34-37)	T R D L
h-FSH-β-(49-52)	K T C T
h-FSH-β-(81-95)-NH_2	Q C H C G K C D S D S T D C Ta
Thioredoxin active site	W C G P C K

Even partial sequences of the α-chain displace h-CG from the receptor to some extent [2.221].

Antibodies against h-FSH-β-(76-118) or h-FSH-β-(1-33) bind h-FSH, LH and TSH [2.222]. A monoclonal antibody directed against h-CG binds in the vicinity of Lys^{45} of the α-chain and Asp^{112} of the β-chain. A peptide in which these two regions are combined produces a specific antibody against h-CG and could be used for the development of a synthetic vaccine [2.223]:

T M L V Q K N V T S H P L T C D D P R F Q

——— h-CG-α-(46-55)——————— ——— h-CG-β-(106-116) ———————

The sugar residues on the peptide chains are of great importance for the biological activity. Although the binding of deglycosylated h-CG, to the receptors is better than that of intact h-CG, it shows no activation of cAMP *in vitro*. In fact, it inhibits the activity of glycosylated h-CG. However, deglycosylated or desialylated h-CG stimulates the synthesis of testosterone *in vivo* (monkeys) [2.224].

2.4.6. Applications [2.225]

Chorionic gonadotropin, h-CG, or h-MG (menopausal gonadotropin, urogonadotropin) and h-FSH (metrodin) are used predominantly in human medicine. The compounds, h-CG (Choragon, Predalon, Pregnesin, Primogonyl) and h-MG (Humegon, Pergonal) are used to treat disturbances of gonadal functions, e.g., primary and secondary amenorrhea, sterility, disturbances of spermatogenesis, delayed puberty, cryptorchidism, oligosperima, and menorrhagia. At present, the main indication is polycystic ovary disease and *in vitro* fertilization [2.225]. A pulsatile therapy with Metrodin [2.226] or Pergonal [2.227] has advantages over daily injections. There is a marked reduction of multiple births in the case of pulsatile application [2.228].

2.5. Thyrotropin [2.229]

2.5.1. Occurrence

Thyrotropin (TSH) is formed in the hypophysis like the structurally related gonadotropins.

2.5.2. Release [2.231]

While the release of TSH is regulated by TRH, high doses of the thyroid hormones, triiodothyronine (T_3) and thyroxine (T_4) inhibit the formation of TSH in the hypophysis by a negative feedback mechanism. In comparison, small doses of T_3 and T_4 stimulate the release of TSH via the activation of TRH. Furthermore, the release of TSH is stimulated by calcitriol (1.25 $(OH)_2D_3$) [2.231], prostaglandins [2.232] and vasopressin [2.233].

Substances that inhibit the release of TSH are neuromedin B [2.234], glucocorticoids [2.235], estrogen, testosterone, β-endorphin [2.236] and somatostatin (e.g., octreotide [2.237].

2.5.3 Receptors

The structures of dog TSH receptor [2.238] and human TSH receptor [2.239]-[2.241] were established via the elucidation of the cDNA. The h-receptor consists of 744 amino acids. There are sequence similarities between these receptors and the LH/h-CG receptors. As far as the function of the thyroid gland is concerned, the TSH receptor does not appear to be adversely affected by repeated administration of TSH. However, repeated doses of TSH elicit a reduced response from adenylate cyclase [2.242].

2.5.4. Biological Effects

TSH stimulates the synthesis and secretion of T_4 and T_3 in the thyroid gland and the formation of thyroglobulin through cAMP [2.243]. TSH enhances the synthesis of DNA and the proliferation of epithelial follicular cells from normal human thyroid tissue [2.244], a process that can be inhibited by somatostatin [2.245]. Lithium salts inhibit the TSH stimulated formation of T_3 and T_4 and produce hypothyreosis in women [2.246].

2.5.5. Structure-Activity Relationships

The sequences 26-46, 31-45 and 21-35 of the α-chain [2.247] and the sequences 101-112, 71-85, 31-45, 41-55, and 1-5 of the h-TSH-β-chain [2.248] inhibit the binding of TSH to human thyroid membrane preparations.

```
                                                                     IC50 (μ M)
h-β-TSH-(101-112)             K T N Y C T K P Q K S Y                     80
     -(71-85)                 V A P Y F S Y P V A L S C K C              104
     -(31-45)       C M T R D I N G K L F L P K Y                        186
     -(41-55)                           F L P K Y A L S Q D V C T Y R    242
     -(1-15)                          F C I P T E Y T M H I E R R E      331
```

2.6. Protein 7B2

A protein (M_r 21000) that was isolated [2.249] from the adenohypophysis of the pig has now been structurally characterized in humans [2.250] and *Xenopus* [2.251] through the cDNA sequence. It has been found that h-7B2 consists of 185 amino acids and has the greatest structural similarity to proinsulin from the chick and the duck.

Similar to FSH, 7B2 is released from the hypophysis by GnRH [2.252] and inhibited by inhibin and testosterone [2.253]. As in the case of the gonadotropic hormones, the level of 7B2 is higher in women after menopause [2.254].

2.7. Inhibins, Activins, and Related Hormones

2.7.1. Common Structural Characteristics

The inhibins (IHB), activins (ATV) and the transforming growth factor (TGF-β) consist of two protein chains which are linked together by disulfide bridges. The inhibins are heterodimers: they consist of an α-chain and a β-chain. The α-chains of the two known inhibins, IHB-A and IHB-B, are identical and there are structural similarities between the α- and β-chains. On the other hand, the activins and TGF-β are the corresponding compounds made of two chains of the β type. The Muellerian inhibiting substance (MIS), bone morphogenetic proteins (BMP-2A, BMP-2B, and BMP-3 and osteogenic protein 1 (OP-1) are high molecular glycoproteins that consist of about 400-500 amino acids. The C-terminal ends of these proteins are very similar to TGF-β and the IHB β-chains. Proteins with homologous structures have also been found in *Drosophila* [2.255] (*Drosophila* decapentaplegic gene = DPP), *Xenopus* [2.256] (Vg-1 gene) and in the mouse [2.257] (Vgr-1 = related to Vg-1), see Table 6.

2.7.2. Inhibins [2.258], [2.259]

Occurrence. Inhibin (IHB) activity is found, above all, in the gonads. Other sites of synthesis are the placenta (in humans) [2.260] and the brain [2.261]. The structure of IHB-A from humans, pigs, cows, and rats and that of IHB-B from humans, pigs, and rats have been elucidated. The α-chain and the β-chain are formed from various mRNAs in the form of prepropeptides.

Release. Inhibin is released primarily by exogenous or endogenous FSH, presumably via an increase in cAMP [2.262]. *In vitro*, h-CG, LH, testosterone [2.263], [2.264] somatomedin C [2.265] and, in the case of hypophysectomized female rats, estradiol stimulate the release of IHB. In comparison, GnRH agonists inhibit both basal as well as PMSG stimulated IHB, while a GnRH antagonist potentiates the stimulating effect of PMSG on the release of IHB [2.266].

The level of both estradiol and IHB is low during the early and mid follicular phase and the FSH level is slightly raised. In the late follicular phase, IHB increases parallel to estradiol, while the levels of FSH and LH decrease and increase respectively. It is evident that both hormones are formed by the maturing follicle. IHB reaches a first maximum when LH and FSH also exhibit a maximum level [2.267]. The activity of IHB is reduced in the early luteal phase of the menstrual cycle (parallels that of estrogen [2.267]) and in the period before menopause [2.268]. In the mid luteal phase, IHB increases parallel to progesterone, while the plasma levels of FSH and LH decrease. For this reason, it is assumed here that IHB and progesterone are formed by the corpus luteum [2.267].

Table 6. Structural comparison of the various inhibins and TGF-β with the C-terminal ends of MIS, BMP-2A, BMP-3 and OP-1

Peptide	Amino acid sequence
	1 … 18
h-IHB-α	S T / P L M S W P W S P S A L R L L Q-
p-IHB-α	S T A P L / P W P W S P A A L R L L Q-
b-IHB-α	S T P P L / P W P W S P A A L R L L Q-
	1 … 15 … 19
h-TGF-β	A L D T N Y C F S S T E / K / N C C V R Q-
h-TGF-β_2	A L D A A Y C F R N V Q / D / N C C L R P-
h-, p-TGF-β_3	A L D T N Y C F R N L E / E / N C C V R P-
g-TGF-β_4	D L D T D Y C F G P G T D E K N C C V R P-
xe-TGF-β_5	G V G Q E Y C F G N N G / P / N C C V K P-
h-MIS-(453-535)	C A L R E-
h-BMP-2A-(296-396)	C K R H P-
h-BMP-2B-(308-408)	C R R H S-
h-BMP-3-(370-472)	C A R R Y-
h-OP-1-(330-431)	C K K H E-
	1 … 11 … 15
h, b, p-IHB-βA	G L E / / / C D G K V N / I / / C C K K Q-
h, p-IHB-βB	G L E / / / C D G R T N / L / / C C R Q Q-
	19 … 34
h-IHB-α	R P P / / / E E P A A H A N / / C H R V A-
p-IHB-α	R P P / / / E E P A V H A D / / C H R A S-
b-IHB-α	R P P / / / E E P A A H A D / / C H R A A-
	20 … 28 … 39
h-TGF-β	L Y I D F R K D L G W / K W I H E P K G Y-
h-TGF-β_2	L Y I D F K R D L G W / K W I H E P K G Y-
h-, p-TGF-β_3	L Y I D F R Q D L G W / K W V H E P K G Y-
g-TGF-β_4	L Y I D F R K D L Q W / K W I H E P K G Y-
xe-TGF-β_5	L Y I N F R K D L G W / K W I H E P K G Y-
h-MIS	L S V D L R A E / / / / R S V L I P E T Y
h-BMP-2A	L Y V D F / S D V G W N D W I V A P P G Y-
h-BMP-2B	L Y V D F / S D V G W N D W I V A P P G Y-
h-BMP-3	L K V D F / A D I G W S E W I I S P K S F-
h-OP-1	L Y V S F / R D L G W Q D W I I A P E G Y-
	16 … 35
h, b, p-IHB-βA	F F V S F / K D I G W N D W I I A P S G Y
h-IHB-βB	F F I D F / R L I G W N D W I I A P T G Y-
p-IHB-βB	F F I D F / R L I G W S D W I I A P T G Y-
	35 … 54
h-IHB-α	L N I S F / Q E L G W E R W I V Y P P S F-
b, p-IHB-α	L N I S F / Q E L G W D R W I V H P P S F-

Peptide	Amino acid sequence
	40 51 58
h-TGF-β	H A N F C L G P C P / Y I W S L D T / / / / Q Y-
h-TGF-β_2	N A N F C A G A C P / Y L W S S D T / / / / Q H-
h, p-TGF-β_3	Y A N F C S G P C P / Y L R S A D T / / / / T H-
g-TGF-β_4	M A N F C M G P C P / Y I W S A D T / / / / Q Y-
xe-TGF-β_5	E A N Y C L G N C P / Y I W S M D T / / / / Q Y-
h-MIS	Q A N N C Q G V C G / W P Q S D R N P R Y G N H-
h-BMP-2A	H A F Y C H G E C P / F P L A D H L / / / / N S-
h-BMP-2B	Q A F Y C H G D C P / F P L A D H L / / / / N S-
h-BMP-3	D A Y Y C S G A C Q / F P M P K S L / / / / K P-
h-OP-1	A A Y Y C E G E C A / F P L N S Y M / / / / N A-
	36 48 59
h, b, p-IHB-βA	H A N Y C E G E C P S H I A G T S G S S L S F H-
h, p-IHB-βB	Y G N Y C E G S C P A Y L A G V P G S A S S F H-
	55 78
h-IHB-α	I F H Y C H G G C G L H I P P N L S L P V P G A-
p-IHB-α	I F H Y C H G G C G L P T L P N L P L S V P G A-
b-IHB-α	I F Y Y C H G G C G L S P P Q D L P L P V P G V-
	59 80
h-TGF-β	S K / / / V L A L Y N Q H N / P G A S A A P C C V P-
h-TGF-β_2	S R / / / V L S L Y N T I N / P E A S A S P C C V S-
h-TGF-β_3	S T / / / V L G L Y N T L N / P E A S A S P C C V P-
p-TGF-β_3	S S / / / V L G L Y N T L N / P E A S A S P C C V P-
g-TGF-β_4	T K / / / V L A L Y N Q H N / P G A S A A P C C V P-
xe-TGF-β_5	S K / / / V L S L Y N Q N N / P G A S I S P C C V P-
h-MIS	/ / / / V V L L L K M Q A R G A A L A R P P C C V P-
h-BMP-2A	T N H A I V Q T L V N S V N / / S K I P K A C C V P-
h-BMP-2B	T N H A I V Q T L V N S V N / / S S I P K A C C V P-
h-BMP-3	S N H A T I Q S I V R A V G V V P G I P E P C C V P-
h-OP-1	T N H A I V Q T L V H F I N / P E T V P K P C C A P-
	60 83
h, b, p-IHB-βA	S T V I N H Y R M R G H S P F / / A N L K S C C V P
h, p-IHB-βB	T A V V N Q Y R M R G L N P / / / G T V N S C C I P-
	79 98
h-IHB-α	P P T P A Q P Y S L L P G A Q / / / / / / P C C A A-
p-IHB-α	P P T P V Q P L L L V P G A Q / / / / / / P C C A A-
b-IHB-α	P P T P V Q P L S L V P G A Q / / / / / / P C C A A-
	81 87 98 100
h-TGF-β	Q A L E P L P I V Y Y / / V / G / R K P K V E Q-
h-TGF-β_2	Q D L E P L T I L Y Y / / I / G / K T P K I E Q-
h-TGF-β_3	Q D L E P L T I L Y Y / / V / G / R T P K V E Q-
p-TGF-β_3	Q D L E P L T I L Y Y / / V / G / R T A K V E Q-
g-TGF-β_4	Q T L D P L P I I Y Y / / V / G / R N V R V E Q-
xe-TGF-β_5	D V L E P L P I I Y Y / / V / G / R T A K V E Q-
h-MIS	T A / Y A G K L L I S / / L S E E R I S / A H H-
h-BMP-2A	T E L S A I S M L Y L / / D / E / N E K V V L K-

Peptide	Amino acid sequence
h-BMP-2B	T E L S A I S M L Y L / / D / E / Y D K V V L K-
h-BMP-3	E K M S S L S I L F F / / D / E / N K N V V L K-
m-Vgr-1	T K L N A I S V L Y F / / D / D / N S N V I L K-
h-OP-1	T Q L N A I S V L Y F / / D / D / S S N V I L K-
	84 104
h, b, p-IHB-βA	T K L R P M S M L Y Y / / D D G / Q N I I K K D-
h, p-IHB-βB	T K L S T M S M L Y F / / D D E / Y N I V K R D-
	99 122
h-IHB-α	L P G T M R P L H V R T T S D G G Y S F K Y E T-
p-IHB-α	L P G T M R S L R V R T T S D G G Y S F K Y E T-
b-IHB-α	L P G T M R P L H V R T T S D G G Y S F K Y E M-
	101 112
h-TGF-β	L S N / M I V R S C K C S.
h-TGF-β_2	L S N / M I V K S C K C S.
h,p-TGF-β_3	L S N / M V V K S C K C S.
g-TGF-β_4	L S N / M V V R A C K C S.
xe-TGF-β_5	L S N / M V V R S C N C S.
h-MIS	V P N / M V A T E C G C R
h-BMP-2A	N Y Q D M V V E G C G C R
h-BMP-2B	N Y Q E M V V E G C G C R
h-BMP-3	V Y P N M T V E S C A C R
h-OP-1	K Y R N M V V R A C G C H
	105 116
h, b, p-IHB-βA	I Q N / M I V E E C G C S
h, p-IHB-βB	V P N / M I V E E C G C A
	123 134
h, b, p-IHB-α	V P N / L L T Q H C A C I

The plasma IHB level in men suffering from idiopathic hypogonadotropic hypogonadism is lowered [2.269]. The level of IHB falls by 70% during treatment with h-CG, which is associated with a complete loss of endogenous FSH [2.270].

Female patients suffering from ovarian granulosa cell tumours [2.271], polycystic ovarian diseases [2.268] and duodenal ulcers [2.272] have high serum inhibin levels.

Biological Effects. The inhibins A and B have the same FSH inhibiting activity. Indeed, IHB specifically inhibits the release of FSH from the hypophysis via a reduction of the FSH-β-mRNA [2.273], without negatively affecting the secretion of lutropin (LH). IHB suppresses the GnRH induced formation of GnRH receptors in cell cultures of rat hypophysis [2.274].

Inhibin also exerts an effect on the gonads. *In vitro*, it inhibits both the formation of progesterone [2.275] and the aromatase activity [2.276].

2.7. 3. Activins [2.258]

Occurrence. Activins are found in the follicular fluid from ovaries and in leukemic cells. Activin A is a homodimer of the β-chain of IHB-A [2.277], [2.278] and activin A-B is a heterodimer consisting of the β-chains of IHB-A and IHB-B [2.279].

Follistatin is an activin binding protein that has been isolated from rat ovaries [2.280].

Biological Effects. In comparison with the inhibins, the activins stimulate the release of FSH. Unlike GnRH, which releases FSH and LH very rapidly, the activins require more than four hours to bring about the secretion of FSH. In hypophyseal cells that have been desensitized by high doses of GnRH, activin is capable of increasing the secretion of FSH by double the amount [2.281]. Activins also potentiate the FSH induced aromatase activity [2.277], [2.279], but stop the synthesis of progesterone [2.282].

Activin that is produced by leukemic cells also acts like erythropoietin (stimulates blood formation) [2.283]. It increases the colony formation and the erythropoietin activity, while inhibin exerts an inhibiting effect [2.284].

2.7.4. Transforming Growth Factor-β [2.283], [2.285]

Occurrence. Transforming growth factor (TGF-β) consists of two identical peptide chains which are linked together by disulfide bridges [2.286]. TGF-β is stored in large amounts in platelets in the form of a high molecular, inactive complex [2.287]. This substance appears to occur widely; indeed, TGF-β-mRNA is found not only in the liver cells of adult animals, but also in all the cell lines that have been studied until now [2.286]. Even blood serum contains large amounts of TGF-β. Intact, biologically active TGF-β is excreted in the urine [2.288].

Apart from h-TGF-β, h-TGF-β_2 [2.289], h- and p-TGF-β_3 [2.290], g-TGF-β_4 and xe-TGF-β_5 [2.291] have been characterized. TGF-β and TGF-β_2 seem to be identical to CIF-A and CIF-B (cofactors for inducing cartilage formation) isolated from the bones of the cow. TGF-$\beta_{1.2}$ is a heterodimer made of the respective chains of TGF-β and TGF-β_2 [2.283].

Release. TGF-β is activated and released from the inactive complex in platelets by thrombin [2.292], acids or urea, i.e., by blood clotting [2.287]. Heparin cleaves the complex between TGF-β and α_2-macroglobulin and, thus, potentiates the effect of TGF-β [2.293]. Bone degrading factors, such as parathyroid hormone (PTH), calcitriol and interleukin-1, release TGF-β activity from the calvaria of rat fetuses and of new born rats. On the other hand, calcitonin reduces the activity of TGF-β [2.294]. Activated osteoclasts release TGF-β from the TGF-β complex formed by bone organ cultures [2.295].

Receptors. TGF-β binds to membrane receptors with high activity. Receptors for TGF-β have been found in practically all cells that have been studied. These receptors appear to be insensitive to a down regulation. Three different receptors exhibiting varying binding characteristics have been observed [2.283]. In osteoblasts, the binding of TGF-β to the receptors is reduced by PTH [2.296].

Biological Effects. On incubation for a longer period of time, h-TGF-β stimulates both the secretion of FSH from rat anterior pituitary cells [2.297] and the aromatase activity in ovarian granaulosa cells [2.276], [2.282], [2.298]. These effects are similar to those produced by the activins. TGF-α inhibits these effects of TGF-β [2.298]. TGF-β has a modulating influence on LH receptors: it potentiates the enhancing effect exerted by FSH on the LH receptors, but inhibits this effect after treatment with insulin [2.299].

TGF-β (10^{-13}-10^{-11} M) inhibits basal and ACTH stimulated steroidogenesis in adrenal cells through the inhibition of cholesterol synthesis [2.300]. The effect is presumably based on the inhibition of LDL metabolism. TGF-β counteracts the binding of LDL to the cell surface [2.301].

In rat pancreas cells, the glucose dependent stimulation of the release of insulin is accomplished to the same extent by TGF-β_1 and TGF-β_2 [2.302].

TGF-β stimulates the expression of endothelin-mRNA in vascular endothelial cells [2.303].

TGF-β is a general regulator of cell growth. Both TGF-β and TGF-α stimulate the growth of mesenchymal cell cultures, independent of anchoring, via the induction of platelet derived growth factor (PDGF) [2.304]. However, the growth of epithelial cells is not stimulated, but inhibited by TGF-β [2.305]. For instance, TGF-β inhibits the growth of many different human cancer cell cultures, independent of anchoring [2.286], [2.306], [2.307]. While TGF-β potentiates PDGF stimulated growth of certain fibroblasts, the same concentrations of TGF-β inhibit the growth of the same cells that is stimulated by the epidermal growth factor (EGF) [2.307]. Endothelial cell growth stimulated by the acidic or basic fibroblast growth factor (FGF) [2.308], [2.309] and the IL-2 activated growth of T cells are also noncompetitively inhibited by TGF-β. TGF-β can exert an inhibitory or a stimulating effect on cell differentiation, depending on the type of cell involved. For instance, it inhibits the differentiation of B lymphocytes and muscle myoblasts [2.310], but stimulates the conversion of 3T3 fibroblasts to adipocytes. The inhibition of EGF or insulin stimulated growth by TGF-β is not the result of the blocking of the receptors for the growth factor [2.311]. Rather, it is due to the reduction of the high affinity of EFG receptors, the number of receptors remaining unchanged [2.312].

The differentiation and proliferation of mesenchymal and epithelial cells are enhanced by TGF-β and these cells exhibit an increased production of fibronectin, various types of collagen and other cell adhesive proteins [2.313], [2.314].

Ascorbic acid potentiates the TGF-β stimulated biosynthesis of collagen types I and III [2.315]. The formation of the corresponding receptors for these cell binding proteins are also enhanced by TGF-β [2.283]. In quiescent MRC-5 human fibroblasts, TGF-β inhibits the collagenase stimulated by other growth factors and stimulates the formation of the specific tissue inhibitor of metalloproteinases (TIMP) [2.316]. In smooth muscle cells from the p-aorta, TGF-β stimulates the production of elastin [2.317]. In the extracellular matrix, TGF-β greatly increases the expression of various forms of chondroitin/dermatan sulfate proteoglycan [2.318] and inhbits the IL-1 induced protease activity and proteoglycan degradation [2.319]. Indeed, TGF-β also plays the part of a regulator in the immune system. It inhibits the basal and the TNF-α

(tumor necrosis factor α) stimulated adhesion of neutrophil blood cells to endothelial cells [2.320], inactivates macrophages [2.321], inhibits *in vitro* the IL-1 stimulated degradation of cartilage [2.322], prevents TNF-α induced cardiac injury [2.323], and increases the survival time of transplanted Langerhans' islets in rats [2.324].

TGF-β appears to be of importance for early bone development and for bone repair mechanisms. In rat fetal calvaria, it increases the synthesis of DNA and collagen [2.325], stimulates the formation of osteopontin in osteosarcoma cells [2.326], and inhibits the production of osteoclast-like cells in cell cultures [2.327]. The effect of TGF-β is inhibited by PTH in osteoblasts [2.296].

In a human erythroleukemia cell line, the synthesis of hemoglobin is stimulated by TGF-β [2.328].

Applications. TGF-β stimulates collagen matrix contraction by fibroblasts, a process that should accelerate the healing of wounds [2.329].

2.7.5. Bone Morphogenetic Proteins and Osteogenic Proteins

Occurrence. M. R. Urist made the observation that decalcified bone matrix stimulates the formation of new bone [2.330]. The substance or the mixture of substances that was isolated from decalcified bones was called osteogenin [2.331], osteopoetin [2.332] or bone morphogenetic proteins (BMP) [2.333]. It has been observed that h-BMP-2A, h-BMP-2B, h-BMP-3, b-osteogenin [2.334] and human osteogenic protein (h-OP-1) [2.335] belong structurally to the TGF-β/IHB family.

The proteins h-BMP-2A, h-BMP-2B, and h-BMP-3 have been found, via the cDNAs, to contain 396, 408 and 427 amino acids respectively [2.333]. The structures of h-BMP-2A and h-BMP-2B are very similar. There is also great similarity between the C-terminal ends of all three glycoproteins and those of the TGF-β/IHB family. Fragments of b-osteogenin that have been sequenced have almost the same amino acid sequence as BMP-3 [2.334]. The structure of the glycoprotein, h-OP-1 was elucidated via its cDNA and found to contain 431 amino acids and a signal sequence of 29 amino acids [2.335]. It corresponds to the proteins m-Vgr-1 and xe-Vg-1 (vegetal protein).

Release. Vitamin D stimulates the formation of the BMPs [2.336]. The production of BMPs is reduced with age [2.337]. Extracts of rat chondrasarcoma and osteosarcoma show no BMP activity. In fact, these extracts have an inhibitory effect on the bone inducing activity of the BMPs [2.338].

Biological Effects. The BMPs initiate the formation of cartilage and result in the production of new bone. Recombinant h-BMP-2A [2.339] and isolated b-osteogenin [2.334] induce the formation of cartilage and bone in rats, which cannot be distinguished histologically from the cartilage and bone formation stimulated by bone extracts.

The activity of demineralized bone matrix is reduced in rats deficient in vitamin D [2.340]. Aluminium salts [2.341] and vitamin A [2.342] have a toxic effect on the chondrogenesis and osteogenesis induced by demineralized bone matrix. In older rats, PDGF potentiates the cartilage and bone formation induced by demineralized bone matrix [2.343].

Applications. Gelatin that contains osteogenin corrects diaphyseal defects in sheep [2.344]. Bone defects in patients with cholesteatomas could be rectified with implants of osteopoetin [2.332].

2.7.6. Muellerian Inhibiting Substance

Occurrence. The glycoprotein, Muellerian inhibiting substance (MIS) is found in large amounts in the Sertoli cells of the testicles of new-born calfs. The structures of h-and b-proMIS have been elucidated via the cDNA sequences [2.345]. h-MIS consists of 535 amino acids and the C-terminal sequence shows similarities to the TGF-β/IHB family.

Biological Effects. MIS inhibits the development of the Muellerian duct during gonadal differentiation in the male embryo. The Muellerian duct develops into Fallopian tubes, uterus, and vagina in the female embryo. Complete virilization is not achieved in the absence of MIS. Preparations of b-MIS inhibit the *in vitro* and *in vivo* growth of tumor cells derived from the ovaries and the endometrium [2.345].

The growth-inhibiting effect of MIS on A-431 cells proceeds via the inhibition of the autophosphorylation of the EGF receptor. The growth-inhibiting effect of MIS on the Muellerian duct can be reversed *in vitro* by EGF [2.346].

Structure-Activity Relationships. A 25 kD C-terminal fragment that is similar to TGF-β, forms a complex with the N-terminal part of MIS. This complex can be made by limited digestion with pepsin and has full biological activity [2.347].

2.8. Follistatin [2.348]

2.8.1. Occurrence

Follistatins are single-chain glycoproteins that were isolated from follicular fluid. Two human follistatin precursors were found via cDNA clones. These precursors contain 344 and 317 amino acids and give rise to two follistatins, each consisting of 315 or 288 amino acids [2.349].

Follistatin is organized in three homologous domains: follistatin-(66-135), follistatin-(139-210), and follistatin-(216-287). The domains exhibit among themselves approximately 52% sequence similarly and have a 51% similarity to the human pancreatic secretory inhibitor (h-PSTI). Follistatin-(309-314) (ISSILE) is identical to a tyrosine kinase domain from the EGF receptor.

	226																									250
h-Follistatin	K	C	L	W	D	F	K	V	G	R	G	R	C	S	L	C	/	D	E	L	C	P	D	S	K	S-
h-PSTI								D	S	L	G	R	E	A	K	C	Y	N	E	L	N	G	C	T	K	I-
	251																									275
h-Follistatin	D	E	P	V	C	A	S	D	N	A	T	Y	A	S	E	C	/	A	M	K	E	A	A	C	S	S-
h-PSTI	Y	D	P	V	C	G	T	D	G	N	T	Y	P	N	E	C	V	L	C	F	E	N	R	K	R	Q-
	276												288												300	
h-Follistatin	G	V	L	L	E	V	K	H	S	G	S	C	N	S	I	S	E	D	T	E	E	E	E	E	D-	
h-PSTI	T	S	I	L	I	Q	K	/	S	G	P	C....														
	301														315											
h-Follistatin	E	D	Q	D	Y	S	F	P	I	S	S	I	L	E	W											

2.8.2. Biological Effects

In hypophyseal cell cultures, both follistatins inhibit, additive to the inhibins and equipotent (1.8 $\times$ 10^{-10} M) with approximately one third of the inhibin activity, the release of FSH. Like the inhibins, the follistatins also inhibit the FSH stimulated production of estrogen. Follistatin is a binding protein for the activins [2.350].

3. Parathyroid Hormone and the Calcitonin Family [3.1]

Parathyroid hormone (PTH) and calcitonin (CT) are both peptides that are secreted from the parathyroid and the thyroid glands, respectively. They regulate the plasma calcium level and bone formation. The level of calcium is increased by PTH and decreased by CT. PTH stimulates the turnover of bone and CT stops it. An important role has also 1α,25-dihydroxyvitamin D_3 (calcitriol, 1,25$(OH)_2D_3$), which is stimulated in the kidneys in response to PTH. Similar to high serum calcium levels, this substance has an inhibitory effect on PTH and stimulates the synthesis of calcitonin. While PTH stimulates osteoblasts to produce new bone and also activates osteoclasts causing increased bone resorption, CT inhibits the mobility and activity of osteoclasts.

Apart from these hormones, the reproductive hormones described in Chapter 2 also strongly affect bone formation. Indeed, TGF-β not only stimulates the release of FSH, like the structurally related activins, but also initiates chondrogenesis and osteogenesis [3.2]. The lack of steroidal sexual hormones (estrogen, testosterone and progesterone), which are, in turn, important for the release of gonadoliberin, leads to osteoporosis. The chaotic release pattern of PTH appears to be dependent on estrogen levels in women [3.3], and very low doses of gonadoliberin agonists have been found to exert a regulating effect on the PTH level in female rats [3.4]. A peptide related to PTH, PTHrP, which is formed primarily in tumors, has properties similar to those of PTH.

In addition to CT, a CT-gene related peptide (CGRP) has been found. CT and CGRP are coded on common genes. There are at least two genes: α-gene and β-gene. While α-CGRP and β-CGRP are formed by these two genes respectively, the mRNA for CT appears to be derived only from the α-gene. CGRP and CT are produced in a large prohormone form by the appropriate splicing of the α-gene product. The alternative production of CGRP or CT is subject to tissue-specific regulation. The mRNA for CT is found almost exclusively in the thyroid gland, whereas the mRNA for CGRP is formed primarily in the nervous system [3.5]. A peptide very similar to CGRP from the β cells of the pancreas is amylin.

3.1. Parathyroid Hormone and PTH-Related Protein [3.6]-[3.8]

3.1.1. Occurrence

Parathyroid hormone (PTH, also called parathyrin according to IUPAC) is formed in the parathyroid gland. It is also found in the brain and in the hypophysis [3.9].

	1																				20
h-α-CGRP	A	C	/	D	T	A	T	C	V	T	H	R	L	A	G	L	L	S	R	S	G-
h-β-CGRP	A	C	/	N	T	A	T	C	V	T	H	R	L	A	G	L	L	S	R	S	G-
h-Amylin	K	C	/	N	T	A	T	C	A	T	Q	R	L	A	N	F	L	V	H	S	S-
h-CT		C	G	N	L	S	T	C	M	L	G	T	Y	T	Q	D	F	N	K	/	/-

	21																37
h-α-CGRP	G	V	V	K	N	N	F	V	P	T	N	V	G	S	K	A	Fa
h-β-CGRP	G	M	V	K	S	N	F	V	P	T	N	V	G	S	K	A	Fa
h-Amylin	N	N	F	G	A	I	L	S	S	T	N	V	G	S	N	T	Ya
h-CT	/	/	F	H	/	T	F	P	Q	T	A	I	G	V	G	A	Pa

A peptide related to PTH (PTHrP) is formed in tumors that are accompanied by hypercalcemia [3.10], in lectin stimulated lymphocytes [3.11], in the T cell line, MT-2 that is infected with HTLV-I [3.12], and in the lactating mammary glands of rats [3.13]. A mRNA corresponding to the h-PTHrP, that is shortened by two amino acids at the C-terminal, was found in the kidney 786-0 cell line [3.14].

The hormones h-, b-, p- and r-PTH consist of 84 amino acids. With 88 amino acids, g-PTH differs from these peptides, which show only slight structural deviations [3.15], see Table 7. The precursor of PTH is preproPTH, a peptide containing 31 additional amino acids at the N-terminal [3.10].

3.1.2. Release of PTH

Low plasma calcium raises the level of PTH, while high plasma calcium, an infusion of calcium, or a calcium diet [3.16] lowers the level of PTH. The circadian rhythm proceeds as follows [3.17]: a calcium peak is observed 2 hours after a PTH peak, a PTH trough is formed 4 hours later followed by a calcium trough 2 hours later, another PTH peak occurs after 4 hours and so on. The secretion of PTH is also stimulated by insulin induced hypoglycemia [3.18], a phosphate rich diet [3.19], [3.20], calcium antagonists [3.21], a NaCl rich diet [3.22], a diet low in vitamin D_3 and calcium [3.23] lithium [3.24], [3.25], stress (induced hypocalcemia) [3.26], β-adrenergic stimulators [3.27], noradrenaline [3.28], histamine, 17-β-estradiol and progesterone (*in vitro*) [3.29], [3.30], dexamethasone [3.31], or citrate infusion [3.32]. In osteoporotic women, the frequency and amplitude of the chaotic pulsatile secretion of PTH are greatly reduced after menopause [3.33]. It has been found that estrogen therapy partly regenerates this pulsatile secretion [3.33].

Apart from calcium, 1,25-dihydroxyvitamin D_3 (calcitriol), which is synthesized in the kidneys in response to stimulation by PTH, is also a physiological inhibitor of the synthesis and release of PTH [3.34]-[3.36].

Magnesium slightly inhibits the release of PTH [3.37]. Dietary Mg correlates positively with the bone mineral density. This may be one of the reasons why vegetarians are less likely to get osteoporosis [3.38]. The release of PTH is also inhibited by calcium canal activators [3.21], cholinergic agonists [3.39], H_2-receptor blockers [3.40], and β-receptor blockers [3.41].

Table 7. PTH and PTHrP of various species

Peptide	Amino acid sequence
	1 … 21
h-PTH	S V S E I Q L M H N L G K H L N S M E R V-
b-PTH	A V S E I Q F M H N L G K H L S S M E R V-
p-PTH	A V S E I Q L M H N L G K H L S S L E R V-
r-PTH	A V S E I Q L M H N L G K H L A S V E R M-
g-PTH	S V S E M Q L M H N L G E H R H T V E R Q-
h-PTHrP	A V S E H Q L L H D K G E S I Q D L R R R-
r-PTHrP	A V S E H Q L L H D K G E S I Q D L R R R-
	22 … 34 … 48
h-PTH	E W L R K K L Q D V H N F V A L G A P L A P R D A G S-
b-PTH	E W L R K K L Q D V H N F V A L G A S I A Y R D G S S-
p-PTH	E W L R K K L Q D V H N F V A L G A S I V H R D G S S-
r-PTH	Q W L R K K L Q D V H N F V S L G V Q M A A R E G S Y-
g-PTH	D W L Q M K L Q D V H S A L E / / / / / / / / / D A R T-
h-PTHrP	F F L H H L I A E I H T A E I R A T S E V S P / N S K-
r-PTHrP	F F L H H L I A E I H T A E I R A T S E V S P / N S K-
	49 … 70
h-PTH	Q R P R K K E D N V L V E S H E K S L G E A / / / / /-
b-PTH	Q R P R K K E D N V L V E S H Q K S L G E A / / / / /-
p-PTH	Q R P R K K E D N V L V E S H Q K S L G E A / / / / /-
r-PTH	Q R P T K K E E N V L V D G N S K S L G E G / / / / /-
g-PTH	Q R P R N K E D I V L G E I R N R R L L P E H L R A A-
h-PTHrP	P S P N T K N H P V R F G G D D E G R Y L T / / / / /-
r-PTHrP	P A P N T K N H P V R F G G D D E G R Y L T / / / / /-
	71 … 84
h-PTH	/ / / / / / / / D K A D V N V L T K A K S Q.
b-PTH	/ / / / / / / / D K A D V D V L I K A K P Q.
p-PTH	/ / / / / / / / D K A A V D V L I K A K P Q.
r-PTH	/ / / / / / / / D K A D V D V L V K A K S Q.
g-PTH	V Q K K S I D L D K A Y M N V L F K T K P.
h-PTHrP	/ / / / / / / / Q E T N K V E T Y K E Q P L K T P G K-
r-PTHrP	/ / / / / / / / Q E T N K V E T Y K E Q P L K T P G K-
	89 … 115
h-PTHrP	K K K G K P G K R K E Q E K K K R R T R S A W L D S G-
r-PTHrP	K K K G K P G K R K E Q E K K K R R T R S A W P G T T-
	116 … 141
h-PTHrP	V T G S G L E G D H L S D T S T T S L E L D S R R H.
r-PTHrp	G S G L L E D P Q P H T S P T S T S L E P S S R T H.

The occurrence of hyperparathyroidism increases with increasing age and this condition is most often encountered in elderly persons over 60 years of age [3.42]. A 60% restriction of food intake results in the inhibition of this age dependent hyperparathyroidism and bone atropy in male rats. Furthermore, in contrast to a casein protein diet, a soybean protein diet was found to have an inhibitory effect on the increase of PTH [3.43]. Raised levels of PTH are also observed in connection with nephrocalcinosis, osteitis fibrosa cystica, chronic kidney failure [3.44], liver diseases [3.45], [3.47] liver cirrhosis [3.47], postnecrotic or alcoholic cirrhosis [3.48], and primary cirrhosis of the gallblader [3.49] myotonic dystrophy [3.50], and obesity [3.51]. Raised plasma levels of PTH were found in about half of the patients suffering from coronary arterial diseases. The plasma PTH level was the highest in patients with myocardial infarction [3.52], [3.53].

In patients suffering from rheumatoid arthritis [3.11] and osteoporosis [3.54], the PTH release to a hypocalcemic signal is reduced.

The release of PTHrP is reduced by glucocorticoids, calcitriol [3.55] and by octreotide [3.56].

3.1.3. Receptors of PTH

The receptor density is reduced by the continuous action of PTH agonists, while PTH antagonists have no effect [3.57]. PTH receptor complexes are internalized by bone cells; this explains the reduced receptor density on the cell surface after treatment with PTH [3.58].

3.1.4. Activity of PTH

The physiological role of PTH is to maintain the extracellular concentration of calcium (1.1-1.3 mM) and to prevent hypo-calcemia. Furthermore, PTH is the principal regulator of bone turnover.

In the cell, PTH stimulates the formation of cAMP and the phosphorylation of intracellular proteins and increases intracellular calcium.

Dexamethasone potentiates the PTH induced formation of cAMP in osteosarcoma cells [3.59]. On the other hand, calcitriol [3.60], 24,25$(OH)_2D_3$ [3.61], 17-β-estradiol [3.62], insulin [3.63] and TNF-α inhibit the PTH stimulated formation of cAMP.

The hormone PTH has a biphasic effect on the bone [3.65]. Low doses of PTH have an osteogenetic effect and high doses result in osteoclasia. Hyperparathyroidism causes osteoporosis, while hypoparathyroidism increases bone formation (ossification of the paravertebral ligaments [3.66].

In cells similar to osteoblasts, PTH stimulates not only DNA synthesis [3.67], but also alkaline phosphatase (low dosed), the Na^+/K^+ pump (K^+ uptake of the cell is increased) [3.68], the incorporation of amino acids into the bone [3.69], the release of mediators (IL-6 [3.70]), osteoclasts to increase bone resorption [3.71]-[3.73], ornithine decarboxylase [3.74]), and the synthesis of osteocalcin mRNA [3.75]. It potentiates the IGF-I stimulated incorporation of thymidine [3.76]. ANF inhibits the PTH stimulated incorporation of [^{3}H] thymidine [3.77] and inhibits the formation of osteopontin, osteopontin mRNA [3.78] as well as (in high doses) alkaline phosphatase. In osteoclasts, PTH stimulates carbonic anhydrase [3.79] and acid phosphatase [3.80], [3.81].

In vivo, PTH also potentiates the IGF-I induced formation of bone [3.76], [3.82]. Although PTH increases the activity of TGF-β, it inhibits the TGF-β stimulated synthesis of DNA, collagen production, decrease in alkaline phosphatase activity and the binding of TGF-β receptors to osteoblasts [3.83]. The PTH stimulated synthesis of PGE_2 and 6-keto-$PGF_1\alpha$ in bone cells is not directly coupled with the effect of PTH [3.84], [3.85]

In the kidney, PTH stimulates the hydroxylation of 25(OH)D_3 to calcitriol. On the other hand, high plasma calcium levels lower the formation of calcitriol [3.85]. Hence, low calcitriol levels are observed in patients who suffer from hyperparathyroidism and have an extremely high calcium level [3.86]. However, the activity of renal 25(OH)-D_3-24-hyroxylase is inhibited by PTH [3.87]. The active enteral resorption of calcium is stimulated by calcitriol. Also b-, h-PTH, and h-PTHrP increase calcitriol receptors in cultivated osteogenic sarcoma cells [3.88]. In the kidneys, physiological doses of PTH separate the calcium phosphate released from the bone. While calcium is returned to the blood, phosphate is excreted in the urine as a function of the blood phosphate levels. In the case of a longer infusion of PTH (more than 12 days), magnesium excretion is increased and in addition, the excretion of calcium is enhanced so that a negative balance for both calcium and magnesium is obtained [3.89]. PTH also increases the concentration of phosphate in the saliva [3.90].

The ability of PTH to lower blood pressure and cause vasodilatation is also interesting [3.91]. Atropin, cyclooxygenase inhibitors and ouabin inhibit the vasodilating effect of PTH-(1-34) [3.92]. Indeed, cAMP also plays a role in the lowering of blood pressure [3.93]. PTH inhibits the entry of calcium into the cell, an effect that may also play a part in the vascular action of PTH [3.94]. PTH-(1-34) causes a marked reduction of the blood pressure of spontaneously hypertensive rats [3.95] and antagonizes the pressor effect of noradrenaline and angiotensin II [3.96]. However, about 40% of the patients suffering from hyperparathyroidism were found to be hypertensive. This is due to the increased level of calcium since PTH potentiates the hypertensive effect of hypercalcemia [3.97].

At high calcium concentrations, PTH, unlike calcitonin [3.98], enhances the arginine induced secretion of glucagon, which can lead to hyperglycemia. Low doses of PTH stimulate the glucose and calcium dependent release of insulin in isolated rat islet cells, whereas high doses have an inhibitory effect. This release is accompanied by an increase in cytosolic calcium and cAMP [3.99]. Both glucose metabolism and the sensitivity to insulin are reduced in hyperparathyroidism [3.100].

In adrenal cells from chicken g-PTH stimulates the secretion of steroid hormones via the PTH receptors, an effect similar to that shown by ACTH [3.101].

Parenteral administration of PTH promotes the development of gastric ulcers, which can be prevented by CT [3.102]. In contrast, cerebroventricular (i.c.v.) application of r-PTH-(1-34) inhibits the secretion of gastric acid and the development of gastric ulcers in rats [3.103].

3.1.5. Activity of PTHrP

The properties of PTHrP are similar to those of PTH [3.104]. It increases the level of calcium and stimulates the formation of calcitriol. The calcium mobilizing activity of PTHrP-(1-34)-amide is comparable with that of b-PTH-(1-34) [3.105]. In

VMR-106 osteosarcoma cells, however, the PTHrP peptide amides were twice as active as b-PTH-(1-34) [3.106]. PTHrP acts as an autocrine growth factor in human kidney cell carcinoma [3.107].

3.1.6. Structure-Activity Relationships

The amino terminal sequence, 1-34, contains all the structural requirements for full biological activity. Both an extension as well as a shortening at the C-terminus results in the loss of biological activity compared to PTH-(1-34). Sequence 1-27 is the shortest fragment possessing activity. The region 25-34 is required for receptor binding, while the two amino acids at the amino end are of great importance for biological activity. In fact, g-PTH-(1-34), which differs structurally from b-PTH, above all, between positions 15 and 27, has only 1/10 of the biological activity (receptor binding to b-PTH receptors, cAMP formation, and relaxing of the rabbit renal artery) [3.108].

Oxidation of the methionine in PTH-(1-34) abolishes the hypotensive and vasodilating effect of this peptide, without affecting its hypercalcemic effect. Oxidation of PTH-(1-34) reduces the ability activating adenylate cyclase in renal micro-vessels, but not in the tubuli [3.109], [3.110].

The activity of b-PTH-(1-34) is increased by the substitution of Tyr or D-Tyr for Phe^{34} and the replacement of the two Met residues by Nle. The blocking of the carboxyl group by an amide function also has a favourable effect on the biological activity. Thus, [$Nle^{8,18}$, D-Tyr^{34}]b-PTH-(1-34)-amide is about 4 times as effective as unsubstituted b-PTH-(1-34).

3.1.7. PTH Antagonists

The peptides PTH-(3-34) and [$Nle^{8,18}$, Tyr^{34}] b-PTH-(3-34)-amide are weak *in vitro* inhibitors. In comparison, [Tyr^{34}] b-PTH-(7-34)-amide inhibits, even *in vivo*, the PTH induced excretion of phosphate in the urine, cAMP formation [3.111] and the increase in plasma calcium levels [3.112]. Intensive cleavage of PTH-(1-34) with cathepsin D yields PTH-(8-34), which also exhibits an antagonistic effect *in vitro* [3.113]. The substitution of D-Trp for Gly^{12} leads to antagonists ([D-Trp^{12}, Tyr^{34}] b-PTH-(7-34)-amide and [$Nle^{8,10}$, D-Trp^{12}, Tyr^{34}]b-PTH-(7-34)-amide) that are 10-30 times more effective [3.114]. The behaviour of the D-α-Nal^{12} and the D-β-Nal^{12} analogs is similar to that of the D-Trp^{12} derivatives [3.115].

3.1.8. Applications

PTH-(1-34) has been used in the treatment of osteoporosis. Alternating therapy with calcitriol [3.116] or sa-calcitriol [3.117] appears to be especially favorable.

PTH-(1-34) can also serve as a diagnosticum to differentiate between pseudo-hypoparathyroidism and hypoparathyroidism [3.118].

3.2. Calcitonin [3.119]

3.2.1. Occurrence

Apart from in the C cells of the thyroid gland, calcitonin (CT)- like material has also been found in the brain, hypothalamus, hypophysis, lungs, thymus, liver, gastrointestinal tract, adrenals, muscles, parathyroid gland, cerebrospinal fluid, seminal fluid [3.120], and in breast milk, in which it is 10-40 times more enriched than in plasma [3.121], [3.122]. This hormone is formed as preproCT. The structures of h-preproCT [3.123] and of g-preproCT [3.124], [3.125] were elucidated via the cDNA. N-terminal of the CT sequence, there are 84 amino acids with a signal peptide containing 25 amino acids. The N-terminal fragment of the propeptide consisting of 57 amino acids, stimulates, like insulin, the proliferation of human osteoblasts [3.126]. C-termial of the CT sequence, a peptide containing 21 amino acids, called PDN-21 (katacalcin), is found adjacent to the Pro-amide forming sequence P G K K R.

The calcitonins are peptide amides containing 32 amino acids. Marked structural differences are observed between species (Table 8).

Table 8. Calcitonin sequences of various species

Peptide	Amino acid sequence
	10 20 30
h	C G N L S T C M L G T Y T Q D F N K F H T F P Q T A I G V G A Pa
r	C G N L S T C M L G T Y T Q D L N K F H T F P Q T S I G V G A Pa
b	C S N L S T C V L S A Y W K D L N N Y H R F S G M G F G P E T Pa
o	C S N L S T C V L S A Y W K D L N N Y H R Y S G M G F G P E T Pa
p	C S N L S T C V L S A Y W R N L N N F H R F S G M G F G P E T Pa
sa1	C S N L S T C V L G K L S Q E L H K L Q T Y P R T N T G S G T Pa
sa2	C S N L S T C V L G K L S Q D L H K L Q T F P R T N T G A G V Pa
sa3	C S N L S T C M L G K L S Q D L H K L Q T F P R T N T G A G V Pa
anq	C S N L S T C V L G K L S Q E L H K L Q T Y P R T D V G A G V Pa
g	C A S L S T C V L G K L S Q E L H K L Q T Y P R T D V G A G T Pa

3.2.2. Release of CT

CT is released from the C cells in response to calcium, strontium, barium, cholinergic peptides, e.g., pentagastrin, cholecystokinin (CCK), CCK-4, CCK-5, CCK-8 [3.127] and caerulein, and also to secretin and glucagon. However, chronic hyper-calcemia has an inhibitory effect on the acute calcium stimulated release of CT [3.128]. VIP potentiates the calcium stimulated release of CT [3.129]. The i.c.v. application of Leu-EK to rats stimulates the secretion of CT through the β-adrenoreceptors [3.130]. Although the effect of 17-β-estradiol and progesterone on the release of CT during menopause is disputed, these substances exert a dose-dependent stimulating effect on the secretion of CT *in vitro* from the C cells of the thyroid glands of eight day old rats [3.131]. Somatostatin and low plasma levels of calcium inhibit the release of CT. Increased intracellular calcium and cAMP appear to be the second messengers for the release of CT [3.132].

Men have a higher plasma CT level than women [3.133]. A raised plasma level of CT is found in cancer of the thyroid gland [3.134], heroin addicts [3.135], and in patients suffering from urticaria pigmentosa [3.136]. The synthesis of substances that are similar to CT, but of higher molecular weight, has been observed in tumors of the liver [3.137] and of the lung [3.138]. A reduced plasma CT level is found in hypothyroidism [3.139], Cushing's syndrom [3.140], Turner's syndrom (low formation of estrogen and high susceptibility to osteoporosis) [3.141], in postmenopausal women [3.142], in hypogonadal osteoporotic men [3.143] and in patients with high levels of PTH [3.144].

3.2 3. Receptors

Receptors for CT in osteoblast like cells are increased by calcitriol [3.145] and decreased via internalization, by CT [3.146].

3.2.4. Biological Effects of CT

Low doses of CT stimulate the formation of cGMP. On the other hand, high doses activate adenylate cyclase and inhibit the release of cGMP [3.147], [3.148].

In mammals, CT leads to a lowering of plasma calcium and phosphate levels via the inhibition of bone resorption and excretion through the kidneys [3.149].

In women, the ovarian steorid hormones potentiate the hypocalcemic effect of CT [3.150]. CT exerts an anti-diuretic and natriuretic effect [3.151], promoting the excretion of uric acid [3.152]. Another effect on the kidneys is the stimulation of 25-hydroxy-vitamin D_3-1α-hydroxylase (calcitriol synthesis) [3.153].

The mobility of osteoclasts, giant cells in the bone marrow that are associated with bone tissue resorption, is inhibited by CT [3.154] via an increased uptake of calcium [3.155] and phosphate [3.156]. In comparison, CT has a stimulating effect on the proliferation of osteoblasts [3.157]. It directly activates the growth of cartilage [3.158].

In humans, CT lowers the insulin level [3.159] and reduces the suppressive effect of glucose on glucagon release. Hence, it leads to an increase in blood sugar levels. On the other hand, CT inhibits the release of glucagon, which is induced by arginine or hypoglycemia, as well as glucagon stimulated glycogenolysis. This explains why the "diabetogenic" CT lowers plasma glucagon levels and the blood sugar concentration in diabetics who are dependent on insulin. These results are, however, disputed [3.160]. On the other hand, the i.c.v. application of sa-CT increases the glucose stimulated release of insulin in rats [3.161].

The synthesis of fatty acids in the liver is increased by CT through the activation of fatty acid synthetase [3.162] and ATP citrate lyase [3.163] in a process that is dependent on calcium and calmodulin [3.164].

CT has both a stimulating and an inhibitory effect on the hypophyseal hormones.

It stimulates the stress hormones ACTH [3.159], [3.165], [3.166] and β-endorphin [3.159], [3.165], [3.167] (analgetic effect) and thus releases cortisol from the adrenals [3.159], [3.165]. CT has an inhibitory effect on the release of somatotropin (STH) [3.168], prolactin [3.169], [3.170], lutropin, follitropin [3.171], thyrotropin [3.172] and, consequently, also on the TRH-stimulated sécretion from the thyroid gland [3.173].

Although CT stimulates the stress hormones, the subcutaneous application of this peptide reduces the stress induced secretion of gastric acid and the formation of ulcers [3.174]-[3.176]. In the gastrointestinal tract, CT releases somatostatin, inhibits the release of gastrin [3.177], and increases motility [3.178]. The secretion of trypsin, amylase and lipase from the pancreas and the contraction of the gallbladder are also reduced by CT (inhibition of CCK) [3.179]. This effect on the gastrointestinal tract could be explained by a central action of CT because the i.c.v. application of much smaller doses, even 1000 times smaller, also results in an inhibition of the secretion of gastric acid [3.180], [3.181] and pepsin [3.180], and of stress induced gastric ulcers, as well as an increase in motility [3.182]. In fact, like the intravenous injection, the i.c.v. application of sa-CT also causes hypocalcemia in rats [3.119].

Furthermore, the central application of CT delays the emptying of the stomach [3.181], [3.183] and reduces the uptake of food and water [3.184]. While the i.c.v. application of CT causes an increase in the blood pressure, heart rate [3.185], and body temperature [3.186], intravenously applied CT has no effect on these functions [3.185].

Similar to CGRP, CT inhibits the SP induced airway resistance in anesthetized guinea pigs [3.187] and the PAF stimulated biosynthesis of the leukotrienes, LTC_4 and LTD_4 [3.188].

3.2.5. Structure-Activity Relationships

The sa- and anq-CTs produce a hypocalcemic effect in mammals that is approximately 10-30 times stronger than that exerted by the mammalian peptides themselves. The C-terminal proline amide is especially important for the biological activity of CT. Indeed, the substitution of the C-terminal carboxamido group or the elimination of the C-terminal proline reduces the activity to less than 1%. On the other hand, the acylation of the N-terminal amino group or its substitution by hydrogen causes a slight increase in the activity. The ring structure at the N-terminal part of the molecule is important for the activity of h-CT, but not for sa-CT. Hence, both cysteines in sa-CT can be replaced by Ala or Cys(Acm) without loss of the hypocalcemic effect. In the case of h-CT, however, these substitutions cause a dramatic decrease in the hypocalcemic effect [3.183]. The disulfide structure or the size of the ring is not essential for the activity of anq-CT. For instance, [$Asu^{1,7}$]anq-CT (elcatonin) still has about 80% of the activity of anq-CT, and Ser^2 or Asn^3 can be deleted without loss of activity [3.190], [3.191]. The amino acids Ser^{29} and Thr^{31} are in part responsible for the high activity of sa-CT; indeed, [Ser^{29}, Thr^{31}]h-CT is about 5 times as active as h-CT. In the sa-CT series, Val^8 can be replaced by Gly without loss of activity. However, this does not apply to Met^8 of h-CT [3.191]. It has been found that [$Arg^{11,18}$, Lys^{14}] sa-CT is as active as sa-CT. The ε-amino group of the Lys^{14} of this compound could be derivatized without loss of activity, e.g., with biotin, fluorescein or 4-azidobenzoate [3.192].

While des-Leu^{16}-sa-CT and des-Phe^{16}-h-CT exhibited only one tenth of the corresponding CT activity in binding tests [3.193], the amino acids Tyr^{22} [3.191], [3.194], Leu^{19}, Ser^{13} [3.195], Gln^{20}, and Thr^{21} [3.194] of sa-CT can be deleted without loss of activity. For instance, des-Ser^{2}, des-Tyr^{22}[Gly^{8}]sa-CT has a somewhat higher biological activity (hypocalcemic test in rats) than sa-CT [3.191].

Both a strongly α-helical structure as well as a less helical component give rise to highly active sa-CT analogs. Model peptides containing substantial variations, but a stable α-helix, exhibit high biological activity. MCT-I [3.196] has about the same activity as b-CT, whereas MCT-II [3.197] is just as active as sa-CT.

sa-CT	C	S	N	L	S	T	C	V	L	G	K	L	S	Q	E	L	H	K	L	Q	T	Y	P	R	....
MCT-II	C	S	M	L	S	T	C	L	L	Q	Q	L	Q	K	L	L	Q	K	L	K	Q	Y	P	R	....
MCT-I	C	G	M	L	S	T	C	L	L	Q	Q	W	Q	K	L	L	Q	K	L	K	Q	L	P	R	....

Highly active analogs with a smaller helical component are [Gly^{8}]- or [Ala^{16}]-sa-CT [3.198] as well as des-Leu^{19}-saCT (almost twice as active as sa-CT) [3.195] or [Gly^{8}] des-Leu^{19}-sa-CT (almost three times as active as sa-CT) [3.199]. An analog exhibiting low helix formation, [Gly^{8}, D-Arg^{24}]des-Leu^{16}-sa-CT, binds and acts (cAMP production) on brain membranes, but not on kidney membranes. The α-helical structure is, therefore, important for the action on the kidney, but not for that on the brain [3.200].

sa-Calcitonins	Hypocalcemic activity (IU/mg)
sa-CT	4 250
[Gly^{8}]sa-CT	6 500
[Gly^{8}] des-Leu^{19}-sa-CT	11 000
[Gly^{8}, Ala^{16}] des-Leu^{19}-sa-CT	2 500
Des-1-Amino-[Ala$^{1.7}$]-des-Leu^{19}-sa-CT	7 400
Des-1-Amino-[Ala$^{1.7}$, Gly^{8}] des-Leu^{19}-sa-CT	5 400

3.2.6. Applications

Subcutaneous, intramuscular, rectal, or intranasal application of CT is possible. Intranasal administration produces the fewest side effects [3.201] and the doses required can be greatly reduced. In fact, the nasal administration of only 50 IU of sa-CT per day inhibit bone resorption without having a hypoglycemic effect or increasing plasma cAMP levels [3.202], [3.203].

The substances applied are sa-CT (calcitonin L, Karil, Calsynar), h-CT (Cibacalcin), p-CT (calcitonin S) and, in Japan, elcatonin (carbocalcitonin, turbocalcitonin) for the indications morbus paget, hypercalcemia, hyperparathyroidism, osteoporosis [3.204], osteolysis [3.205], Sudeck syndrom, acute pancreatitis, chronic polyarthritis [3.206] and tumorosteolysis. CT is also used as a centrally acting analgesic [3.207], [3.208].

3. 3. Calcitonin Gene Related Peptides [3.209]

3. 3.1. Occurrence

The mRNA for calcitonin gene related peptides (CGRP) is found primarily in the central and peripheral nervous system. The CGRPmRNA is highly expressed in trigeminal ganglia, in medullary thyroid carcinoma, in lung tumor cells, and in the muscle layers of the pylorus [3.210]. CGRP has been found together with substance P in the ends of central and peripheral sensorial nerves and in the nerves of the bone [3.211]. Circulating CGRP is formed primarily by the perivascular and cardiac nerves.

Two different human and rat CGRPs have been observed: α-CGRP coded in the α-gene, which is also responsible for the formation of calcitonin (CT), and β-CGRP coded in the β-gene [3.212]. p-CGRP [3.213] and g-CGRP [3.214] differ only slightly from h-α-CGRP.

	1																			20
h-α-CGRP	A	C	D	T	A	T	C	V	T	H	R	L	A	G	L	L	S	R	S	G-
h-β-CGRP	A	C	N	T	A	T	C	V	T	H	R	L	A	G	L	L	S	R	S	G-
r-α-CGRP	S	C	N	T	A	T	C	V	T	H	R	L	A	G	L	L	S	R	S	G-
r-β-CGRP	S	C	N	T	A	T	C	V	T	H	R	L	A	G	L	L	S	R	S	G-
p-CGRP	S	C	N	T	A	T	C	V	T	H	R	L	A	G	L	L	S	R	S	G-
g-CGRP	A	C	N	T	A	T	C	V	T	H	R	L	A	D	F	L	S	R	S	G-

	21																37
h-α-CGRP	G	V	V	K	N	N	F	V	P	T	N	V	G	S	K	A	Fa
h-β-CGRP	G	M	V	K	S	N	F	V	P	T	N	V	G	S	K	A	Fa
r-α-CGRP	G	V	V	K	D	N	F	V	P	T	N	V	G	S	E	A	Fa
r-β-CGRP	G	V	V	K	D	N	F	V	P	T	N	V	G	S	K	A	Fa
p-CGRP	G	M	V	K	S	N	F	V	P	T	D	V	G	S	E	A	Fa
g-CGRP	G	V	G	K	N	N	F	V	P	T	N	V	G	S	K	A	Fa

3. 3.2. Release

The peptide, CGRP is released by capsaicin [3.215] and by glucocorticoids [3.216]. Testosterone [3.217] and δ- and μ-opioid receptor agonists [3.218] inhibit the rlease of CGRP. The level of immuno reactive (i.r.)-CGRP is lowered in the ulcerated duodenum [3.219].

3. 3. 3. Receptors

Various types of receptors have been found. For instance, the linear analog, $[Cys(Acm)^{2,7}]$ h-α-CGRP has a high agonistic activity in the rat vas deferens bioassay, but has no effect on guinea pig heart preparations. On the other hand, the fragment, h-α-CGRP-(12-37) has a weak antagonistic effect on guinea pig heart preparations, but has no activity in the rat vas deferens bioassay [3.220].

3. 3.4. Biological Effects

CGRP is involved in sensory, motory and autonomous functions. It stimulates the formation of cAMP.

Intravenous injections result in a rapid decrease in blood pressure and an increase in the heart rate. It appears that CGRP is of physiological importance for the prevention of cerebral ischemia caused by excessive vasoconstriction and for the control of the peripheral resistance [3.221]. CGRP has a proliferative effect on human endothelial cells. Therefore, CGRP may be important for the formation of new vessels, e.g., in ischemia, inflammations and in the healing of wounds [3.222]. CGRP stimulates the release of ANP in the isolated rat atrium [3.223] and the synthesis of prostaglandins [3.224] in a dose dependent process. The release of noradrenaline in mice is stimulated by i.c.v. application (2.2 nmol), increasing the arterial blood pressure and resulting in tachycardia.

In the lungs, CGRP exerts a higher bronchoconstricting effect than substance P [3.225].

In comparison with SP, the intradermal administration of CGRP causes a more pronounced reddening of the skin (erythema with pseudopodia surrounded by a pallor edge) [3.226]. In some pathological cases, circulating CGRP could be responsible for producing hyperemia. It has been suggested that CGRP potentiates the release of SP from primary afferent terminals and promotes the transmission of nociceptive information induced by mechanical noxious stimuli [3.227]. Similar to sa-CT, the i.c.v. application of CGRP, however, has an antinociceptive effect [3.228].

A large variety of stimuli lead to an increase in the secretion of somatotropin (STH), which is inhibited by the i.c.v. administration of CGRP to rats, presumably via the release of SRIF [3.229]. Although the stress induced release of prolactin (PRL) is reduced by the intraperitoneal or subcutaneous application of CGRP [3.230], it is stimulated *in vitro* [3.231].

In the adrenals, CGRP increases the secretion of cortisol and aldosterone and inhibits the release of Met-enkephalin [3.232], [3.233].

CGRP appears to be a central mediator of ingestive behaviour. The central application of this peptide suppresses the uptake of food. Similar to CT, both the intravenous and the central application of CGRP results in the inhibition of gastric acid secretion [3.234], [3.235]; however, the gastrin level is not lowered, but is raised instead [3.236]. It appears that β-CGRP is, above all, important for the lowering of gastric acid [3.237], [3.238]. The inhibition of gastric acid may be a direct effect or/and mediated through the release of somatostatin [3.239]-[3.241]. Thus, the subcutaneous application of β-CGRP to rats causes the inhibition of gastric ulcers induced by alcohol, aspirin or indomethacin [3.242], [3.243]. CGRP relaxes the intestinal muscle [3.244] and inhibits the emptying of the stomach in rats on either central or intraperitoneal application [3.245], [3.246].

Low doses of both α- as well as β-CGRP have an anti-secretory effect on the epithelium of the colon. High doses stimulate the secretion in each case [3.247]. α-CGRP may be involved in the pathogenesis of secretory diarrhoea in medullary thyroid carcinoma [3.248].

CGRP has an inhibitory effect on the secretion of insulin [3.249]-[3.251] and on the insulin stimulated formation of glycogen [3.252]-[3.254] and promotes glycogenolysis [3.254]. Therefore, CGRP (and/or amylin) may be responsible for the etiology of insulin resistance in type 2 diabetes mellitus. In healthy men, intravenously applied CGRP inhibits the secretin or CCK-8 stimulated release of trypsin and amylase, but has no effect on the arginine induced release of insulin and glucagon [3.255]. In dogs, CGRP and CT inhibit synergistically the secretin induced secretion from the pancreas [3.241], [3.256].

CGRP also inhibits, in a dose dependent process (10^{-10}-10^{-7} M), the proliferation of cells from the lymph nodes and the spleen of mice [3.257].

The effect on the plasma calcium level and on bone metabolism is similar, but less potent, than that of CT.

3. 3.5. Structure-Activity Relationships

α- and β-CGRP cause an equivalent elevation in blood flow in the rabbit skin and enhance the contractility of the rat atrium. They also identically inhibit osteoclastic bone resorption. g-CGRP exerts a stronger effect for a longer period of time than h-CGRP [3.214].

The complete structure appears to be required for full biological activity. Acetylation of the lysine in positions 24 and 35 or of the N-terminal amino group leads to a significant reduction of the biological activity [3.258]. The N-terminal part of the molecule appears to be important for activity and the C-terminal part for receptor binding.

In binding tests with membranes of the CNS or the spleen, [Tyr°]h-α-CGRP was found to be more active than h-α-CGRP. Like h-α-CGRP, [Ala$^1\varphi$(CH_2NH)Cys^2]h-α-CGRP also binds in these tests. The peptides, [Cys(Acm)^{2,7}]h-α-CGRP and cyclo^{2,7}-[Asp^2, Lys^7]h-α-CGRP exhibit slightly weaker binding. The binding of h-α-CGRP-(12-37) was about ten times less than that of h-α-CGRP Cyclo^{2,7}-h-α-CGRP-(1-7) does not bind at all [3.220].

The N-terminal fragments (1-12), (1-15) and (1-22) of CGRP reduce the blood pressure of rats anesthetized with urethane in a dose dependent process. Among these compounds, CGRP-(1-15) exerts the strongest effect (about 100 times weaker than α-CGRP [3.259].

h-α-CGRP-(12-37) was found to have a weak antagonistic effect in guinea pig heart preparations [3.220]. [Tyr°]CGRP-(28-37) antagonizes the relaxing effect of h-α-and β-CGRP on opossum internal anal sphincter [3.260].

3.4. Amylin [3.261]

Occurrence. Amylin (islet amyloid polypeptide, IAPP), a peptide amide containing

37 amino acids, was isolated from amyloid deposits in the pancreatic islet cells of patients suffering from type II diabetes (non-insulin-dependent) [3.262], [3.263]. There is a 46% structural similarity between amylin and CGRP. The structures of h-pro-amylin [3.263], [3.264] and h-prepro-amylin (89 amino acids) [3.265] were also elucidated via the DNA sequence. Amylin has also been isolated and characterized from rats, mice, guinea pigs, cats, and degu (Octodon degus) [3.266].

Amylin also occurs in normal β-cells of various mammals. Not only patients suffering from type II diabetes, but also healthy persons have i.r.-amylin in their islet β-cells. However, the amyloid deposits are only found in type II diabetics and not in healthy persons, indicating a pathogenic effect of amylin [3.267].

Release. Amylin, like insulin, is released from the rat pancreas during glucose or arginine stimulation [3.268], [3.269].

Streptozotocin inhibits the release of amylin more than that of insulin [3.269]. The human plasma level of amylin is reduced in insulin-dependent diabetes [3.270].

Biological Effects. Amylin, like CGRP, inhibits the basal and insulin-stimulated uptake of glucose and the synthesis of glycogen in the skeletal muscle at a concentration of 10^{-9} M [3.271], [3.272]. Therefore, amylin (and/or CGRP) may be responsible for the etiology of insulin resistance in type II diabetes mellitus.

Amylin and CT exert similar effects: it inhibits the osteoclastic resorption of bone at low levels of plasma calcium in rats and rabbits and lowers the plasma calcium level in rats [3.273].

Structure-Activity Relationships. The disulfide ring and the C-terminal amide are important for full biological activity [3.264].

	1																			20
h-Amylin	K	C	N	T	A	T	C	A	T	Q	R	L	A	N	F	L	V	H	S	S-
f-Amylin	K	C	N	T	A	T	C	A	T	Q	R	L	A	N	F	L	I	R	S	S-
r-Amylin	K	C	N	T	A	T	C	A	T	Q	R	L	A	N	F	L	V	R	S	S-
m-Amylin	K	C	N	T	A	T	C	A	T	Q	R	L	A	N	F	L	V	R	S	S-
cav-Amylin	K	C	N	T	A	T	C	A	T	Q	R	L	T	N	F	L	V	R	S	S-
oc-Amylin	K	C	N	T	A	T	C	A	T	Q	R	L	T	N	F	L	V	R	S	S-
h-α-CGRP	A	C	D	T	A	T	C	V	T	H	R	L	A	G	L	L	S	R	S	G-

	21																37
h-Amylin	N	N	F	G	A	I	L	S	S	T	N	V	G	S	N	T	Ya.
f-Amylin	N	N	L	G	A	I	L	S	P	T	N	V	G	S	N	T	Ya.
r-Amylin	N	N	L	G	P	V	L	P	P	T	N	V	G	S	N	T	Ya.
m-Amylin	N	N	L	G	P	V	L	P	P	T	N	V	G	S	N	T	Ya.
cav-Amylin	H	N	L	G	A	A	L	L	P	T	D	V	G	S	N	T	Ya.
oc-Amylin	H	N	L	G	A	A	L	P	P	T	K	V	G	S	N	T	Ya.
h-α-CGRP	G	V	V	K	N	N	F	V	P	T	N	V	G	S	K	A	Fa.

4. Corticoliberin – Proopiomelanocortin Cascade [4.1]

Corticoliberin (corticotropin releasing hormone or factor, CRH or CRF) and the peptides of proopiomelanocortin (POMC) are released in response to stress and are called the stress hormomes. Through the stimulation of cAMP and a cAMP dependent protein kinase [4.2], the hypothalamic factor, CRH, causes the secretion of POMC from the corticotrophic cells of the adenohypophysis, the endocrine cells of the pars intermedia, and from neurons that extend from the nucleus arcuatus of the hypothalamus to various regions of the brain. The prohormone, POMC, is then enzymatically cleaved to form POMC-(1-108)-amide (HP-N-108), corticotropin (ACTH) and β-lipotropin (β-LPH), see Figure 2. Further degradation of HP-N-108 gives rise to POMC-(1-76) and HP-N-30 [4.3] (POMC-(79-108)-amide, amidated joining peptide, JP-N [4.4]), POMC-(1-28) [4.5] (increased the mitotic activity of the adrenal cortex [4.6]), δ-melanotropin (δ-MSH) and γ-MSH (4.7). ACTH can be cleaved to form α-MSH and the corticotropin like intermediate lobe peptide (CLIP). Cleavage of the N-terminal tetrapeptide of CLIP can give rise to β-cell-tropin (β-CTP), and β-LPH yields γ-LPH, β-MSH, β-endorphin (β-EP) and the melanotropin potentiating factor (MPF). After the proteolytic hydrolysis of POMC, the peptide

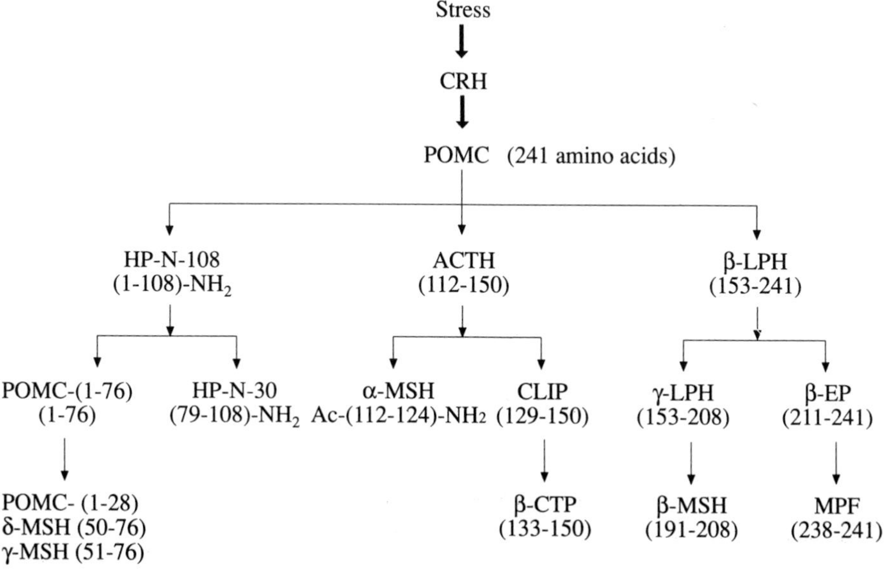

Figure 2. Degradation of proopiomelanocortin (POMC)

sequences of α-MSH and β-EP can be acetylated by opiomelanotropin acetyltransferase. Acetylation causes, e.g., a strong increase in the biological activity of α-MSH, whereas the activity of ACTH decreases. Acetylation abolishes the opiate activity of β-EP.

In the adrenal cortex, ACTH stimulates primarily the formation of the glucocorticoids and α-MSH releases aldosterone in the presence of angiotensin II (AT II), see Figure 3. The glucocorticoids and the mineralocorticoids play an important role in stressful situations. The glucocorticoids enhance liver gluconeogenesis and exhibit immunosuppressive and antiinflammatory activity. This antiinflammatory activity may be mediated by the lipocortins. The release of ACTH and β-EP is inhibited by the glucocorticoids by a negative feedback mechanism. Aldosterone exerts an antidiuretic and antinatriuretic effect, which is of importance during blood loss.

POMC and its cleavage products have also been found in the pineal gland, placenta, ovaries [4.8], testes (Leydig cells) [4.9], [4.10], epididymis [4.11], gastrointestinal tract, human leukocytes and in mouse macrophages [4.12].

Both CRH and POMC-derived stress hormones usually inhibit the reproductive hormones [4.13]-[4.15] and, like the blood pressure raising substance endothelin, enhance the release of the atrial natriuretic factor (ANF). On the other hand ANF down regulates the stress hormones and stimulates the reproductive hormones. Therefore, ANF is presumably very important as a regulator in coping with stress. While the androgens and gonadotropins stimulate the mRNA for POMC, [4.16], [4.17], estrogen and testosterone inhibit the formation of the stress hormones [4.18], [4.19]. However, estrogen also has a stimulating effect on the synthesis of both cortisol and the corticosteroid binding globulin. For this reason, estrogen has an immunosuppressive character and increases the plasma level of cortisol in pregnant women [4.20]. The D_2 dopamine agonists, which exhibit a GnRH releasing activity, inhibit the synthesis of POMC [4.21], [4.22]. The expression of the POMC gene in the testes

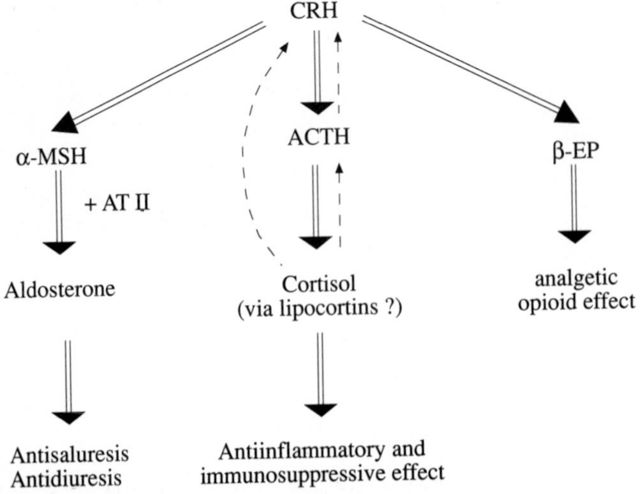

Figure 3. Corticoliberin, mediator of stress-respond

correlates closely with the maturation of the Leydig cells. In contrast to the adenohypophysis, the POMC gene in the testes is not regulated by the glucocorticoids [4.17].

Permanent stress probably desensitizes the immunosuppressive ACTH/glucocorticoid system. This condition is characterized by chronic fatigue and an increased formation of antibodies against the body's own tissues, a preliminary stage of autoimmune diseases [4.23].

The structures of h-[4.24], r-[4.25], b-[4.26], sa-, and xe-POMC [4.27] were elaborated via their DNA sequences (Table 9).

Table 9. Structures of different POMCs

Peptide	Amino acid sequence
	1 31
h-POMC	W C L E S S Q C Q D L T T E S N L L E C I R A C K P D L S A E-
r-POMC	W C L E S S Q C Q D L T T E S N L L A C I R A C R L D L S A E-
b-POMC	W C L E S S Q C Q D L T T E S N L L A C I R A C K P D L S A E-
	32 40 51 61
h-POMC	T P M F P G N G D E Q P L T E N P R K Y V M G H F R W D R F-
r-POMC	T P V F P G N G D E Q P L T E N P R K Y V M G H F R W D R F-
b-POMC	T P V F P G N G D E Q P L T E N P R K Y V M G H F R W D R F-
xe-POMC	L P V F P G N G H L Q P L S E S I R K Y V M T H F R W N K F-
	62 65 76 79 89
h-POMC	G R R N / S S S S G S S G A G Q K R E D / V S A G E D C G P-
r-POMC	G P R N / S S S / / A G G S A Q R R A E E E T A G G / / / /-
b-POMC	G R R N G S S S S G V G G A A Q K R E E E V A V G E G P G P-
xe-POMC	G R R N / S T G N D G S N T G Y K R E D I S S Y P V F S L F-
	90 108 112 116
h-POMC	L P E G G P E P R S D G A K / P G P R / E G K R / S Y S M E-
r-POMC	/ / / / / / / / / / D G R P E P S P R / E G K R / S Y S M E-
b-POMC	R / / / / / / / / G D D A E / T G P R / E D K R / S Y S M E-
xe-POMC	P L S D Q N A P G D N M E E E P L D R Q E N K R / A Y S M E-
sa-POMC	H S D K R H S Y S M E-
	117 124 130 133 145
h-POMC	H F R W G K P V G K K R R P V K V Y P N G A E D / E S A E A-
r-POMC	H F R W G K P V G K K R R P V K V Y P N V A E N / E S A E A-
b-POMC	H F R W G K P V G K K R R P V K V Y P N G A E D / E S A Q A-
xe-POMC	H F R W G K P V G R K R R P I K V Y P N G V E E / E S A Q S-
sa-POMC	H F R W G K P I G H K R R P I K V Y A S S L E G G D S S Q G-
	146 150 153 174
h-POMC	/ F P L E F K R E L T G Q R L R E G D G P D G P A D D G A G-
r-POMC	/ F P L E F K R E L E G E / / / Q / / / P D G / / / / / / /-
b-POMC	/ F P L E F K R E L T G E R L E Q A R G P E A Q A E S A A A-
xe-POMC	/ Y P M E L R R E L S / / / L / E L D Y P E I / / / / / D L-
sa-POMC	T F P L Q A R R Q L G S W E D E M V G A L G N Q G A K A E T-

Peptide	Amino acid sequence
	175 191 198
h-POMC	A Q A D L E H S L L V / / / / / / A A E K K D E G P Y R M E-
r-POMC	/ / / / L E Q V L E P / / / / / / D T E K A D / G P Y R V E-
b-POMC	R A / E L E Y G L / V A E A E A E A A E K K D S G P Y K M E-
xe-POMC	D E D I E D N E V K S / / / / / / A L T K K N / G N Y R M H-
sa-POMC	K V V P R T L T V T G / / / / / / L Q D K K D / G S Y R M G-
	199 208 211 226
h-POMC	H F R W G S P P K D K R Y G G F M T S E K S Q / / T P L V T-
r-POMC	H F R W G N P P K D K R Y G G F M T S E K S Q / / T P L V T-
b-POMC	H F R W G S P P K D K R Y G G F M T S E K S Q / / T P L V T-
xe-POMC	H F R W G S P P K D K R Y G G F M T P E R S Q / / T P L M T-
sa-POMC	H F R W G S P T A I K R Y G G F M K P Y T K Q S H K P L I T-
	227 238 241
h-POMC	L F K N A I I K N A Y K K G E.
r-POMC	L F K N A I I K N V H K K G Q.
b-POMC	L F K N A I I K N A H K K G Q.
xe-POMC	L F K N A I I K N S H K K G Q.
sa-POMC	L L K H I T L K N E Q.

4.1. Corticoliberin [4.28]

4.1.1. Occurrence

Corticoliberin also known as corticotropin releasing hormone or factor (CRH or CRF), is found in nerve fibers all over the brain. In the brain, CRH can occur with substance P (SP), acetylcholine esterase [4.29], oxytocin [4.30] or with vasopressin [4.31]. It has also been found in the pancreas [4.32], adrenal medulla [4.33], placenta [4.34], stomach, testes [4.35] and in the tissues of ACTH producing tumors [4.36]. The structurally related peptide sauvagin (SV) and urotensin I (UT I) were isolated from the skin of *Phyllomedusa sauvagii* and from the *urophysis* of the white sucker, *Catostomus commersoni.*

It has been observed that h-CRH is identical to r-CRH. However, the prepro sequences of these two species differ considerably [4.37]. Also, cap-[4.38] and o-CRH have identical sequences, but they differ from the b-sequence by a single amino acid [4.39]. On the other hand, p-CRH [4.40] is very similar to h- and r-CRH.

There is some structural similarity between the antiarrhythmic peptide (AAP) from the bovine atrium and SV. AAP has an inhibitory effect on experimentally induced thromobosis and arrythmia [4.41].

	1									10										20								28
h-, r-CRH	S	E	E	P	P	I	S	L	D	L	T	F	H	L	L	R	E	V	L	E	M	A	R	A	E	Q	L	A-
p-CRH	S	E	E	P	P	I	S	L	D	L	T	F	H	L	L	R	E	V	L	E	M	A	R	A	E	Q	L	A-
o-, cap-CRH		*Q	E	P	P	I	S	L	D	L	T	F	H	L	L	R	E	V	L	E	M	T	K	A	D	Q	L	A-
b-CRH		*Q	E	P	P	I	S	L	D	L	T	F	H	L	L	R	E	V	L	E	M	T	K	A	D	Q	L	A-
cat-UT-I	N	D	D	P	P	I	S	I	D	L	T	F	H	L	L	R	N	M	I	E	M	A	R	I	E	N	E	R-
cy-UT-I	N	D	D	P	P	I	S	I	D	L	T	F	H	L	L	R	N	M	I	E	M	A	R	N	E	N	Q	R-
SV		*Q	G	P	P	I	S	I	D	L	S	L	E	L	L	R	K	M	I	E	I	E	K	Q	E	K	E	K-
α-hel-CRH	S	Q	E	P	P	I	S	L	D	L	T	F	H	L	L	R	E	M	L	E	M	A	K	A	E	Q	E	A-
CRH-Inhib.									D	L	T	F	H	L	L	R	E	M	L	E	M	A	K	A	E	Q	E	A-
b-AAP			G	P	Ph	G	A	G																				

	29												41
h-, r-CRH	Q	Q	A	H	S	N	R	K	L	M	E	I	Ia
p-CRH	Q	Q	A	H	S	N	R	K	L	M	E	N	Fa.
o-, cap-CRH	Q	Q	A	H	S	N	R	K	L	L	D	I	Aa.
b-CRH	Q	Q	A	H	N	N	R	K	L	L	D	I	Aa.
cat-UT-I	E	Q	A	G	L	N	R	K	Y	L	D	E	Va.
cy-UT-I	E	Q	A	G	L	N	R	K	Y	L	D	E	Va.
SV	Q	Q	A	A	N	N	R	L	Y	L	D	T	Ia.
α-hel-CRH	E	Q	A	A	L	N	R	L	L	L	E	E	Aa.
CRH-Inhib.	E	Q	A	A	L	N	R	L	L	L	E	E	Aa.

4.1.2. Release

There is a circadian rhythm in the secretion of CRH. Similar to ACTH, the plasma level of CRH in humans peaks at 6.00 in the morning and has its lowest values in the evening at 18.00 and 22.00 [4.42]. Otherwise CRH is, in general, released in response to stress, e.g., hunger [4.43], loss of blood [4.33], peripheral increase in noradrenaline [4.44], osmotic stimulation [4.45] or insulin induced hypoglycemia [4.46]. Very small doses of catecholamines have a central stimulating effect and large doses exert an inhibitory effect on the release of CRH [4.47]. Noradrenaline inhibits the formation of CRH mainly via the β-receptors and partly also via α-adrenergic receptors [4.48], [4.49].

The pulse frequency of CRH is greatly increased by the chronic administration of alcohol to rats [4.50]. The release of CRH from the hypothalamus is also stimulated by morphine [4.51], acetylcholine, [4.52], scorpion poison, K^+ [4.53], IL-1β, IL-1α [4.54], [4.55] and IL-6 [4.54].

Endogenous glucocorticoids appear to be important for the synthesis and release of CRH. As a rule, dexamethasone has been found to inhibit the release of CRH [4.56], [4.57] and CRH-mRNA [4.58], [4.59].

The release of CRH from the hypothalamus is inhibited by a negative feedback mechanism by ACTH [4.60], [4.61] and endogenous opioids [4.62] via μ- and κ-receptors [4.63]. However, very small doses [0.1-1-nM) of β-endorphin stimulate the release of CRH *in vitro*, but higher doses (0.1 μM) inhibit both the basal as well as the acetylcholine stimulated release of CRH [4.64].

Oxytocin and the prostaglandins stimulate the release of i.r.-CRH and i.r.-ACTH from placenta cell cultures [4.65]. Thus, CRH is involved in the paracrine formation of ACTH in the placenta. Indeed, plasma CRH levels increase continually during pregnancy from the second trimester up to birth. The level of CRH in the plasma falls rapidly after delivery [4.66]. This CRH originates from the placenta [4.67], [4.68].

In Alzheimer's disease, a reduced CRH level in the brain [4.69] and in bone marrow fluid [4.70] is observed. In comparison, patients suffering from anorexia nervosa [4.71], [4.72] and depression [4.73], [4.74] exhibit enhanced release of CRH.

4.1.3. Receptors

CRH receptors are found in the adenohypophysis and in regions of the brain that are related to the limbic system and to the autonomic nervous system [4.75].

In spite of the increased plasma ACTH level in adrenalectomized rats, a decrease in the CRH receptors in the hypophysis has been found [4.76]. This down regulation of the CRH receptors in the hypophysis after adrenalectomy is caused by the enhanced secretion of vasopressin [4.77]. However, the CRH receptors on membranes of the liver, spleen, testes, prostate, and of the pancreas [4.36] increase in number after adrenalectomy, and are reduced by high doses of dexamethasone [4.78], [4.79].

4.1.4. Biological Effects

Both CRH as well as SV and UT I stimulate the synthesis and release of proopiomelanocortin and its degradation in the hypophysis and placenta [4.65], [4.80] to form, e.g., ACTH, α-melanotropin (α-MSH) [4.81], γ_3-MSH [4.82], and β-endorphin. Another effect is an increase in the plasma levels of corticosterone, aldosterone [4.83] and 18-hydroxycorticosterone [4.84]. However, CRH does not only act through the formation of ACTH, but it also has a direct potentiating effect on the ACTH stimulated synthesis of corticosterone [4.85]. The physiological importance of CRH is emphasized by the fact that CRH antigens cause atrophy of the adrenal glands and lower the plasma levels of ACTH and β-EP.

The activity of CRH depends on calcium. Extracellular Ca^{2+}, the entry of Ca^{2+} into the cell, calmodulin [4.86], cAMP dependent protein kinase [4.87] all play a role in the CRH stimulated release of ACTH. Stimulators of protein kinase C enhance the CRH stimulated formation of cAMP ten fold [4.88].

It has been found that intravenously applied CRH raises the plasma level of oxytocin (OT) and vasopressin (VP) [4.89], [4.90], which potentiate the activity of CRH [4.91]. Angiotensin II (AT II) [4.92] and gastrin releasing peptide-(14-27) [4.93] also potentiate the ACTH releasing effect of CRH.

Somatostatin (SRIF), melanophore concentrating hormone (MCH) [4.94], δ-sleep inducing peptide (DSIP) [4.95], glucocorticoids, morphine [4.96], endogenous opioides and synthetic enkephalins, e.g., DAMME [4.97], all inhibit the ACTH secreting effect of CRH.

In vitro, CRH stimulates the release of both β-endorphin as well as dynorphin from the rat hypophysis [4.98]. Both intravenous dynorphin A (1-13) as well as intravenous or intradermal CRH inhibit the inflammatory response (protein extravasation, edema formation) to scalding in rats [4.99]. Intravenous sauvagin is in this case about ten times and intravenous urotensin I is about four times as active as intravenous h- or r-CRH [4.100]. This activity of CRH is also found in hypophysectomized and adrenalectomized animals [4.101]. In the rat paw edema model, systemically applied CRH reduces hyperalgesia, formation of edema and hyperthermia. The antinociceptive effect is abolished neither by hypophysectomy nor by adrenalectomy. On the other hand, the antiinflammatory effect is not dependent on the hypophysis, but on adrenal function [4.102].

CRH has a direct [4.103]-[4.105] and an inhibitory effect, via the release of β-EP [4.106], on the reproductive functions of both sexes [4.107], [4.108]. The release of GnRH to the hypophysis from the hypothalamus is evidently prevented by CRH [4.109]. The prolactin (PRL) dependent inhibition of GnRH is also partly caused by the activation of CRH containing neurons [4.110]. CRH also centrally inhibits the release of somatotropin (STH) [4.111] through the stimulation of SRIF in the hypothalamus [4.112]. On the other hand, intravenous CRH raises plasma levels of PRL [4.113].

Intravenous doses of the three peptides (CRH, SV and UT I) that are higher than those required for the ACTH secreting effect cause a selective dilation of the blood vessels of the mesenterium, which is responsible for the decline in peripheral resistance and in the systemic arterial blood pressure. UT I and SV are about five to ten times as effective as o-CRH. Similar to the kinins, UT I or CRH [4.114] activates the synthesis of the prostaglandins (6-keto-$PGF_1\alpha$, PGE_2). Here, too, the prostaglandins, in turn, inhibit the hypotensive effect of UT I and CRH by a negative feedback mechanism [4.114]. In human beings, only doses higher than 200 μg of h-CRH cause a reduction of diastolic blood pressure, which is accompanied by tachycardia and increased pulse rate [4.115]. The central application of CRH raises blood pressure and pulse rate by increasing the levels of adrenaline and noradrenaline.

Intravenous CRH increase plasma pancreatic polypeptide (PP) [4.116] and the secretion of HCO_3^- from the pancreas. *In vivo*, it potentiates the cholecystokinin induced pancreatic protein secretion and the secretin induced secretion of HCO_3^- [4.117]. The stress and CRH induced secretion of HCO_3^- proceeds via the release of β-EP [4.118].

The intravenous application of CRH results in a strong and persistent lowering of the level of pentagastrin stimulated gastric acid, while the plasma gastrin level increases [4.119], [4.120]. Intraperitoneal CRH reduces stress induced gastric ulcers in rats [4.121]. CRH inhibits the emptying of the stomach [4.122]-[4.124] in a vagus dependent process [4.125], reduces the transit in the small intestine, increases the transit in the large intestine and causes diarrhea [4.126]. Restraint stress leads to reduced food uptake due to the secretion of CRH [4.127].

In humans, intravenous CRH increases the breath-time volume, breath frequency [4.128], the attentivenss and improves the mood and well-being of depressive patients [4.129].

CRH exerts a central antipyretic effect that may be due to the central release of ACTH or α-MSH, substances which also have antipyretic activity (4.130). The behavior induced by the central application of CRH to experimental animals contrasts sharply with that produced by tranquilizers and anxiolytic compounds [4.131]. Indeed, locomotoric activity and emotionalism are increased [4.132], cleaning behavior is enhanced, food uptake is reduced [4.134], aggressiveness is reduced and defensive behavior is enhanced [4.134].

4.1.5. Structure-Activity Relationships

The three amino acids at the amino terminus of o-CRH and UT I are of no importance for the biological activity. UT I-(4-41)-amide and o-CRH-(4-41)-amide retain their full vasodilating and ACTH releasing activity. However, the activity falls to below 0.1% on further deletion of amino acids at the N-terminal end, e.g., in the case of o-CRH-(10-41) amide. Indeed, SV-(17-40) has practically no influence on the endocrine glands and on the circulating system. A maximally α helical CRH with structural characteristics of the known CRHs was twice as active as o-CRH. The C-terminal fragments of this α helical CRH (8-41, 9-41, and 10-41) acted as competitive inhibitors *in vitro* and *in vivo* [4.135].

The C-terminal region is of extreme importance for CRH activity. Both o-CRH-(1-41) with a free C-terminal carboxyl group and o-CRH-(1- 39)-amide have less than 0.1% of the biological CRH activity. However, the C-terminal end of the molecule is not essential for the vasodilating activity: UT I-(4-28) still has about 5% of UT I activity, which roughly corresponds to the vasodilating activity of o-CRH. Met^{21} is of some significance for activity because $[Met(O)^{21}]$o-CRH exhibits only 10% of the biological activity. However, norleucine and norvaline can be substituted for this amino acid with practically no loss of activity.

4.1.6. Applications

Application by means of subcutaneous injection is advisable and is at least just as effective as intravenous administration. In comparison, the intranasal activity is only about 1% of that obtained on parenteral administration [4.136].

A pulsatile injection of CRH leads to the normalization of ACTH and cortisol secretion after glucocorticoid therapy (secondary adrenal insufficiency) [4.137], [4.138] and apparently does not result in the desensitization of the hypophysis [4.139]. A pulsatile application (6.25 μg every 30 minutes) is preferable to a longterm infusion of 50 or 100 μg/4 hours [4.140].

CRH can be used as a diagnostic aid for hypophyseal function [4.141]. After 100 μg of CRH (i.v.), the plasma ACTH and cortisol levels in patients suffering from Cushing's syndrom increased to a higher extent than in healthy persons [4.142]. In depressive patients with a high level of cortisol, CRH produces a lower ACTH response than in normal persons [4.143]. An abnormal increase in somatotropin after application of CRH is observed in patients suffering from congenital thyrotropin deficiency [4.144].

4.2. Corticotropin [4.145], [4.146]

4.2.1. Occurrence

Table 10. ACTH sequences from various species [4.149]

Source	Amino acid sequence																		
	1																		20
h, o, b, p, whale	S	Y	S	M	E	H	F	R	W	G	K	P	V	G	K	K	R	R	P	V-
m, r	S	Y	S	M	E	H	F	R	W	G	K	P	V	G	K	K	R	R	P	V-
cu	S	Y	S	M	E	H	F	R	W	G	K	P	V	G	K	K	R	R	P	V-
guinea pig	S	Y	S	M	E	H	F	R	W	G	K	P	V	G	K	K	R	R	P	V-
st	S	Y	S	M	E	H	F	R	W	G	K	P	V	G	R	K	R	R	P	V-
g	S	Y	S	M	E	H	F	R	W	G	K	P	V	G	R	R	K	R	P	I-
α-MSH	Ac-S	Y	S	M	E	H	F	R	W	G	K	P	Va							
h-CLIP																		R	P	V-
h-PTH-(17-25)			S	M	E	R	V	E	W	L	R									

Source																			
	21																		39
h, whale	K	V	Y	P	N	G	A	E	D	E	S	A	E	A	F	P	L	E	F.
o, b	K	V	Y	P	N	G	A	E	D	E	S	A	Q	A	F	P	L	E	F.
p	K	V	Y	P	N	G	A	E	D	E	L	A	E	A	F	P	L	E	F.
m, r	K	V	Y	P	N	V	A	E	N	E	S	A	E	A	F	P	L	E	F.
cu	K	V	Y	P	N	G	A	E	N	E	S	A	E	A	F	P	V	E	F.
guinea pig	K	V	Y	A	N	G	A	E	E	E	S	A	E	A	F	P	L	E	F.
st	K	V	Y	A	D	G	V	Q	E	E	T	S	E	A	F	P	L	E	F.
g	K	V	Y	A	D	G	S	V	B	Z	Z	Q	A	S	Y	P	V	E	F.
h-CLIP	K	V	Y	P	N	G	A	E	D	E	S	A	E	A	F	P	L	E	F.
h-β-CTP		V	Y	P	N	G	A	E	D	E	S	A	E	A	F	P	L	E	F.

Corticotropin (ACTH) is formed in the hypophysis as a metabolite of POMC. ACTH is also found along with β-endorphin-(β-EP-)like material in the interstitial fluid of the testes [4.147] and in the pancreas [4.148]. Enzymatic degradation of ACTH gives rise to α-melanotropin (α-MSH), the corticotropin like intermediate lobe peptide (CLIP) and β-cell-tropin (β-CTP). A certain structural similarity is observed between the N-terminal part of ACTH and the h-parathyroid hormone (17-25).

ACTH sequences from various species are shown in Table 10.

4.2.2. Release

ACTH is formed along with the other POMC cleavage products in response to stress, CRH, and to all the stimuli that release CRH and potentiate its activity. The formation of ACTH is accompanied by an increase in cAMP (see 4.1.) [4.146]. ACTH is secreted in micropulses of a few minutes. The frequency remains the same throughout the day, only the amplitude changes [4.150]. The amplitude is higher in the morning than in the afternoon and in the night, and it is higher in men than in women [4.151]. Serotonin increases the formation of ACTH and cortisol by stimulating CRH. It appears to be jointly responsible for the circadian increase in cortisol [4.152].

In vivo, corticosteroids inhibit the secretion of ACTH [4.146] in the hypophysis and cerebrospinal fluid. On the other hand, higher concentrations of ACTH are found in the hypothalamus after treatment with dexamethasone [4.153]. Glucocorticoids have no effect *in vitro* on the CRH stimulated release of ACTH from the hypophysis [4.154]. GABA appears to reduce the release of ACTH via the inhibition of CRH [4.152]. An increased intake of salt reduces the formation of ACTH in response to stress, VP or CRH. This effect is probably due to the release of the atrial natriuretic factor (ANF) [4.155].

Endogenous opioids appear to have an inhibiting as well as a stimulating effect on the release of ACTH. Thus, both the peripheral application of naloxone and μ- and κ-agonists stimulate the release of ACTH [4.156].

ACE inhibitors, e.g., p.o. (*per os*) Enalapril (reduces the insulin stimulated release of ACTH in patients suffering from essential high blood pressure [4.157]), inhibit the release of ACTH via the reduced formation of angiotensin II.

While ACTH secretion is increased in Cushing's syndrom, it is lowered in Sheehan's syndrom. The circadian ACTH rhythm is disturbed in patients suffering from depression. In fact, hypersecretion of ACTH, cortisol and STH is observed in this group of people [4.158], [4.159]. A deficiency of ACTH linked with high levels of PRL, estrogen, and LH has been found in gynaecomastia (breast development in men) [4.160].

4.2.3. Receptors

An ACTH receptor from mouse adrenal cells has a molecular weight of 225 kDa. It consists of four subunits with molecular weights of 83 kDa, 64 kDa, 52 kDa and 22 kDa. The 83 and 52 kDa units are linked together by a disulfide bond and are not covalently associated with the 64 and 22 kDa subunits [4.161].

In humans, ACTH receptors have also been found on splenocytes and peripheral mononuclear leukocytes [4.162].

IGF 1 potentiates the ACTH stimulated increase in ACTH receptors [4.163] and TGF-β reduces the ACTH receptors on adrenocortical cells [4.164].

4.2.4. Biological Effects

In the adrenal cortex, ACTH stimulates the synthesis of glucocorticoids and mineralocorticoids in a calcium dependent process [4.165] and via the activation of adenylate cyclase. Both aldosterone and 19-hydroxy-androstenedione, which act synergistically, are increased [4.166]. ACTH must be administered in the morning to achieve the maximum stimulation of the glucocorticoids. The sensitivity of adrenal cells to ACTH is reduced when these cells are exposed to ACTH for longer periods [4.167].

The 16 K fragment of POMC (HP-N-108), γ_3-MSH, and β-LPH, but not β-EP, potentiate the ACTH stimulated synthesis of glucocorticoids and aldosterone [4.168]. The ACTH stimulated secretion of aldosterone probably proceeds via angiotensin II [4.169]. ACTH also enhances the release of adrenaline and noradrenaline in the adrenal gland [4.170] and, in a calcium dependent process, potentiates the noradrenaline induced contraction of strips of rat aorta [4.171]. ACTH exerts a trophic effect on the adrenocorticotropic cells of the adrenal gland. Adrenal hyperplasia is observed as a result of the hypersecretion of ACTH (Cushing's disease). The adrenal glands shrink as a result of ACTH insufficiency of the hypopyhsis (Sheehan's syndrom).

Corticostatin (CS) from granulocytes inhibits the ACTH stimulated formation of corticosterone [4.172]. The substance first isolated from rabbits, cu-CS (CS I), is a little more active than the corresponding h-CS (HP4) [4.173], [4.174].

	1																			
cu-CS (CS I)	G	I	C	A	C	R	R	R	F	C	P	N	S	E	R	F	S	G	Y	C-
h-CS (HP4)		V	C	S	C	R	L	V	F	C	R	R	T	E	L	R	V	G	N	C-

	21													34
cu-CS (CS I)	R	V	N	G	A	R	Y	V	R	C	C	S	R	R.
h-CS (HP4)	L	I	G	G	V	S	F	T	Y	C	C	T	R	V.

Independent of the gluco-and mineralocorticoids, ACTH also has a curative effect in cases of shock [4.175]. Cholinergic mechanisms in the central nervous system play an important role here [4.176], [4.177]. This anti-shock effect is also exhibited by N-terminal fragments of ACTH, e.g., ACTH-(4-10). Alsactide is not as effective as ACTH (1-24) and ACTH-(4-10) [4.178].

While ACTH has no effect on the gonadoliberin (GnRH) stimulated secretion of lutropin (LH) in hypophyseal cells, cortisol increas the basal LH secretion but inhibits the GnRH stimulated LH secretion [4.179]. Cortisol also reduces the formation of testosterone in Leydig cells [4.180]. However, ACTH stimulates the synthesis of testosterone in Leydig cells [4.181], [4.182].

ACTH-(1-24) stimulates the pancreatic secretion of $NaHCO_3^-$ and protein by means of cholinergic mechanisms [4.183]. Similar to β-cell-tropin, the entire ACTH molecule has, in addition, insulinotropic and hypoglycemic activity [4.148]. However, it also potentiates the diabetogenic activity of growth hormone induced by hypoglycemia [4.184]. ACTH lowers the plasma glucose level in slim mice, while it raises the plasma level of glucose in fat mice [4.185]. In rats, corticosterone inhibits the activity of lipoprotein lipase in adipose tissue. This inhibitory effect of corticosterone is abolished by the simultaneous administration of ACTH-(1-24) [4.186].

In mice, ACTH prevents morphine induced tolerance [4.187], constipation and reduced bile flow [4.188].

ACTH binds specifically to splenocytes, inhibiting the production of antibodies and interferon (direct immunosuppressive activity) [4.162].

Sequence 4-10 accounts for the influence of ACTH on behavior and on learning processes (see α-melanotropin). For instance, this fragment increases attentiveness, sexual activity, and the memory capacity of rats, at the same time reducing anxiety.

Cholinergic neurons and muscarinic receptors play a role in ACTH dependent behavior patterns [4.189].

The central application of ACTH, ACTH-(1-24) as well as α-MSH results in the inhibition of food uptake [4.190], [4.191]. For instance, the effect of κ-opiate receptor agonists in stimulating food uptake is inhibited [4.191]. Similar to CRH, centrally applied ACTH raises blood pressure and the heart rate. While CRH primarily increases motory activity, centrally applied ACTH induces cleaning behavior. These central effects (behavior and circulatory effect) of ACTH are abolished by pretreatment with centrally applied naloxone [4.192]. Like CRH, centrally applied ACTH prevents persons from falling asleep, while des-acetyl-α-MSH and CLIP intensify slow wave sleep and paradoxical sleep [4.193].

It is assumed that the antiinflammatory activity of cortisol is mediated by the lipocortins (LPC), vasocortin [4.194], endonexins [4.195], [4.196] or uteroglobulin [4.197]. The LPCs are proteins from monocytes or cells of the adrenal medulla. For instance, the structures of h-LPC I with 346 amino acids [4.198] and of h-LPC II with 338 amino acids [4.199] have been elucidated. LPC inhibits phospholipase A_2 (PLA_2). Various fragments have been obtained by the limited proteolysis of recombinant h-LPC I with elastase. The region common to all active fragments is the sequence h-LPC I-(97-178) [4.200].

4.2.5. Structure-Activity Relationships

A fragment as small as ACTH-(4-7) (10^{-4} M) exhibits corticosterone forming activity *in vitro*. A dramatic increase in activity occurs with the addition of Met^3. However, the full activity of the N-terminal sequence is attained with the appearance of a free amino group at the N-terminal Ser^1. Thus, ACTH-(1-10) is 100 times as active as ACTH-(5-10), and ACTH-(1-18) has the full activity of native ACTH.

With regard to the corticosteroid releasing activity, ACTH-(1-24) (tetracosactide) is more active than ACTH *in vitro*. The C- terminal part of ACTH (ACTH-(25-39)) increases the antigenicity and the duration of action of ACTH *in vivo*. Thus, ACTH can be subdivided into at least four segments. Sequence 11-18 is very important for receptor binding and has been called the “address” by R. Schwyzer. Sequence 4-10, which is responsible for the corticotropic effect, contains the “message” and the N-terminal tripeptide is the “amplifier”. The C-terminal fragment, ACTH-(25-39), which is responsible for the antigenicity and safe transport, can be considered to be the “envelope”. ACTH-(11-24), which contains the address but not the message, acts as a competitive antagonist for the corticotropic activity of ACTH. Even ACTH-(7-38), which was isolated from the human hypophysis and contains only a mutilated message, inhibits the corticosteroid releasing activity of ACTH. The inhibitory effect of ACTH-(11-24) can be increased 1000 fold by using lysine amide to link two peptide chains [4.201]. An identical linkage of ACTH-(1-24) by lysine amide also gives rise to an antagonist, which is, in fact, more effective than the dimer of ACTH-(11-24) [4.202]. [Phe^9] ACTH-(1-24) also has an inhibiting effect on glycolysis and steroidogenesis.

ACTH from guinea pigs in which Pro^{24} is replaced by Ala has a higher aldosterone releasing activity than h-ACTH or ACTH-(1-24)-amide [4.149]. The compound

[Cys(carboxamidomethyl)^{25}]ACTH-(1-26) has three to four times the activity of ACTH in stimulating the secretion of aldosterone, but a lower corticosterone stimulating activity than ACTH.

The substitution of D-Ser or β-Ala for Ser^1 and of Lys for Arg^{17} and lysine amide or 1,4-diaminobutylamine for Arg^{18} gives long-acting analogs that have about up to eight times the ACTH activity.

Of the short, highly active ACTH peptides, [β-Ala^1, Lys^{17}] ACTH-(1-17)-4-amino-n-butylamide (ACTH 1-17, alsactide) has been developed. This ACTH derivative is stable to aminopeptidases and carboxypeptidases and has a long duration of action.

Similar to β-CTP, which is isolated from the pars intermedia of genetically fat mice, CLIP exhibits insulin releasing activity. Shortening the molecule at the N-terminal end or the acetylation of β-CTP results in a reduction of the insulin relasing activity [4.203]. Indeed, short N-terminal fragments of β-CTP are devoid of insulin releasing activity [4.203]. CLIP also releases pancreatic amylase *in vitro* [4.204]. Even ACTH has this insulinotropic activity, but its N-terminal fragments, e.g., ACTH-(1-24), are devoid of it. CLIP, like α-MSH and des-acetyl-α-MSH, lowers the β-endorphin stimulated secretion of prolactin in rats [4.205].

ACTH (1-13)-amide (des-acetyl-α-MSH), which has mostly lost the cortisol and aldosterone releasing activity has a higher melanophore stimulating activity than ACTH.

At a concentration of 0.1 nM/l, human parathyroid hormone, which contains structural elements of the ACTH message sequence, has *in vitro* in the adrenal cells of mice eleven times the steroidogenic activity of ACTH [4.206].

4.2.6. Applications

Natural corticotropin from the pig is on the market under the name Acethropan. The synthetic ACTH derivatives, tetracosactide (Synacthen) and alsactide (Synchrodyn 1-17) are being applied. ACTH preparations are used as a diagnostic aid for the functioning of the adrenal cortex and as a therapeutic agent for insufficient functioning of the adrenal cortex, in multiple sclerosis, inflammatory rheumatic diseases, collagen diseases, acute gout, radicular pain syndrom, severe allergic skin diseases and in collitis ulcerosa. ACTH has also an antiemetic effect and is administered to cancer patients who are being treated with Cisplatin [4.207].

4.3. α-, β-, γ- and δ-Melanotropin (MSH) [4.208]

4.3.1. Occurrence

α-MSH, a metabolite of POMC and ACTH, is found primarily in the hypophyseal pars intermedia and in the hypothalamus.

The central element of α-MSH, the tetrapeptide His-Phe-Arg-Trp, also occurs in two other regions of POMC. γ- and δ-MSH arise from POMC-(1-28) and β-MSH from γ-LPH.

β-MSH from vertebrates was isolated from the hypophyseal pars intermedia and h-β-MSH from the hypothalamus [4.209]. In many species, two different β-MSH peptides are found.

The peptide, b-δ-MSH was isolated from the bovine hypophysis. Peptides which can be present in the glycosylated form [4.208], i.r.-γ_3-MSH peptides and i.r.-γ_1-MSH peptides, have been found in the bovine [4.212]-[4.212] and human [4.211] hypophysis and in the brain.

		1				5					10			13		
Mammalian α-MSH	Ac-	S	Y	S	M	E	H	F	R	W	G	K	P	Va		
sa-α-MSH I		S	Y	S	M	E	H	F	R	W	G	K	P	Va		
sa-α-MSH II	Ac-	S	Y	S	M	E	H	F	R	W	G	K	P	I	G	H
um-α-MSH		S	Y	S	M	E	H	F	R	W	G	K	P	Ma		
um-α-MSH		S	Y	S	M	E	H	F	R	W	G	K	P	M		
ra-α-MSH [4.209]		S	Y	S	M	E	H	F	R	W	G	K	P	V		
xe-α-MSH [4.209]		A	Y	S	M	E	H	F	R	W	G	K	P	V		

	1				5					10					15			18
h-β-MSH	D	E	G	P	Y	R	M	E	H	F	R	W	G	S	P	P	K	D
si-β-MSH	D	E	G	P	Y	R	M	E	H	F	R	W	G	S	P	P	K	D
e-β-MSH	D	E	G	P	Y	K	M	E	H	F	R	W	G	S	P	P	K	D
		E	G	P	Y	K	M	E	H	F	R	W	G	S	P	P	K	D
o-, p-β-MSH	D	E	G	P	Y	K	M	E	H	F	R	W	G	S	P	P	K	D
b-, o-β-MSH	D	S	G	P	Y	K	M	E	H	F	R	W	G	S	P	P	K	D
cam-β-MSH	D	G	G	P	A	K	M	E	H	F	R	W	G	S	P	P	K	D
	D	G	G	P	Y	K	M	Q	H	F	R	W	G	S	P	P	K	D
um-β-MSH	D	G	I	D	Y	K	M	G	H	F	R	W	G	A	P	M	D.	K
	D	G	D	D	Y	K	F	G	H	F	R	W	S	V	P	L		
sa-β-MSH	D	G	G	S	Y	K	M	N	H	F	R	W	S	G	P	P	A	S
	D	G	G	S	Y	R	M	G	H	F	R	W	G	S	P	T	A	I
st-β-MSH	D	G	G	S	Y	R	M	R	H	F	R	W	Q	A	P	L	K	D
xe-β-MSH [4.210]		X	G	N	Y	R	M	H	H	F	R	W	G	S	P	P	K	

b-δ-MSH K Y V M G H F R W D R F G R R N G S S S-
S G V G G A A Q
b-γ_3-MSH Y V M G H F R W D R F G R R N G S S S-
S G V G G A A Q
b-γ_2-MSH Y V M G H F R W D R F G
(also γ-MSH)
b-γ_1-MSH Y V M G H F R W D R Fa

4.3.2. Release

Release of α-MSH is probably regulated through the enzymatic degradation of oxytocin as the prohormone for melanostatin (MIF) and melanoliberin [MRF]. Stress [4.213], corticoliberin, sauvagin, and urotensin I increase the i.r. plasma α-MSH level and the α-MSH level in the median eminence. β-Endorphin (β-EP) is also capable of centrally releasing α-MSH. Estradiol appears to stimulate the formation of α-MSH in rats [4.214]. α-MSH is released in the septum in response to fever induced by interleukin 1, especially when the temperature is rising (shivering phase) [4.215].

The release of α-MSH from the pars intermedia of the frog hypophysis is inhibited by NPY, PYY, a-PP [4.216] and by γ-MSH. Dopamine, which inhibits the formation of all the POMC-containing peptides in *Rana ridibunda*, exerts a still higher inhibiting effect on the acetylation of des-acetyl-α-MSH [4.217]. Ascorbic acid deficiency leads to a reduced formation of α-MSH amide *in vivo*, while the N-terminal acetylation remains unchanged [4.218].

Increased levels of γ-MSH are observed in physiological stress, cardiovascular distress, blood loss [4.219], and in cardiac arrest [4.220].

4. 3. 3. Biological Effects of α-MSH

In melanocytes, α-MSH stimulates by activating a tyrosinase the synthesis of the melanophoric pigment melanin (melanogenesis) and the transport of granula that contain pigment (melanosomes). A dark coloring of the skin is caused by the dispersion of the melanosomes through the numerous dendrites of the melanocytes. In comparison, the skin is lightened by a concentration of melanosomes. Human skin becomes darker within 24 hours after the administration of α-MSH. A darkening of the skin is observed in Cushing's, Addison's or Nelson's disease, in kidney damage, diseases in which the plasma α-MSH level is increased (absence of feedback control by corticosterone) or in ACTH secreting tumors. On the other hand, the skin becomes lighter in cases of hypophyseal insufficiency. The C-terminal tetrapeptide of β-EP, the melanotropin potentiating factor (MPF), enhances the melanotropic activity of α-MSH.

The second messenger of α-MSH appears to by cAMP. Inhibitors of α-MSH, e.g., melatonin, inhibit the α-MSH stimulated formation of cAMP and increase the concentration of cGMP instead. Melatonin, a hormone present in the pineal body, stimulates the aggregation of melanosomes.

Apart from this effect on melanocytes, α-MSH has a large number of other activities in mammals [4.221]:

increase in testicular Sertoli cell cAMP,
secretion of estradiol and plasminogen activator,
increase in lipogenesis and sebum production in the skin,
increase in lipolysis in adipose tissue,
increase in plasma free fatty acid levels,
increase in the adrenal steroidogenesis related to fetal growth and development,
increase in aldosterone synthesis and secretion by the adrenal zona glomerulosa (angiotensin II seems to be important [4.222]),
increase in pineal serotonin levels, decrease in pineal melatonin levels,
increase in somatotropin secretion by the pituitary gland,
inhibits stress and β-endorphin stimulated prolactin release [4.223]-[4.225],
increase in plasma luteinizing hormone [4.226],

increase in plasma glucagon and insulin levels, decrease in blood pressure,
decrease in plasma levels of calcium, reduction of bone resorption,
decrease in immunomodulatory and inflammatory activities of IL-1 [4.227],
reduction of body temperature following i.c.v. or parenteral administration [4.228], [4.229],
modulation of opioide peptide induced analgesia (nociceptive activity),
increase in sex attractant odours (female rat),
modulation of sexual behaviour (female rat),
increase in behaviour (arousal, attention, learning, memory retention, sexuality),
increase in acetylcholine in the brain,
decrease in food uptake in the case of central application [4.224],
increase in liquid uptake in the case of central application [4.230],
increase in nerve regeneration, and increase in forelimb regeneration.

4. 3.4. Biological Effects of β-MSH

h-β-MSH exhibits the same biological activity as α-MSH in the Anolis skin test. β-MSH sequences from other species also show high melanotropic activity, which is usually slightly lower than that of α-MSH. Similar to α-MSH, β-MSH also raises the plasma levels of glucose, glucagon, insulin and free fatty acids in the rabbit. In comparison, γ-MSH is inactive [4.231].

4. 3.5. Biological Effects of γ-MSH

Of the γ-MSH peptides, acetyl-γ_1-MSH, which still has less than 0.1% of the α-MSH activity, exerts the highest melanotropic effect. In behaviour tests, γ_2-MSH acts more like an antagonist of α-MSH and ACTH peptides (impairs learning). γ_2-MSH also inhibits the release of α-MSH from the hypophysis and reduces the β-endorphin induced antinociceptive and hypothermic effect. The direct infusion of γ-MSH into the renal artery results in prompt natriuresis and kaliuresis of the infused person, but not of the contralateral kidney [4.232]. The ACTH stimulated synthesis of glucocorticoids and aldosterone is potentiated by γ_3-MSH [4.233]. The intraventricular application of γ-MSH leads to a long term increase in the blood pressure [4.234]. The actual function of POMC-(1-108) and its fragments is probably based on the growth promoting effect on the adrenal cortex and on the hypertensive effect.

4. 3.6. Structure-Activity Relationships

The acetyl group is of considerable importance for the melanotropic activity of α-MSH.

α-MSH can be subdivided into three sequences: the classical messenger sequence, His-Phe-Arg-Trp (α-MSH-(6-9)), the C-terminal tetrapeptide, Gly-Lys-Pro-Val-amide which has seven times the melanotropic activity of α-MSH (6-9), and the N-terminal sequence, Ac-Ser-Tyr-Ser-Met-Glu which acts as a potentiator. Within the classical messenger sequence, -Phe-Arg- appears to be the crucial point for melanotropic activity. The minimum effective sequence is represented by Ac-His-Phe-Arg-Trp-amide [4.235]. The amino acid Lys in the C-terminal tetrapeptide is important for activity. In fact, even Ac-Lys-amide exhibits weak MSH activity. It is not the basic charge of Lys^{11} that is important for α-MSH activity because $[Lys(For)^{11}]$-, $[Lys(Msc)^{11}]$- and $[Lys(Tos)^{11}]$-α-MSH have almost the same melanophore stimulating activity as α-MSH in the frog skin test. In comparison, amino acids with shorter side chains exhibit a weaker activity. The significance of the C-terminal tetrapeptide for melanophore stimulating activity is demonstrated by ACTH-(7-38), which contains this tetrapeptide sequence. Although it has no corticosterone releasing activity, it still has 30% of the MSH activity of ACTH. Met^4, Gly^{10}, and Pro^{12} are also important for MSH activity [4.235]. If Met^4 and Gly^{10} are replaced by a cysteine disulfide bridge ($[Cys^4,Cys^{10}]\alpha$-MSH), an isosteric cyclopeptide is obtained which contains the two important messenger sequences. This peptide is a superagonist and is 10000 times as active as α-MSH in the frog skin test (*Rana pipens*). The cyclic Ac-$[Cys^4,Cys^{10}]\alpha$-MSH-(4-10)-amide is less active than α-MSH. But Ac-$[Cys^4,Cys^{10}]\alpha$-MSH-(4-13)-amide is again superactive and has the same activity as cyclo-$[Cys^4,Cys^{10}]\alpha$-MSH in the frog skin test. This again shows that the C-terminal sequence is of great significance for MSH activity. The disulfide bridge is important for biological activity. It is possible to incorporate -S-CH_2- or -S(O)-CH_2- bridges, instead of the disulfide bridge, with a slight loss of activity. The reduction of the disulfide bridge results in a 1000-10000 fold decrease in the biological activity [4.236].

Met^4, which is sensitive to oxidation, can be replaced by Nle without loss of activity. Treatment of α-MSH with hot alkali prolongs the duration of action. In this procedure, Met^4 and Phe^7 undergo partial racemization. It has been shown with synthetic peptides that $[Nle^4$. D-$Phe^7]$ αMSH is 60 times as active as α-MSH in the frog skin test. The corresponding (4-10)-fragment was found to be even more active than α-MSH in the lizard skin test and in the melanoma adenylate cyclase test. In mice, $[Nle^4$, D-$Phe^7]$-(4-11) or -(4-10)-α-MSH has 100 times the activity of α-MSH [4.237]. Even Ac-$[Nle^4$, D-$Phe^7]$ α-MSH-(4-9)-amide which has one amino acid less, is still ten times as active as α-MSH in the melanoma tyrosinase and in the lizard skin test, but it is ten times less active than α-MSH in the frog skin test. The activity of this hexapeptide was very prolonged compared to that of α-MSH [4.238]. The central application of $[Nle^4$, D-$Phe^7]$ α-MSH produces an antipyretic effect that is ten times greater than that of α-MSH. The antipyretic effect after intravenous application is, however, not pronounced. The C-terminal tripeptide, which also exhibits activity, appears to be very important for the antipyretic effect of α-MSH [4.228].

Cyclic compounds with D-Phe^7, which contain an acidic amino acid in position 5 and a basic amino acid in position 10, are more active than α-MSH. The two most active and long acting compounds in the lizard skin bioassay are [4.239]:

Ac-$[Nle^4,Asp^5$,D-$Phe^7,Lys^{10}]$ α-MSH-(4-10)-amide (90 times the activity of α-MSH) and

Ac-$[Nle^4,Asp^5$,D-$Phe^7,Lys^{10},Gly^{11}]$ α-MSH-(4-13)-amide (100 times the activity of α-MSH).

The activity of cyclo-[Cys^4,D-Phe^7,Cys^{10}]α-MSH on melanocytes is about as high as that of cyclo-[Cys^4,Cys^{10}]α-MSH, but it acts considerably longer. Cyclo-[Cys^4,D-Phe^7,Cys^{10}]-α-MSH-(1-12)-amide obtained by deleting the C-terminal valine has the same potency as cyclo-[Cys^4,D-Phe^7,Cys^{10}]α-MSH. This indicates that Val^{13} is not required for the melanotropic activity [4.240]. These analogs have little effect on the glomerulosa cells of the adrenal cortex: [Nle^4,D-Phe^7]α-MSH has no activity and cyclo-[Cys^4,Cys^{10}]α-MSH is only as active as α-MSH itself [4.241].

An analog of α-MSH-(4-9), H-Met(O_2)-Glu-His-Phe-D-Lys-Phe-OH (ORG-2766) [4.242] is, depending on the test, about 100 to 1000 times as active as α-MSH-(4-9) in promoting learning. However, high doses have the opposite effect. The subcutaneously or orally applied peptide has an antiamnestic effect, antagonizes pentobarbital anesthesia and reduces the uptake of morphine by the brain. ORG-2766 reduces cramps in rats [4.243], improves the function and regeneration of the nervous ischiadicus [4.244]-[4.246], reduces the cytotoxic effect of Cisplatin, and inhibits the tyramine stimulated increase in blood pressure in streptozocin diabetic rats [4.247]. The application of Z-D-Lys-Phe-NH-$(CH_2)_8$-NH_2 in the same doses as ORG-2766, also results in the stimulation of the acetylocholine level in the striatum of rat. H-Met(O_2)-Glu-His-Phe-D-Lys-Phe-NH-$(CH_2)_8$-NH_2 (HOE 427, Ebiratide) [4.248] and H-Met(O)-Glu-His-Phe-D-Lys-Phe-NH-$(CH_2)_8$-NH_2 have 100 times the ORG-2766 activity in this test [4.249]. In learning tests with rats, HOE 427 is about 500 times as active as ORG 2766 [4.250]-[4.252]. The sequence, Phe-D-Lys-Phe appears to be especially important for the activity of ORG-2766 and HOE 427 [4.253].

4. 3.7. Antagonists [4.221]

Ac-α-MSH-(7-10)-amide is a weak and selective α-MSH antagonist in the lizard skin bioassay. A variation of the structure is represented by Ac-[D-Trp^7,D-Phe^{10}]α-MSH-(7-10)-amide, a moderately strong specific competitive inhibitor of α-MSH in the frog and lizard skin bioassay [4.254]. A strong antagonist that is specific for frog skins is the uncyclized Ac-Nle-Asp-Trp-D-Phe-Nle-Trp-Lys-amide; the cyclic lactam of this compound is a full agonist.

Other antagonists are the growth hormone releasing peptide, His- D-Trp-Ala-Trp-D-Phe-Lys-amide and its analogs.

4. 3.8. Applications

The peptide, [Nle^4, D-Phe^7] α-MSH (intermedin alpha) can be applied topically to the skin. This compound is resorbed through the skin and systemically causes increased pigmentation in the yellow mouse [4.255].

α-MSH-(6-9) was found to exert a positive transdermal effect (better mood, less anxiety, less pain, less spasticity and less muscular weakness) in 24 of 44 patients suffering from multiple sclerosis [4.256].

In clinical studies, ORG-2766 positively altered the mood and level of performance of patients without influencing their sleep. Even in elderly, mentally weak patients,

ORG-2766 increased attentiveness and induced social behaviour [4.257]. At present, Ebiratide is being tested on patients suffering from Alzheimer's disease.

4.4. Melanotropin-Like Peptides from Crustacea and Fish

[4.258]

The pigment dispersing hormone (PDH), also called light adapting distal retinal pigment hormone (DPRH), is found in crustacea (*Pandalus borealis* and *Uca pugilator* [4.259]. An almost identical PDH has been isolated from the heads of the grasshopper *Romalea microptera* [4.260]. The PDHs have faint resemblance to the C-terminal tetrapeptide of α-MSH, which, like PDH, has melanotropic activity.

	1					6			9									
Pandalus PDH	N	S	G	M	I	N	S	I	L	G	I	P	R	V	M	T	E	Aa
Uca PDH	N	S	E	L	I	N	S	I	L	G	L	P	K	V	M	N	D	Aa
Romalea PDH	N	S	E	I	I	N	S	L	L	G	L	P	K	L	L	N	D	Aa
α-MSH(10-13)-amide										G	K	P	/	Va				

The melanophore concentrating hormones, which cause the lightening of the skin, are the melanophore concentrating hormone from salmom (sa-MCH) [4.261] and the red pigment concentrating hormone (RPCH) from crustacea. RPCH belongs to the family of adipokinetic hormones, which are, in turn, structurally assigned to the Bombesin family.

There is a definite structural similarity between sa-MCH and ACTH-(13-19), and sa-MCH has been shown to inhibit the CRH stimulated release of ACTH and formation of cortisol [4.262]. There is also a slight structural similarity between RPCH and sa-MCH.

I.r. MCH has also been found in the brain and, above all, in the hypothalamus of rats. This is, however, a peptide that differs from sa-MCH.

sa-MCH	D	T	M	R	C	M	V	G	R	V	Y	R	P	C	W	E	V
ACTH-(13-19)							V	G	K	K	R	R	P.....				
RPCH								*Q	L	N	F	S	P	G	Wa		

4.5. Opioid Peptides [4.263]

4.5.1. Introduction

Until now, the numerous peptides isolated, that have an effect on opioid receptors are formed from various prohormones. The activity of β-endorphin (β-EP), which arises from proopiomelanocortin (POMC), is based on the N-terminal pentapeptide sequence Tyr-Gly-Gly-Phe-Met. This sequence is identical to that of Met-enkephalin (Met-EK), which is produced primarily from prepro-EK A of the adrenal medulla. A series of extended Met-EKs that have structures related to prepro-EK A were isolated from the adrenal medulla. The extended Leu-EKs isolated from the hypophysis and hypothalamus, e.g., dynorphins (DP), α- and β-neoendorphin (neo-EP), and rimorphin, are derived from a common precursor, prepro-EK B (prepro-β-neo-EP/DP). Prepro-EK A and B are structurally very similar and are probably derived from a common early gene.

The C-terminal tetrapeptide sequence of prepro-EK A is, apart from the missing amide group, identical to that of femarfarm amide (FMRF amide), which was isolated from molluscs. FMRF amide is not an opiate. FMRF amide is not derived from prepro-EK A since the C-terminal glycine important for the formation of the amide group is missing. In molluscs, it is formed from a prepropeptide which contains 21 copies of the precursor FMRFG sequence.

Kyotorpin (KP), a dipeptide with analgetic activity, was isolated from the bovine brain. Although the structure Tyr-Arg occurs in β-MSH (positions 5-6) and in prepro-EK A (positions 35-36), it is definitely not derived from these peptides because neo-KP, also isolated from the bovine brain, is the obvious procompound. Apart from the structure Tyr-Arg, there is absolutely no similarity between neo-KP and prepro-EK or β-MSH; indeed, neo-KP is more similar to positions 7-11 of glucagon and somatoliberin.

The dermorphins (DM) and deltorphins (DT) are heptapeptide amides from the skin of the South American frogs *Phyllomedusa sauvagil, Phyllomedusa rhodei,* and *Phyllomedusa bicolor.* They contain D-alanine or D-methionine in position 2.

The exorphins are another group of opiate peptides which are formed by the enzymatic degradation of food. Examples are α- and β-casomorphins (CM).

Modified morphines, e.g., naloxone or certain EK analogs, are antagonists for endogenous opioids and exogenous opiates. In a definite dosage range, cholecystokinin-8 (CCK-8) also acts as an endogenous selective antagonist for the analgetic activity of opioids.

4.5.2. Opioid Receptors [4.263]

Prior to the discovery of endogenous opioids, three receptors were postulated which were named according to the prototypes for these receptors:

1. the μ-receptor (according to morphine),
2. the κ-receptor (according to ketocyclazocin), and
3. the σ-receptors (according to SKF 10 047).

The discovery of the enkephalins (EK) led to the introduction of the δ-receptors. β-EP prompted the postulation of the ε-receptors and the naloxone sensitive effect on thermoregulation gave rise to γ-receptors. Furthermore, iota-(ι-) receptors have been postulated in the dog and rabbit intestine, and λ-receptors with a high affinity for 4,5-epoxymorphinan have been described. The μ- and κ-receptors have already been divided into subunits (μ_1-1, μ_2- and 3 κ-receptors [4.264] [4.265]).

All the endogenous opioids known until now are ligands for the μ-(or μ_1-), δ-, κ- and ε-receptors. Met-EK, Leu-EK and β-EP are the main ligands for the μ- and δ-receptors. The dermorphins are specific for the μ-receptors, the deltorphins for the δ-receptors, and the dynorphins are specific for the κ-receptors. β-EP blinds primairly to the ε-receptors. Opiates, e.g., morphine, bind to the μ_2-receptors. The δ-receptors in rat brain and mouse vas deferens appear to be nonidentical [4.266]. In a model postulated by R. Schwyzer [4.267], the δ-site is exposed to the aqueous compartment surrounding the target cell at a distance comparable to or greater than the Deby-Hückel length and is in a cationic environment. The μ-site is exposed to the anionic fixed-charge compartment of the membrane in aqueous surroundings. The κ-site is buried in a more hydrophobic membrane compartment close to the fixed-charge compartment.

In vitro tests for the various receptors are the inhibition of electrically stimulated tissue samples from:

a) Guinea pig ileum for μ- and κ-receptors,
b) Mouse vas deferens for μ-, δ- and κ-receptors,
c) Rat vas deferens for ε-receptors,
d) Rabbit vas deferens for κ-receptors, and
e) Hamster vas deferns for δ-receptors [4.268].

Selective Ligands for the μ-Receptor [4.269]

[D-Ala^2, MePhe^4, Gly-ol^5]EK (DAGO, DAMGO),
[DELTA-Ala^2] EK [4.270],
[D-Nle^2, Glu^5]EK-amide [4.270]
Tyr-D-Met-Gly-NH-CH(CH_3)-CH_2-CH(CH_3)$_2$,
H-Tyr-D-Glu-Gly-Phe-Lys-amide [4.271],
[D-Ala^2] Leu-EK-chloromethylketone (DALECK), irreversible reagent [4.272], [4.273],
Tyr-D-Ala-Gly-MePhe-chloromethylketone (DAMK), irreversible reagent [4.274],
DM-(1-6)-amide, more selective than DAGO [4.275],
[D-Met(O)^2]DM-(1-4)-OMe [4.276],
[D-Arg^2, Lys^4] DM-(1-4)-amide (DALDA), more selective than DAGO [4.277],
H-Tyr-D-Glu-Phe-Lys-amide [4.271],
H-Tyr-D-Orn-Phe-Asp-amide [4.269], [4.270], [4.278],
H-Tyr-D-Asp-Phe-Orn-amide [4.278],
Morphiceptin,
D-Phe-Cys-Phe(p-N_3)-D-Trp-Lys-Thr-Pen-Thr-amide
(somatostatin analog) for photo affinity labeling [4.279].

Selective Ligands for the δ-Receptor [4.269]

[D-Ala^2, D-Leu^5] EK (DADLE),
[D-Ala^2, MeMet^5] EK-amide, (Metkephamide) [4.280],
[D-Ala^2, Pya^5] EK (Pya = Pyrenylalanine, fluoresces, 5 times as selictive as DADLE [4.281],
[D-Thr^2] Leu-EK-Thr (DTLET),
[D-Ser^2] Leu-EK-Thr [DSLET],
[D-Ser (tBu)^2] Leu-EK-Thr(DSTBULET) [4.282],
[D-Ser^2] Leu-EK-Thr(tBu) (DSLETBU) [4.282],
[D-Ser (tBu)^2] Leu-EK-Thr (tBu) (BUBU), strongest selective ligand for the δ-receptor until now [4.282],
[DELTA-Ala^3] EK [4.269],
[Tyr (m-SO_3H)^1]Leu-EK [4.283],
[D-Ser^2] β-EP-(1-6),
[D-Pen^2,Pen^5]EK (DPLPE) [4.284]-[4.286]
[D-Pen^2,D-Pen^5]EK (DPDPE), very selective [4.285], [4.286],
[(S,S)-β-Me-pNO_2Phe^4] DPDPE 5 times as effective as DPDPE [4.287],
[pFPhe^4] DPDPE and [pClPhe^4]DPDPE (more selective and about 7 times as effective as DPDPE [4.288]),
[pN_3Phe^4] DPDPE, for affinity labeling [4.279],
Dimers of EKs which are linked via the carboxyl groups with -NH-(CH_2)n-NH-bridges (n = 2-12) ([D-Ala^2]EK-(1-4)-tetrapeptide and [D-Ala^2]Leu-EK) [4.289],
[D-Ala^2, E-(2R,3S)-cyclopropylphenylalanine^4]Leu-EK (CP-OH), selective for rat brain δ-receptors [4.266],
Deltorphins (= Dermenkephalins) [4-290], [4.291].

Peptide Ligands for the κ-Receptor

DP-A-13 (also binds to μ-receptor) [4.292],
DP-A-8 (also has high affinity for δ- and μ-receptors [4.292],
[D-Pro^{10}] DP-A-11 (DPDYN) [4.293],
Pepties E, BAM-12 and BAM-22 [4.294].

Peptide Ligands for the ε-Receptor

β-Endorphin > BAM 22 > Peptide E > Peptide F > BAM 12 [4.295]

4.5.3. Opioid Actions [4.296]

μ_1 mediated:

Supraspinal analgesia,
Favourable effect on central receptors in states of trauma and shock caused by hemorrhage [4.297],
Dosis dependent stimulation (low dose) or inhibition (high dose) of acetylcholine

release from guinea pig ileum [4.297],
Reduced acetylcholine turnover,
Inhibition of norepinephrine release from guinea pig ileum [4.298],
Small, but significant decrease in dopamin levels in striatum,
Catalepsy,
Hypothermia,
Development of anorexia,
Stimulation of prolactin release [4.299],
Stimulation of ACTH-release in rats [4.300],
Inhibition of gastrointestinal transit and motility (i.c.v., i.t., s.c., i.v.) [4.301],
Inhibition of gastric acid secretion [4.275], [4.301],
Gastric acid releasing effect on pheripheral receptors [4.302],
Inhibition of VIP-induced secretion in the jejunum of the rat [4.303],
Stimulation of contractions of the canine small intestine *in vivo* [4.304], [4.305],
Suppression of experimentally induced diarrhea [4.301],
Pressor response and tachycardia (nucleus tractus solitarius, nucleus ambiguus) [4.306]

δ-mediated:
Spinal analgesia,
High dopamin turnover [4.307],
Inhibition of acetylcholin release from rat corpus striatum [4.308],
Impairment of avoidance learning in rats [4.309],
Reversal of endotoxic shock,
Inhibition of SP-stimulated plasma extravasation and vasodilation [4.310],
Inhibition of gastrointestinal transit (only after i.t. application) [4.301],
Reduction of stress ulcers and basal gastric acid [4.280],
Contraction of the canine small intestine *in vivo* [4.304].

κ-mediated:
Analgesia,
Anticonvulsants,
Increased food uptake [4.311],
Motor disfunction after spinal cord injury [4.312],
Inhibition of the release of acetylcholine in placenta villi [4.313] or guinea pig ileum [4.298],
Inhibition of dopamin-release from rat corpus striatum [4.308],
Decrease in blood pressure (nucleus ambiguus) [4.306],
Cardioaccelaration (nucleus tractus solitarius) [4.306],
Contractions of gastric reticulum in sheep [4.314],
ACTH-release in rats [4.300],
Diuresis (mediated by the adrenal medulla) [4.315].

ε-mediated:
Supraspinal analgesia.

δ-mediated:
Psychometric actions and behavioural patterns.

4.5.4. β-Endorphin [4.316]

Occurrence [4.317]. β-Endorphin contains 31 amino acids and is derived from proopiomelanocortin (POMC). It has been isolated primarily from the hypophysis. However, β-EP is also found in human placenta cell cultures, human ovaries [4.318], human sperm [4.319], in the endometrium during the secretory phase [4.320], in the gallbladder [4.321], pancreas, and in the small intestine.

Apart from β-EP and β-EP-(1-27) (C fragment or δ-EP), acetyl-β-EP, acetyl-β-EP-(1-27) [4.322] and β-EP-(1-18) [4.323] have also been found in the anterior lobe of the rat hypophysis. Ac-β-EP-(1-18) is found primarily in the posterior lobe of the rat hypophysis [4.323], [4.324]. In the hypophyseal pars intermedia of rats, Ac-β-EP-(1-16) (Ac-α-EP) and Ac-β-EP-(1-17) (Ac-γ-EP) [4.325] and, in particular, β-EP are detected after exposure to stress [4.326]. In rats, the degree of acetylation of the hypothalamic endorphins increases with age [4.327].

The C-terminal tetrapeptide of β-EP is the melanotropin potentiating factor (MPF).

	1 10 20 31
h-β-EP	Y G G F M T S E K S Q T P L V T L F K N A I I K N A Y K K G E
b, o, cam-β-EP	Y G G F M T S E K S Q T P L V T L F K N A I I K N A H K K G Q
p-β-EP	Y G G F M T S E K S Q T P L V T L F K N A I V K N A H K K G Q
g-β-EP	Y G G F M T S E H S Q M P L L T L F K N A I V K N A Y K K G E
st-β-EP	Y G G F M S S E R G R A P L V T L F K N A I V K N A Y K K G Q

A few deletions clearly show the similarity between shark [4.328], the two salmon endorphins and human endorphin.

		1																13
h-β-EP		Y	G	G	F	M	/	T	/	S	E	K	S	Q	T	/	/	P-
sa-EP I	Ac-	Y	G	G	F	M	K	P	Y	T	/	K	/	Q	S	H	K	P-
sa-EP II	Ac-	Y	G	G	F	M	K	S	W	N	E	R	S	Q	/	/	K	P-
pr-EP		Y	G	G	F	M	K	S	W	D	E	R	G	Q	/	/	K	P-

	14																	31
h-β-EP	L	V	T	L	F	K	N	A	I	I	K	N	A	Y	K	K	G	E
sa-EP I	L	I	T	L	L	K	H	I	T	L	K	N	/	E	Q.			
sa-EP II	L	L	T	L	F	K	N	V	I	I	K	D	G	Q	Q.			
pr-EP	L	L	T	L	F	R	N	V	I	V	K	D	G	E	H.			

Release. β-EP is secreted in the hypophysis in response to stress, corticoliberin, angiotensin II, lipoxygenase or epoxygenase products [4.329], insulin induced hypoglycemia (via cholinergic mechanisms [4.330], adrenaline, food uptake, and chronic alcohol consumption [4.331]. β-EP can be released from the mucosa of the small intestine by gastric acid or bile acid and from the mucosa of the gallbladder by CCK-8 [4.321]. Estrogen and testosterone increase the plasma level of β-EP [4.332]-[4.334]. Both h-CG and PMSG stimulate the formation of β-EP in the ovaries [4.335] and in the testicular interstitial fluid [4.336]. In the latter case, testosterone and the chronic application of LH-RH agonists [4.337] have a negative effect on the formation of β-EP.

Dexamethasone inhibits the release of β-EP by blocking CRH. In humans, somatostatin and oxytocin have no effect on the basal plasma β-EP level. However, they lower the levels of β-EP, β-LPH and cortisol that are raised after insulin induced hypoglycemia [4.338].

In humans, there is a circadian rhythm in the release of β-EP which is similar to that of cortisol: a high level in the morning and a trough at night [4.339]. The basal β-EP level is lower in people who bear a high risk of becoming alcoholics compared to those who do not. On the other hand, the plasma level of β-EP increases with alcohol consumption in the high risk group, but not in the low risk group [4.340]. The plasma levels of β-EP are significantly raised in depressive patients [4.341] and in obesity (children and adults) [4.342].

Biological Effects. β-EP has analgetic and lipolytic activity [4.343]. The i.c.v. application of CCK-8 [4.344] and α-MSH results in the inhibition of the analgetic activity of β-EP. β-EP lowers phosphodiesterase activity, somatostatin secretion in the isolated pancreas of the dog, the release of VIP and TRH [4.345], the release of gonadoliberin in the mesencephal central grey substance, plasma LH, the release of oxytocin induced by sucking, isoprenalin induced increase in plasma vasopressin [4.346], and, in high doses, the formation of CRH, ACTH, LPH and cortisol. On the other hand, very low doses release CRH in rat hypothalami [4.347].

Both the intraventricular and the intravenous application of β-EP causes the release of STH (via SRF [4.348]) and PRL. The β-EP stimulated release of PRL can be inhibited by α-MSH or CLIP [4.349]. β-EP increases plasma insulin, glucagon, and glucose, especially in fat persons [4.350], while contradictory results have been obtained in experimental animals.

In humans, intravenously applied β-EP raises the phasic pressure of the pylorus, reduces the activity of the antrum and causes an activity in the small intestine that is similar to fasting phase III [4.351]. In rats, intravenous β-EP stimulates the secretion of HCO_3^- in the duodenum [4.352] and reduces cysteamine induced duodenal ulcers.

In pharmacological doses, β-EP reduces the left ventricular systolic and diastolic pressure in the isolated rat heart [4.353].

Depending on the dose and on environmental influences, the i.c.v. application of β-EP results in hyperthermia or hypothermia. Other central effects of β-EP are the release of catecholamines, Met-EK [4.354], thyrotropin, vasopressin, and α-MSH as well as an increase in food uptake.

Structure-Activity Relationships. β-EP has a highly specific opiate recognition sequence at the N-terminal end (positions 1-5) which is bound to an amphiphilic helix (positions 13-31) by a hydrophilic link (positions 6-12) [4.355]. The N-terminal

Tyr-Gly bond of β-EP is more protected against aminopeptidases than that of the EKs. Hence, β-EP has a prolonged duration of action compared to the EKs.

Acetylated sa-β-EPs have no activity in an opiate receptor assay, while des-acetyl-sa-β-Ep I is 2.5 times as active as β-EP from warm-blooded animals. In contrast to β-EP from warm-blooded animals, des-acetyl-sa-β-EP has no affinity for the ε-receptors of rat vas deferens, but binds more strongly to δ- and κ-receptors.

Synthetic derivatives show that Gln or Arg in position 8 doubles the analgetic effect and with Trp in position 27, the analgetic effect is four times that of β-EP. [Gln^8, Trp^{27}]h-β-EP has almost 8 times the receptor binding ability of h-β-EP, while its analgetic effect corresponds to that of h-β-EP. The replacement of Glu^8 appears to be of great importance for receptor binding and may find application for the design of β-EP antagonists [4.356]. [Gln^8, Gly^{31}]h-β-EP-Gly-Gly-amide is really a strong antagonist for β-EP induced analgesia. It is 200 times as active as naloxone [4.357].

The substitutions, D-Ala^2 and Me-Phe^4 which are favourable to EK activity change the β-EPs in the direction of ligands for δ-receptors. An increasing hydrophobicity in position 5 correlates with decreasing analgetic activity [4.358]. [D-Ala^2]-cam-β-EP is still equipotent with cam-β-EP, but [D-Ala^2,D- Leu^5]h-β-EP has only 8% of the h-β-EP activity.

If the dermorphin sequence is substituted for the seven N-terminal amino acids of β-EP, a highly analgetic peptide is obtained, which is 4.4 times as active as h-β-EP and about as active as dermorphin [4.358].

Analogs of β-EP, which have no homology in the 12 C-terminal amino acids but imitate the helical structure of the C-terminus of β-EP, are as active or more active than β-EP.

Cysteine bridges between positions 14 and 26, 15 and 26, 16 and 26 as well as 17 and 26 are tolerated. These derivates containing Phe^{27} and Gly^{31} all exhibit stronger receptor binding than h- β-EP [4.359].

The peptide, β-EP-(1-27), which was found along with β-EP-(1-26) in the rat brain, has only 0.2% of the β-EP activity in the opiate receptor displacement assay. β-EP-(1-27) inhibits the β-EP induced analgesia and release of STH [4.360], and the β-EP induced hypothermia in the mouse [4.361], but not the β-EP induced release of PRL [4.361].

The N-terminal Tyr is very important for the analgetic activity of β-EP. The binding of des-Tyr^1-β-EP to opiate receptors is absolutely nil. h-β-EP-(6-31) is an inhibitor of β-EP induced analgesia. It also inhibits the β-EP stimulated release of PRL, but has no effect on the release of TSH [4.362]. h-β-EP-(6-31), like h-β-EP-(28-31) (Lys-Lys-Gly-Glu, h-MPF) and h-β-EP-(30-31), inhibits the α-MSH induced grooming, stretching and yawning syndrom as well as the β-EP induced grooming and catatonia [4.363]. In amphibians, h-β-EP and h-MPF have the ability to regenerate limbs. However, the corresponding b-β-EP-(28-31) (Lys-Lys-Gly-Gln), the C-terminal dipeptide Gly-Glu or the N-terminal β-EP sequences no longer exhibit this property [4.364]. The C-terminal end is of prime importance for the lipolytic activity of β-EP. Derivatives in which the two amino acids at the C-terminus are deleted have no activity. [4.365]. MPF, which has no lipolytic activity, inhibits the lipolytic activity of β-LPH [4.366].

4.5.5. Peptides of Preproenkephalins A and B [4.367]

There is considerable structural similarity between the two preproenkephalins (prepro-EK) that were isolated from endocrine and nerve tissue. The structures of h-, b-, r-[4.368], xe-[4.369] preproenkephalin A and h- and b-preproenkephalin B have been elucidated. Prepro-EK A is primarily processed to give Met-EK and Met-EK containing longer peptides, e.g., peptide E and adrenorphin. On the other hand, prepro-EK B gives rise to Leu-EK, the neo-endorphins and the dynorphins.

Prepro-EKs are widely distributed in the body (e.g., in the brain, gastrointestinal tract, adrenal cortex, corpus luteum [4.370], heart [4.371], and germ cells [4.372] and are processed differently according to their location [4.373].

Table 11. Partial sequences isolated from prepro-EKA [4.375]

Part of sequence	Amino acid sequence
b-(104-137) = Peptide F	Y G G F M K K M D E L Y P L E V E E E A-
	N G G E V L G K R Y G G F M.
b-(104-129) = Amidor-phin [4.376]	Y G G F M K K M D E L Y P L E V E E E A-
	N G G E V La.
b-(111-129)-amide [4.377]	M D E L Y P L E V E E E A N G G E V La.
b-(104-109)	Y G G F M K
Met-EK	Y G G F M
b-(140-189)	D A E E D D G L G N S S N L L K E L L G-
	A G D Q R E G S L H Q E G S D A E D V S-
	K R Y G G F M R G L
b-(182-189) = MERGL	Y G G F M R G L.
b-(192-230) = Peptide I	S P T L E D E H K E L Q K R Y G G F M R-
	R V G R P E W W M D Y Q K R Y G G F L.
b-(206-230) = Peptide E	Y G G F M R R V G R P E W W M D Y Q K R-
	Y G G F L.
b-(206-277) = BAM-22	Y G G F M R R V G R P E W W M D Y Q K R-
	Y G.
b-(206-225) = BAM-20	Y G G F M R R V G R P E W W M D Y Q K R.
b-(206-217) = BAM-12	Y G G F M R R V G R P E.
b-(206-213) = Adrenor-phin = Metorphamide	Y G G F M R R Va.
b-(206-212)	Y G G F M R R.
b-(206-211)	Y G G F M R.
b-(226-230) = Leu-EK	Y G G F L.
b-(233-263) = Peptide B	F A E P L P S E E E G E S Y S K E V P E-
	M E K R Y G G F M R F.
b-(233-247) [4.378]	F A E P L P S E E E G E S Y S.
b-(257-263) = MERF	Y G G F M R F.

4.5.5.1. Peptides Derived from Preproenkephalin A

Occurrence. Numerous fragments of prepro-EK A have been isolated from the adrenal medulla which can, in turn, be regarded as precursors of Met-and Leu-EK, see Table 11. Plasma proteins also appear to be a pool for peptides related to Met-EK.

Plasma proteins treated with pepsin generate peptides related to Met-EK [4.374].

Release. The release of the enkephalins is stimulated by GABAergic mechanisms, insulin induced hypoglycemia [4.379], [4.380], endotoxin shock [4.381], electroshock [4.382], intraventricular application of β-endorphin [4.383], and glucocorticoids [4.384], [4.385]. On the other hand, dopamine exerts a more inhibitory effect on the formation of Met-EK [4.386], [4.387]. The mRNA for prepro-EK A is stimulated by stress [4.388], AT II [4.389] (in adrenal chromaffin cells), FSH and cAMP (in Sertoli cells) [4.390]. Chronic administration of glucocorticoids inhibits the formation of the mRNA for prepro-EK A [4.388].

The plasma level of Met-EK increases considerably in marathon runners [4.391]. There is an increase in the plasma peptide F level in young men in the resting phase after exercise [4.392].

Met- and Leu-EK are rapidly degraded by enzymes. Apart from the aminopeptidases, the EKs are hydrolyzed primarily by enkephalinase, which cleaves the C-terminal dipeptide.

Biological Effects [4.393]. For a description of the activity of the enkephalins, see also Chap. 4.5.3.: μ_1-, δ- and κ-mediated opioid actions.

Enkephalins inhibit transmitter release from presynaptic nerve ends in the central and peripheral nervous system by blocking the type 2 calcium channels [4.394].

Met-and Leu-EK bind, above all, to δ-receptors and have an analgetic effect only if applied centrally. Analogs that have a preference for the μ- and δ-receptor exhibit an analgetic effect even on peripheral application. The actions of morphine with respect to tachyphylaxis and substance dependency are also observed after chronic application of the highly active EK analogs. Withdrawal symptoms in morphine-dependent dogs can be suppressed, e.g., with DAMME.

The endocrine and exocrine pancreatic and gastrointestinal secretion is modulated by the enkephalins [4.395]-[4.402].

Low doses of CCK have a stimulating effect and high doses an inhibitory effect on the opioid stimulated release of insulin [4.403].

Depending on the location, enkephalins exert a contracting or relaxing effect on the muscles of the gastrointestinal tract. Enkephalins contract the lower esophagus and the pyloric sphincter [4.404], delay gastrointestinal transit [4.405], [4.406], generate the interdigestive migrating motor complex in the duodenum [4.407], [4.408] (inhibition of the VIPergic system [4.409], cause, like CCK-8, gallbladder contraction *in vitro* [4.410], relax the sphincter of Oddi [4.411], cause vasoconstriction in the lungs [4.412], and inhibit the motility of the bladder on central application [4.413]. Depending on the dosage and species, enkephalins (above all when applied centrally) exert a hypotensive [4.414] or hypertensive effect [4.415].

FMRF-amide acts as an antiopioid substance in the central modulation of cardiovascular functions [4.416].

The i.c.v. application of μ- and δ-agonists to rats results in the inhibition of the release of oxytocin [4.417] (role in parturition [4.418]) and of vasopressin (diuresis).

The enkephalins modulate the release of ACTH and cortisol via extrahypophyseal mechanisms [4.419]. Enkephalins potentiate the ACTH stimulated release of corticosterone [4.420] and inhibit the CRH induced increase in the plasma levels of ACTH, β-EP and cortisol [4.421].

The enkephalins raise the level of prolactin (PRL) in the plasma by inhibiting the release of dopamine and stimulate the release of somatotropin through the somatotropin releasing factor. The increase in plasma PRL is potentiated by estrogen and testosterone.

The enkephalins inhibit ovulation and the release of lutropin by inhibiting the synthesis of gonadoliberin, which is induced by VIP, dopamine and adrenergic mechanisms. The inhibitory effects of endogenous opioids on the secretion of the gonadotropins depends on the gonadal steroids. The FSH stimulated formation of progesterone is promoted by the enkephalins [4.422], while the secretion of testosterone from the testes of immature rats is suppressed [4.423].

In experimental animals, endogenous opioids inhibit the release of TRH and, thus, the formation of TSH [4.424]. On the other hand, treatment with naloxone lowers the plasma TSH level in humans and DAMME potentiates the TSH releasing effect of TRH.

The endogenous opioids act also as immunomodulators. They may act as immunostimulants [4.425] and as immunosuppressants [4.426]; in fact, endogenous opioids also acquire a role in the stress induced growth of tumors [4.427], [4.428].

Structure-Activity Relationships. For structure-activity relationships, see also Chap. 4.5.2.: selective ligands for opioid receptors.

Tyr^1 is essential for opiate activity. However, the N-terminal amino group can be methylated, guanylated or extended by amino acids, e.g., Phe or Tyr, without especially affecting the activity. The N-allyl-EKs (particularly N-allyl-Met-EK) are, analogous to naloxone, antagonists for morphine and EK.

Gly^2 can be replaced by α-aminoisobutyric acids or by D-amino acids. These compounds are considerably more active and more resistant to enzymatic degradation. The more lipophilic the substitution, the stronger the binding to the μ-receptors and, therefore, the analgetic activity. In the case of hydrophilic substitution, the affinity for the δ-receptors increases.

Gly^3 is important for the biological activity of the EKs. Only the substitution of Aza-Gly leads to a more active compound.

The replacement of Phe^4 by other amino acids usually lead to loss of activity. However, N-methylation or substitution of Aza-Phe leads to an increase in the analgetic activity. The substitution of p-nitrophenylalanine ($Phe(NO_2)$), pFPhe [4.429] or carboranylalanine (Car) for Phe^4 results in an increase in activity in the GPI test, which may be caused by the electrophilic effect of these substituents.

The more lipophilic the amino acid in position 5, the stronger the analgetic effect. For instance, derivatives with Pro-amide and thiazolidine carboxylic acid amide in position 5 exert a specially strong analgetic effect. The i.c.v. application of the O-galactosyl derivative of [D-Met^2, Hyp^5] EK amide produces an analgetic effect that is 50 000 times that of morphine [4.430]. In the GPI assay, the replacement of the amide structure by isosteric formyl amides results in roughly another 7-8 fold increase in activity. Position 5 does not necessarily have to be occupied by amino acids. The position can contain any residue, but preferably a lipophilic one. For instance, methioninol sulfoxide (Met(O)-ol), the thiolactone of homocysteine and substituted hydrazides proved to be favourable.

Further shortening of the chain leads to the tripeptidyl-N-methyl-phenethylamides or -2-amino-4-methyl-pentane-amides, which have a still higher analgetic activity. Another simplification results in H-Tyr-N(Me)-$(CH_2)_4$CO-N(Me)-C_2H_4-phenyl, which shows high analgetic activity when applied intravenously [4.431].

An interesting EK analog that is cyclized via a disulfide bridge is [D-Cys^2,D-Cys^5] EK-amide, which shows about twice the DAMME activity in the GPI assay. All these highly active compounds are more stable to enzymatic degradation than the rapidly degradable natural EKs, see Table 12.

The two hydrophilic EK analogs, [D-Arg^2, $Phe(NO_2)^4$, Pro^5]EK-amide (BW 942C) [4.431a], [4.431b] and [D-$Met(O)^2$, $Phe(NO_2)^4$, Pro^5]EK-amide (nifaltide) =CAS-98311-64-9 and CAS-73385-60-1> [4.431c] with peripheral selectivity [4.431a] are effective against diarrhea.

Table 12. Enkephalin analogs with high analgetic activity

Enkephalin analogs	Relative analgetic activity	
Morphine (in comparison)	1.0	
Tyr-D-Ala-Gly-Phe-Met-amide (DAME, DAMEA, DALAMID, Dalamet) [4.432]	0.05	(i.v.)
Tyr-D-Lys(For)-Gly-Phe-homocysteinthiolacton (Ociltide)	0.5	(i.v.)
Tyr-D-Ala-Gly-MePhe-Met (O)-ol (DAMME, FK 33-824 [4.433]	4.0	(i.v.)
Tyr-D-Met-Gly-Phe-Pro-amide (DMPEA)	3.0	(i.v.)
Tyr-D-Met (O)-Gly-MePhe-NH-NH-CO-C_2H_5	5-6	(i.v.)
Tyr-D-Met (O)-Gly-N-methyl-phenethylamide (SD-33)	1.0	(s.c.)
MeTyr-D-Met (O)-Gly-N-methyl-phenethylamide	1.7	(s.c.)
Tyr-D-Met (O)-Gly-MePhe-ol (SD 25, Syndyphalin)	6.6	(s.c.)

4.5.5.2. Peptides Derived from Preproenkephalin B

Occurrence. Leu-enkephalin is formed primarily from preproenkephalin B [4.434]. The longer dynorphins (DPs) and neoendorphins (neo-EP), which are formed from preproenkephalin B, were isolated first from the hypothalamus, hypophysis, and from the duodenum of the pig. Later, DPs were found in the adrenal medulla and in isolated chromaffin cells from the bovine adrenal cortex [4.435], in guinea pig heart [4.436] and in the rat duodenum [4.437]. Big-DP contains the sequences of DP-A-17 and DP-B-13.

Release. During dehydration, immunoreactive DP is raised in the hypothalamus in the night and during the day and is lowered in the hypophysis.

Leu-Enkephalin	Y	G	G	F	L														
p-(175-183), β-Neo-EP	Y	G	G	F	L	R	K	Y	P										
p-(175-184), α-Neo-EP	Y	G	G	F	L	R	K	Y	P	K									
p-(209-256), Big-DP	Y	G	G	F	L	R	R	I	R	P	K	L	K	W	D	N	Q	K	R-
	Y	G	G	F	L	R	R	Q	F	K	V	V	T	R	S	Q	E	D	P-
										N	A	Y	Y	E	E	L	F	D	V
p-(209-216), DP-A-8	Y	G	G	F	L	R	R	I											
p-(209-219), DP-A-11	Y	G	G	F	L	R	R	I	R	P	K								
p-(209-221), DP-A-13	Y	G	G	F	L	R	R	I	R	P	K	L	K						
p-(209-225), DP-A-17	Y	G	G	F	L	R	R	I	R	P	K	L	K	W	D	N	Q		
p-(228-240), DP-B-13 (Rimorphin)	Y	G	G	F	L	R	R	Q	F	K	V	V	T						
p-(228-256), DP-B-29	Y	G	G	F	L	R	R	Q	F	K	V	V	T	R	S	Q	E	D	P-
(Leumorphin)							N	A	Y	Y	E	E	L	F	D	V			

In ovariectomized rats, the level of immunoreactive dynorphinlike material increases in the anterior lobe of the hypophysis, an increase that is prevented by estrogen [4.438]. There is an accumulation of immunoreactive DP-8 in the hippocampus and frontal cortex of old rats which appears to be accompanied by a reduced learning ability [4.439]. After traumatic injuries, i.r. dynorphin is raised in the spinal cord of the rat. This appears to be responsible for hind limb paralysis, an attendant symptom of traumatic injuries [4.440].

The i.r.-DP levels are lowered in the spinal cord of schizophrenic patients [4.441].

Biological Effects. See Chap. 4.5.3. for κ-mediated opioid actions. DPs potentiate the glucose or amino acid stimulated release of insulin [4.442]. They inhibit the release of somatostatin [4.442], TRH [4.443], pancreatic polypeptide [4.442], oxytocin [4.444], vasopressin [4.445], [4.446] and counteract morphine tolerance [4.447]. DP-A or des-Tyr-DP-A inhibits the binding of the growth hormone releasing hexapeptide, His-D-Trp-Ala-Trp-D-Phe-Lys-amide (SK&F 110679) via non-opioid mechanisms [4.448]. Presumably due to the Arg-Pro sequence, DP-A-13 and DP-A-17 release histamine in rat peritoneal mast cells [4.449] and lead to the extravasation of plasma in the rat skin [4.450].

The central application of DPs results in a decrease in the body temperature, increase in the uptake of food [4.451] and water [4.452], cataleptic activity [4.453], the suppression of motory activity [4.454], the reduction of the response to acoustic signals [4.454], and the stimulation of large amplitude slow wave activity in the EEG of rats, and also leads to hindlimb paralysis similar to that observed after spinal cord injury [4.455].

In cerebral ischemia, DP-A-13 increases the survival rate [4.456].

Structure-Activity Relationships. See also Chap. 4.5.2.: peptide ligands for the κ-receptor. DPs are specific ligands for the κ-receptors. The affinity for the δ-receptors increases with decreasing chain length [4.457], see Table 13.

DP-A-17 and DP-A-13 are more stable to peptidases than DP-A-8 and DP-A-9. Therefore, short chain DPs have a considerably shorter biological activity. For this reason, it is assumed that DP-A-8 and DP-A-9 have a neural transmitter or modulator function at the κ-receptors, and the more stable DP-A-13 and DP-A-17 have a more hormonal function.

Table 12. Activity of dynorphin-like peptides in the GPI assay

Peptide	Amino acid sequence																	Relative activity in GPI assay
	1	2	3	4	5	6	7	8	9		11		13					
p-DP-A-17	Y	G	G	F	L	R	R	I	R	P	K	L	K	W	D	N	Q	100
p-DP-A-13 [4.458]	Y	G	G	F	L	R	R	I	R	P	K	L	K					100
p-DP-A-8	Y	G	G	F	L	R	R	I										3.1
Rimorphin	Y	G	G	F	L	R	R	Q	F	K	V	V	T					88
p-α-neo-EP	Y	G	G	F	L	R	K	Y	P	K								4.5
p-β-neo-EP	Y	G	G	F	L	R	K	Y	P									1.5
Leu-EK	Y	G	G	F	L													0.2

The C-terminal carboxyl group appears to be unfavourable since the corresponding methyl esters have a much higher potency in the GPI assay. Similar to other opioids, the N-terminal tyrosine is essential for opiate activity.

The basic amino acids are particularly important for the biological activity of DP-A-13. Indeed, Arg^6 and Arg^7 are of special significance [4.459]. The substitution of Ala for Ile^8 in DP-A-13 leads to a more active (2-9 times) compound [4.459]. Cyclo-[D-Cys^2, Cys^5, N-$MeArg^7$, D-Leu^8]DP-(1-8)-ethylamide is extremely active. It is 3.4 times as analgetically active as DP-A, but binds more strongly to the μ- and δ-receptors than to the κ-receptors [4.460].

The substitution of D-Trp in DP-A-11 gives rise to weak and not very selective antagonists: [D-$Trp^{2,8}$, D-Pro^{10}]-, [D-$Trp^{5,8}$, D- Pro^{10}]- and [D-$Trp^{2,4,8}$, D-Pro^{10}]DP-A-11 [4.461]. The compounds, N-di-allyl-[D-Pro^{10}]DP-A-11 and N-di-allyl-[$Aib^{2,3}$, D-Pro^{10}]DP-A-11 are also opioid antagonists with weak selectivity for κ-receptors [4.462]. Des-Tyr^1-rimorphin also inhibits morphine induced activities [4.463].

Applications. The intrathecal (i.t.) application of 15 μg of DP-A-13 to cancer patients produces a nociceptive effect that lasts for more than 4 hours [4.464].

4.5.6. Dermorphin [4.464] and Deltorphins

Occurrence. Dermorphin (DM), [6-hydroxyproline]DM, deltorphin (DT, dermenkephalin) [4.465] and [D-Ala^2] deltorphin I and II [4.466] were until now the only naturally occurring regulatory peptides that contain a D-amino acid. They were isolated from the skin of the South American frogs, *Phyllomedusa sauvagii, Phyllomedusa rhodei* and *Phyllomedusa bicolor*. A precursor of DM contains five copies of a sequence of 35 amino acids, which, in turn, contains the dermorphin sequence at the C-terminal end [4.467]. Another procompound contains only four copies of DM and, in addition, a copy of deltorphin [4.465]. A propeptide from *Phyllomedusa bicolor* contains three copies of [D-Ala^2]deltorphin I and one copy of [D-Ala^2]deltorphin II [4.466]. The precursors still contain alanine and methionine in the L form. Thus, the conversion to the D configuration is a post-translational reaction.

DM-precursor-(162-196):

	1 ... 10 ... 20 ... 30 ... 35
	I K R E S E E E N E N E E N H E E G S E M K R Y A F G Y P S G E A K K
Dermorphin	Y a F G Y P Sa
Deltorphin (DT Dermenkephalin)	Y m F H L M Da
[D-Ala^2]Deltorphin I	Y a F D V V Ga
[D-Ala^2] Deltorphin II	Y a F E V V Ga

Immunoreactive DM has also been found in the brain, adrenal glands and in the gastrointestinal tract of the rat [4.468].

Biological Effects of Dermorphin. DM binds preferably to the μ-receptors. In the GPI test, DM is 57 times as active as Met-EK and, when given intravenously, has on a molar basis about 10 times the analgetic activity of morphine.

DM inhibits both in small doses (i.c.v.) as well as in large doses (s.c. or i.p.) the secretion of gastric acid, the emptying of the stomach, pancreatic secretion and intestinal motility. In this process, the plasma level of gastrin rises and that of PP and secretin, falls [4.469]. DM raises (i.v.) the plasma prolactin level (via the central nervous system) [4.470], somatotropin [4.471], thyrotropin [4.472], somatostatin, plasma glucagon [4.473], and the blood pressure and heart rate. On the other hand, DM inhibits the release of ACTH, β-LPH, β-EP [4.474] and LH [4.475].

The N-terminal tetrapeptide, DM-(1-4)-amide stimulates the basal as well as the pentagastrin or histamine induced gastrointestinal secretion in a process that can be reversed by naloxone. It also stimulates phase III of the migratory motorial complex (MMC) in the duodenum/jejunum and increases contraction in the proximal and distal parts of the colon [4.476].

Biological Effects of the Deltorphins. In contrast to DM, which binds preferably to the μ-receptors, DT favours the δ-receptors. Unlike other δ-ligands, deltorphin when given i.c.v. improves the memory of mice [4.477].

Structure-Activity Relationships of Dermorphin. The rules that applied to the enkephalins are not applicable to dermorphin. Position 3 is occupied by Phe instead of Gly, which was assumed to be essential, and the important aromatic Phe in position 4 is replaced by Gly.

The C-terminal amide group is important for the biological activity of DM [4.478]. In the GPI assay, [Tyr^7] DM has twice the DM activity and 1.4 times the analgetic activity [4.479]. The substitution of Hyp for Pro^6 gives rise to a compound with the same analgetic and gastrointestinal activity (inhibition of the gastrointestinal transit) and a higher prolactin releasing activity [4.478]. [Tyr(OMe)^5]DM has a higher affinity for the μ-receptors and [Phe^5] DM binds more strongly than DM to the δ-receptors [4.480].

[D-Arg^2]DM has about the same analgetic activity as DM [4.481], [4.482]. Shortening the [D-Arg^2]dermorphins at the C-terminal end results, especially in the case of the hexapeptide and the tetrapeptide, in an enhancement of the activity. The tetrapeptide amide is more than twice as active as [D-Arg^2]DM. The tripeptide amide still has about 10% of the [D-Arg^2]DM activity [4.482].

N-terminal guanylated DM exerts a higher analgetic effect and clearly inhibits the gastrointestinal transit [4.483].

[D-Phe^5]- and [phenylglycine^5]DM-(1-5)-amide [4.484] or [Met(O)^5]DM-(1-5)-amide [4,483] are about as active as DM.

In the case of the tetrapeptide analogs, amides with a bulky side chain, N-terminal guanylated compounds, and the [Sar^4], [D-Arg^2], and [D-Met(O)^2] substitutions all have a higher activity. The N-guanylated DM-(1-4)-adamantyl amide is 160 times as active as DM-(1-4)-amide [4.485], and the N-guanylated [Sar^4]DM-(1-4)-D-NH-CH(CH_3) C_6H_5 has 16-28 times the potency of DM [4.486]. [D-Arg^2] DM-(1-4)-amide when applied subcutaneously to mice exerts a higher analgetic effect than DM [4.487] and it also has an antinociceptive effect in morphine tolerant mice [4.488]. Depending on the assay, H-Tyr-D-Arg-Phe-Sar-OH is 21-58 times as active as morphine [4.489] and is, consequently, more active than DM [4.481]. When given subcutaneously to mice, [D-Arg^2, β-Ala^4]DM-(1-4)-OH is 32.7 times more effective than morphine [4.490]. The i.c.v. application of [D-Met(O)^2]DM-(1-4)-amide to mice produces 1500 times the effect of morphine and the subcutaneous application produces 17 times the effect of morphine [4.491]. In this series, the subcutaneously most active compounds are [D-Met(O)^2, Sar]DM-(1-4)-amide and [D-Met(O)^2, D-Ala^4]DM-(1-4)-OH (22 and 30 times the potency of morphine respectively) [4.492]. One of the best μ-selective ligands was found to be [D-Met(O)]-DM-(1-4)-OMe [4.492].

Although DM-(1-3)-amide, which is shortened to three amino acids, still has 15 times the analgetic activity of Leu-EK, it has only 1/4 of the activity in the GPI assay. The N-terminal dipeptidyl-3-phenyl-1-propylamide is three times as active as DM-(1-3)-amide in the GPI assay. Extremely analgetic (3-4 times the potency of morphine) subcutaneously active compounds are Tyr-D-Met(O)-Phe-hydrazide (SD-49) and Tyr-D-Met(O)-Phe-Gly-OEt (SD-48).

Among the various [D-Arg^2]DM-(1-3)-amides, the methylamide is 2.8 times, the dimethylamide 5.3 times, the glycine extended 4.8 times and the β-Ala extended compound 32.7 times as active as morphine when each is administered subcutaneously [4.493].

Structure-Activity Relationships of Deltorphin. It has been shown by ^{1}H-NMR studies that the conformational preference of the N-terminal sequences of both peptides, DM and DT, is the same. The varying selectivity for the opioid receptors is explained by charge effects and by the hydrophobicity of the C-terminal part of the molecule [4.494]. DT-(1-4)-amide practically binds only to the μ-receptors. Met^6, Asp^7 and Leu^5 are important for the binding to the δ-receptors. Met^6 can be oxidized or replaced by D-Met [4.495].

4.5.7. Kyotorphin [4.496], [4.497]

Occurrence. Kyotorphin (KP), a dipeptide with the structure Tyr-Arg, was isolated from the bovine brain. It is presumably formed from neo-KP (Thr-Ser-Lys-Tyr-Arg), which is also isolated from the bovine brain. This sequence corresponds to the

C-terminal end of the h- and b-hemoglobin α-chain. ACE cleaves neo-KP producing KP [4.498].

Biological Effects. When applied centrally, KP exhibits analgetic activity in mouse tail pinch, hot plate and acidic acid writhing tests. It releases Met-EK.

Structure-Activity Relationships. D-Tyr-Lys [4.499] is about twice and Tyr-D-Arg [4.500] is four times as active as KP. Similar peptides which on i.c.v. application exhibit antinociceptive activity, reversible with naloxone, are H-Trp(2-Nps)-Arg-OH and H-Trp(2-Nps)-Lys-OH and their methylester hydrochlorides, which have about 50 times the potency of KP [4.501]. A retro-compound, Lys-Trp(2-Nps) is also about half as active as [D-Ala^2]Met-EK. Orn-Trp(2-Nps) and Arg-Trp(2-Nps) are more active than Lys-Trp(2-Nps). These compounds do not bind to opitate receptors [4.500].

In Neo-KP, the replacement of lysine and arginine by the corresponding D amino acids produces an increase in activity. D-Ser or Pro can also be substituted for Ser^2 with a gain in activity. The most active compounds were found to be [D-Ser^2]neo-KP and [Pro^2]neo-KP, which had about 10 times the potency of neo-KP [4.502].

4.5.8. Exorphins [4.503]

The term exorphins refers to peptides with opiate activity that are produced from food during digestion.

Occurrence. β-Casomorphin-7 (β-CM-7) was isolated from casein peptone and corresponds to b- or o-β-casein-(60-66) [4.504]. Even the more active N-terminal β-CM-tetrapeptide-amide, morphiceptin, which was already synthesized prior to isolation, could be isolated from enzymatically digested milk proteins [4.505].

α-Casomorphins were isolated from α-casein by hydrolysis with pepsin. They correspond to the sequence 90-96 and 90-95 of α-casein. An exorphin-like sequence, α-gliadin-(43-49), was found in α-gliadin [4.506].

A fragment (F 12-5) of the fraction F 12 from cobra poison is structurally similar to the casomorphins and has analgetic activity [4.507].

b-β-Casein-(57-69)	S	L	V	Y	P	F	P	G	P	I	P	N	S
o-β-Casein-(57-69)	S	L	V	Y	P	F	P	G	P	I	A	N	S
h-β-Casein-(48-60)	P	L	I	Y	P	F	V	E	P	I	P	Y	G
β-CM-7				Y	P	F	P	G	P	I			
β-CM-5				Y	P	F	P	G					
β-CM-4				Y	P	F	P						
Morphiceptin [4.508]				Y	P	F	Pa						
h-β-CM-5				Y	P	F	V	E					
h-β-CM-4				Y	P	F	V						
α-Gliadin-(43-49)				Y	P	Q	P	Q	P	F			
α-CM-7			R	Y	L	G	Y	L	E				
α-CM-6			R	Y	L	G	Y	L					
F 12-5				V	P	R	P	Y					

Biological Effects The β-casomorphins have only slight opiate activity. Intragastric β-CMs cause the release of somatostatin. In a process that can be reversed by naloxone, both digested gluten as well as β-CM (oral) stimulate the release of insulin and glucagon in dogs after a test meal of liver extract and sucrose [4.509].

From the serosal side, the casomorphins exert an opioid effect on intestinal electrolyte transport [4.510]. When applied parenterally, they stimulate the postprandial release of PP and the amino acid or glucose stimulated release of insulin [4.511], increase the plasma level of PRL [4.512], inhibit the release of somatostatin and TRH, and reduce the level of TSH [4.513].

Structure-Activity Relationships. The N-terminal tripeptide has no activity. Morphiceptin, the N-terminal tetrapeptide amide, has the same activity as Met-EK in the GPI test and, like dermorphin, is a specific ligand for the μ-receptors. The peptides, h-β-CM-(1-4)-amide (valmuceptin) and particularly h-[D-Val^4] β-CM-(1-4)-amide (devalmuceptin) bind more tightly to the μ opiate receptors than morphiceptin [4.514].

The substitutions of D-Pro or D-pipecolic acid for Pro^4 lead to [D-Pro^4]CM-5 (deprolorphin), [D-Pro4] morphiceptin (deproceptin, Wellcome PL 017) [4.515], and [D-pipecolic acid^4] β-CM-5 (depilorphin), which are more active than morphine both *in vivo* and *in vitro*. These compounds have a high analgetic activity in the rat on either i.c.v. or intravenous application [4.516]. The substitution of D-Phe for Phe^3 increases, although to a lower extent, the antinociceptive activity [4.516]. While the replacement of D-Pro^2 leads to loss of opioid activity, [D-pipecolic acid^2]-β-CM-5 exerts a greater long-lasting analgetic effect [4.516]. On i.c.v. and i.v. administration, deprolorphin develops the highest activity with 28 and 2 times the analgetic activity of morphine respectively [4.516]. It has on i.c.v. application 645 times the activity of β-CM-5 [4.516]. Both the substitution of D-Pro for Pro^4 as well as D-Phe for Phe^3 promotes the binding to the μ-receptors. In organ preparations, [D-pipecolic acid^2]-β-CM-5 binds best to the δ-receptors, but in brain membranes it binds best to the μ-receptors. There appears to be a difference between central and peripheral μ-receptors [4.517].

4.5.9. Femarfarmamide and Related Structures [4.518]

Occurrence. Femarfarmamide (FMRF-amide) was isolated from molluscs, e.g., from the mussel *Macrocallista nimbosa*. FMRF-amide is not derived from prepro-EK A, which contains FMRF at the C-terminal end, but is coded in a gene, e.g., in *Aplysia,* that contains 21 copies of the FMRF sequence [4.519].

With the help of immunocytochemical methods, FMRF-amide like material was discovered in the pancreas of chicks, in the ileum of the dog, and in brain nerve cells from the frog and the rat. The first FMRF-amide like peptide from vertebrates, LPLRF-amide, was isolated from chick brain. Two other peptides, which cross reacted with antibodies to FMRF-amide, were isolated from bovine brain extracts (b-A-19-F-amide and b-F-8-F-amide) [4.520].

The following compounds are structurally related to FMRF amide: ly-SDPFLRF-amide and ly-GDPFLRF-amide [4.521] from the water snail *Lymnaea stagnali*, sip-GDPFLRF-amide and sip-NDPFLRF-amide from the snail *Siphonaria pectinata* [4.522], the small cardioactive peptides SCPa and SCPb [4.523], peptide II [4.524] from *Aplysia* and other gastropod molluscs, antho-RF-amide from *Helix aspera* [4.521], sch-FLRH-amide from the locust *Schistocera gregaria* [4.525], leucomyosuppressin from the cockroach *Leucophaea maderae* [4.525], ma-FLRF-amide from the hawkmoth *Manduca sexta* [4.526], and antho-RW-amide I and II from the sea anemone *Anthopleura elegantissima* [4.527] [4.527] or the pennatulid *Renilla köllikeri* [4.528].

Two further peptides which cross react with FMRF-amide antibodies were isolated from brain extracts of cattle (b-A-19-F-amide and b-F-8-F-amide) [4.529].

h-Prepro-EK-(259-267)										...K	R	Y	G	G	F	M	R	F.
FMRF-amide															F	M	R	Fa.
ly-SDPFLRF-amide												S	D	P	F	L	R	Fa.
ly-, sip-GDPFLRF-amide												G	D	P	F	L	R	Fa.
sip-NDPFLRF-amide												N	D	P	F	L	R	Fa.
sch-FLRF-amide									P	D	V	D	H	V	F	L	R	Fa.
Leucomyosuppressin									*Q	D	V	D	H	V	F	L	R	Fa.
ma-FLRF-amide									*Q	D	V	V	H	S	F	L	R	Fa.
SCPa								A	R	P	G	Y	L	A	F	P	R	Ma.
SCPb										M	N	Y	L	A	F	P	R	Ma.
Helix-Peptide													E	N	F	I	R	Fa.
Antho-RF-amide															*Q	G	R	Fa.
Antho-RW-amide I														*Q	S	L	R	Wa.
Antho-RW-amide II														*Q	G	L	R	Wa.
Peptide II *Aplysia*									E	A	E	E	P	S	F	M	T	RL.
g-LPLRF-amide														L	P	L	R	Fa.
b-A-19-F-amide [4.529]	A	G	E	G	L	S	S	P	F	W	S	L	A	A	P	Q	R	Fa.
b-F-8-F-amide [4.529]											F	L	F	Q	P	Q	R	Fa.

Biological Effects [4.530]. In molluscs, FMRF-amide was found to exert both a stimulating as well as an inhibitory effect on the heart. Electrophysiological studies indicate a hyperpolarizing effect on the neurons in the snail *Helix pomatia*. In rats, both FMRF-amide and LPLRF-amide [4.531] (intravenously and centrally applied) as well as intravenously applied b-A-19-F-amide and b-F-8-F-amide [4.520] increase the arterial blood pressure.

FMRF-amide has an inhibitory effect on the spontaneous or acetylcholine induced contraction of the anterior gizzard of *Aplysia california* and the stomach of *Navanax* [4.532]. It also inhibits the colon motility of mammals [4.530].

FMRF-amide has absolutely no opiate activity, but appears to be an opiate antagonist. While analogs of enkephalin relax the isolated rectum of *Mercenaria mercenaria* and *Macrocallista nimbosa*, FMRF-amide exerts a contracting effect [4.533]. On i.c.v. application, FMRF-amide, like naloxone, inhibits not only the defence induced analgesia, food uptake and aggression [4.534], but also the food uptake in response to κ-agonists (U-50, 488H) [4.535]. However, the weak binding of FMRF-amide to the κ- or μ-receptors allows the postulation of other mechanisms for the antagonistic activity of FMRF-amide [4.536]. Among the FMRF-amide like peptides isolated from the cow, the intraventricular application of b-F-8-F-amide reduces the morphine induced analgesia in rats [4.529] and could be responsible for the abstinence syndrome in morphine addiction [4.537].

On i.c.v. application in rats, FMRF-amide exhibits amnesic activity [4.538] and increases the plasma level of somatotropin after 15 minutes, without affecting the plasma level of LH, FSH and PRL [4.539].

The cardioactive peptides SCPa and SCPb from *Aplysia* stimulate the heart of molluscs and contract their intestine.

Structure-Activity Relationships. The C-terminal amide group and the full length of the molecule are important for the biological activity of FMRF-amide. Acetylation and benzoylation increase the contractile effect of FMRF-amide. A hydrophobic amino acid, e.g., Nle and Phe, is required in position 2, while shorter chained amino acids are not favourable for activity [4.540].

5. Blood Pressure Regulating Peptides

This chapter deals with the angiotensin-kinin system, the neurokinins and tachykinins, vasopressin and the structurally related peptides, oxytocin and vasotocin, the endothelins, the atrial natriuretic factor (ANF), neuropeptide Y (NPY) and peptides of similar structure.

These peptide hormones influence blood pressure and evidently function as modulators of the reproductive hormones (Chap. 2) and of the stress induced proopiomelanocortin cascade (Chap. 4). Angiotensin II (AT II) and, above all, the kinins appear to play a role in ovulation and in the formation of semen. On the one hand, AT II stimulates the proopiomelanocortin (POMC) cascade, presumably via corticoliberin (CRH) and the formation of gluco- and mineralocorticoids, but, on the other hand, it raises the level of lutropin (LH) when applied centrally. Vasopressin has a peripheral stimulating effect and inhibits the POMC cascade when administered centrally, while oxytocin plays a role during mating as well as during and after birth. Endothelin releases LH from the hypophysis and ANF, which can be released by endothelin, inhibits the formation of the stress hormone cascade and, above all, the synthesis of aldosterone. NPY also exerts an effect on the secretion of LH through the release of GnRH from the hypothalamus.

5.1. The Angiotensin-Kinin System [5.1]

The kinins as well as the angiotensins are formed in biological fluids by the enzymatic cleavage of proteins. Kallikreins cleave kininogens (KG) to yield kinins, which are blood pressure lowering peptides. The species specific enzyme renin, which can also be generated by kallikrein from inactive prorenin, is responsible for the formation of angiotensin I (AT I) from α-globulin angiotensinogen (ATG). AT I which has no effect on the blood pressure, is split by the membrane bound angiotensin converting enzyme (ACE, kininase II), a zinc containing carboxy dipeptidase, to form angiotensin II (AT II), a vasopressor. On the other hand, the hypotensive kinins are degraded by ACE. Therefore, the inhibition of ACE results in a double hypotensive effect because both the formation of the blood pressure raising AT II as well as the degradation of the blood pressure lowering kinins are inhibited. The first inhibitors of ACE were isolated from snake venom and called bradykinin potentiating peptides. Orally active inhibitors of ACE are, e.g., captopril, enalapril and ramipril. Inhibitors of renin also exert a hypotensive effect. Potent inhibitors of renin are ATG analogs in which the peptide bond between positions 10 and 11 is reduced or replaced by statin, an amino acid occurring in pepstatin, or by similar synthetic variants.

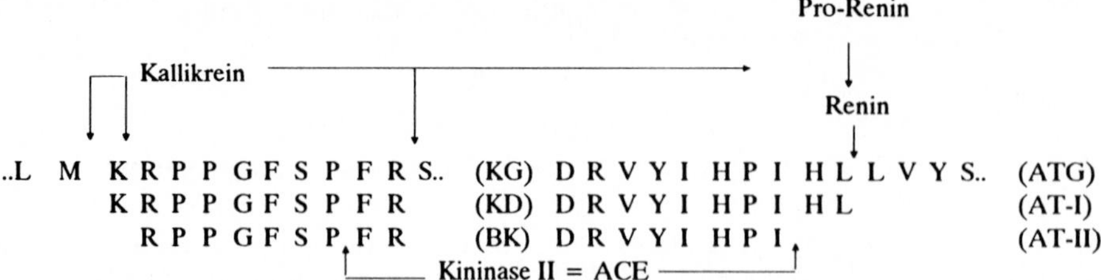

In each case, AT I is located at the N-terminal end of ATG. The structures of r-prepro-ATG and h-Prepro-ATG have been elucidated [5.2].

	1									10															
e-ATG-(1-14)	D	R	V	Y	I	H	P	F	H	L	L	V	Y	S	.	.	.	.,							
h-ATG-(1-25)	D	R	V	Y	I	H	P	F	H	L	V	I	H	N	E	S	T	C	E	Q	L	A	K	A	N .
r-ATG-(1-25)	D	R	V	Y	I	H	P	F	H	L	L	Y	Y	S	K	S	T	C	A	Q	L	E	N	P	S .
h, r, e-AT I	D	R	V	Y	I	H	P	F	H	L															

Glucocorticoids stimulate the formation of ATG [5.3], [5.4]. IL-6 potentiates the glucocorticoid stimulated synthesis of ATG [5.4], and AT II or [Sar[1]] AT II inhibits the production of ATG in hepatocytes [5.5].

5.1.1. Angiotensins

Occurrence. Angiotensins (ATs) are found both in the periphery and in the brain. Plasma AT II is formed primarily in the lungs. Other sites of production of AT II and AT III are the juxtaglomerula cells of the kidney [5.6], the adrenal cortex [5.7], the gonadotropic cells of the hypophysis [5.8], the hypothalamus [5.9], the ovarian follicle [5.10] and the Leydig cells of the testes.

A rapid accumulation of the AT-(1-7)-heptapeptide was observed in the dog brain after administration of AT I. This effect could also be of physiological importance [5.11]. Roughly equimolar amounts of At II-(1-7), AT II and At I have also been found in the rat hypothalamus [5.9].

Contrary to previous results, position 5 of o-AT I is occupied by Ile and not by Val [5.12].

				AT-1 (1–10)									
				AT-II (1–8)									
					AT-III (2–8)								
				1	2	3	4	5	6	7	8	9	10
h, ca, p, o, m, r, cu, e				D	R	V	Y	I	H	P	F	H	L
b				D	R	V	Y	V	H	P	F	H	L
jap. goosefish				N	R	V	Y	V	H	P	F	H	L
g				D	R	V	Y	V	H	P	F	S	L
snake, amphibian			X-	B	R	V	Y	V	H	P	F	Y	L
Crinia georgiana	A	P	G	D	R	I	Y	V	H	P	F		

Release. AT I is formed by the action of renin on α-globulin angiotensinogen

(ATG). AT I is in turn cleaved by the angiotensin converting enzyme (ACE), which is identical to kininase II, or carboxypeptidases to form AT II [5.13]. This effect is potentiated in the pulmonary arteries by the platelet activating factor (PAF) [5.14]. AT III is formed from AT I via AT II (aminopeptidase A) or de-Asp^1-AT I (ACE or carboxypeptidases), see Figure 4. Large amounts of aminopeptidase A were found in the adrenal cortex.

Estradiol or a combination of estradiol and progesterone increase the plasma level of AT I and II in rats, while progesterone alone exerts no such effect [5.15]. A low NaCl diet raises and a high NaCl diet lowers the level of AT II in the serum [5.16]. However, the i.c.v. application of NaCl appears to result in a pressor effect through the secretion of AT II [5.17]. The plasma AT II level is increased in cases of thirst [5.18], dehydration, loss of blood [5.19] and endotoxin shock [5.20].

Receptors. It is possible to distinguish between two AT II receptors in the adrenal gland.

The AT1 receptor (adrenal cortex) is sensitive to DuP-753 [5.21].

The AT2 receptor (adrenal medulla) is sensitive to PD 123 319 ([5.21], Exp-655, WL-19 [5.22]).

AT1 and AT2 receptor like binding sites have also been found in the brain [5.21.]-[5.23.]. AT III has its own subclass of receptor in the vascular smooth muscles. Here, [Sar^1, IIe^7] AT III acts as a selective antagonist for AT III [5.24].

The density of the AT II receptors can be increased by dehydration [5.25], stress [5.26] or IGF I [5.27]. Spontaneously hypertensive rats have a higher AT II receptor density in the nucleus tractus solitarius [5.28]. Treatment with AT II leads to the down regulation of the receptors in rat hepatocytes [5.29]. The density of the AT II receptors on the glomeruli of the kidney is reduced by a low intake of sodium chloride [5.18], in diabetes mellitus [5.30] and by ACTH-(1-24) [5.31]. Chronic treatment with estrogen reduces the AT II receptor density in the pituitary gland, adrenal cortex

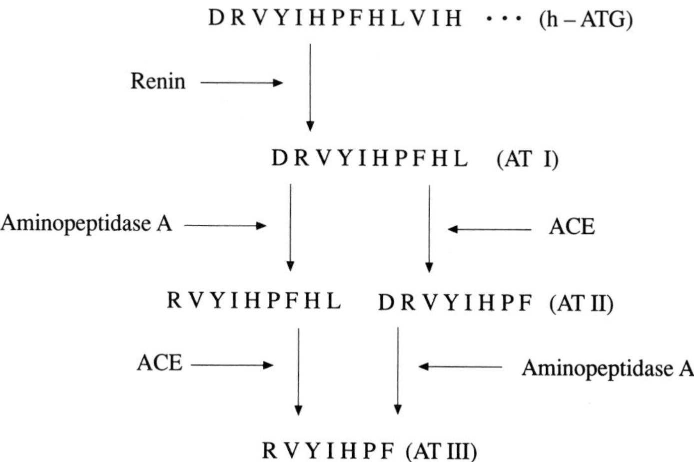

Figure 4. Enzymatic release of angiotensins

[5.32] and in the rat placenta [5.15].

Biological Effect. In the cells, AT II and AT III raise intracellular levels of calcium, phosphatidic acid and phosphoinositides [5.33], stimulate cGMP [5.34], and increase the formation of PGE_2 and PGI_2 by activating phospholipase A_2 [5.35].

In the kidney, AT II inhibits the activity of renin [5.36] and the mRNA for renin by a negative feedback effect through a lipoxygenase product, while an increase in the mRNA for angiotensinogen in the liver is observed [5.37].

AT II exerts a contracting effect on the vascular smooth muscles of the uterus, intestine, aorta, myocardium and kidney, resulting in a potent vasoconstricting and blood pressure increasing activity. The renal vasoconstricting effect of AT II is enhanced by indomethacin. Prostaglandins counteract the activity of AT II in the kidney [5.38].

Other effects of AT II. It increases plasma vasopressin levels [5.39], [5.40]; reduces receptors for endothelin in vascular smooth muscle cells [5.41]; exerts a positive inotropic effect on the myocardium [5.42]; reduces the glomerular filtration rate and, in physiological doses, the excretion of Na^+ and water [5.43], but it exerts a natriuretic and diuretic effect when given in high doses [5.44]; reduces the water absorption in the jejunum through the synthesis of prostaglandins [5.45]; stimulates the formation of both aldosterone and cortisterone in the adrenal glands on short term administration, but only of aldosterone on chronic application [5.46]. AT II also stimulates proliferation in the fasciculata zone of the adrenal gland [5.47] and the synthesis of IGF I receptors on bovine adrenocortical cells [5.27], PDGF-BB receptors on rat aortic smooth muscle [5.48] and B-FGF mRNA in b-luteal cells [5.49]. It might possibly play a role in ovulation [5.50], [5.51]. It stimulates the secretion of both androgen and estrogen from rat ovaries [5.10] and reduces the LH stimulated synthesis of progesterone in bovine luteal cells [5.49]; stimulates the release of the atrial natriuretic factor (ANF) [5.52], [5.53], prolactin [5.54], somatotropin [5.55], ACTH [5.55] (presumably via CRH [5.55]), and β-endorphin [5.56], [5.57]; increases the feeling of thirst, plasma levels of aldosterone, vasopressin, and oxytocin, LH [5.59], the locomotor and exploratory behavior and acquisition of active avoidance when applied centrally [5.60]; and it reduces morphine induced analgesia [5.61] and the alcohol effects on passive and active avoidance responses when applied centrally to mice [5.62].

Other Effects of AT III. Similar to AT II, AT III stimulates the release of aldosterone and prostaglandin.

Structure-Activity Relationships. Both the C-terminal carboxyl group and all the aromatic amino acids are of importance for the actvity of AT II [5.63]. The N-terminal sequence is responsible for the specificity, intensity and duration of the biological activity of AT II. However, AT II-(1-7), like AT II, releases vasopressin and, probably plays a physiological role in the formation of vasopressin [5.11].

The peptides, h-[Sar^1]AT II, h-[Aib^1]AT II and [D-Ser^1] AT II [5.64] are more stable to aminopeptidases and more potent than AT II. The activity is lowered if Asn occupies position 1 [5.65].

The N-terminal amino group is not important for the pressor activity of AT II. However, acylation of the N-terminal amino group causes a decrease in the myotropic and aldosterone stimulating effect.

Tyr^4 is very important for biological activity and fragments without Tyr^4 are devoid of activity [5.63]. The activity depends on the electronegativity of the aromatic ring in position 4. Increasingly electronegative substituents lower activity. On the other hand, [Sar^1, 3-amino-Tyr^4]AT II [5.66] and monoiodo-AT II [5.67] are more active than [Sar^1] AT II and AT II. A β-branched amino acid in position 5 is important for the agonistic activity. In fact, O-methyl threonine proved to be more effective than Ile in position 5 [5.68].

If Pro^7 in AT I or AT II is replaced by Ala, cardioselective substances are formed which have a positive inotropic activity and a smaller pressor effect [5.42].

β, β-Diphenylalanine (Dip) appears to be better than Phe^8 since it has 284% of the activity of [Asn^1]AT II [5.69]. A cyclic analog with about the same activity as AT II is cyclo-[Cys^3, Cys^5] AT II [5.70].

AT II Antagonists. Positions 4 and 8 are responsible for the agonistic or antagonistic action of a AT II peptide.

Competitive antagonists are formed when the side chain of Phe^8 is extended or shortened by a methylene group or when aliphatic or branched aromatic [5.69], [5.71] amino acids are substituted for Phe^8. One of the most active antagonists of this series is [Sar^1, Bip^8] AT II [5.69] (Bip, biphenylalanine). The presence of D-amino acids, such as D-phenylglycine (D-Phg), D-phenethylglycine (D-Peg), D-Phe-amide, or D-(αMe)Phe, in position 8 give rise to active antagonists, which are not degradable by carboxypeptidases [5.72]. The peptide, [Sar^1, Val^5, Ala^8]AT II (Saralasin) has been developed. Apart from Saralasin, [Sar^1, Ile^5, Thr^8]AT II (DE-3489, Sarthran) has also been examined more closely. Peptides that are more active than Saralesin are, e.g., [Sar^1, $Thr(Me)^{5,8}$] AT II and [Sar^1, Chg^5, Lac^8]AT II [5.73] (Chg = L-cyclohexylglycine, Lac = L-lactic acid). Unlike the agonists, the antagonists do not require a β-branched amino acid in position 5. For instance, [Sar^1, Tyr^5, Ile^8]AT II has three times the antagonistic activity (pA^2 = 9.6) of [Sar^1, $Ile^{5,8}$] AT II (pA^2=9.1) [5.68].

Alkylation of the phenolic hydroxyl group of Tyr^4 [5.74] or the substitution of Phe [5.75] or pCl-Phe [5.76], [5.77] for Tyr^4 also gives rise to competitive AT II antagonists. However, [Sar^1, $Tyr(Me)^4$]AT II (Sarmesin) is less active than Saralesin.

Shortening the molecule at the C-terminus also leads to AT II antagonists. Saralesin or Sarmesin is about six times as active as [Sar^1]AT II-(1-7)-amide [5.78]. This compound has no partial agonistic properties and does not inhibit the central dipsogenic activity of AT II [5.79].

Non-peptide AT II receptor antagonists are imidazole derivatives, e.g., methyl-2-n-butyl-4-chloro-1-[4-(2-carboxybenzamido)-benzyl]-imidazole-5-acetate sodium salt (X-6803) (IC_{50} = 0.12 μM) [5.80], 2-n-butyl-4-chloro-5-methoxymethyl-1-[O-sulfo-benzoyl-(4-aminophenyl)-methyl-imidazole (IC_{50} = 0.012 μM) [5.81], 2-n-butyl-4-chloro-5-hydroxylmethyl-1-([2'-(1H-tetrazole-5-yl)-biphenyl-4-yl]methyl)imidazole potassium salt (DuP 753) [5.21], 1-(4-dimethyl-amino-3-methyl-phenyl)methyl-5-diphenylacetyl-4,5,6,7-tetra-hydro-1H-imidazolo-(4,5-c)pyridine-6-carboxylic acid (PD 123 319) [5.21].

AT III Analogs. Substitution of D-N-methyl alanine for Arg^1 in AT III produces a more potent analog. [D-N-methyl-Ala^1, Ile^7]AT III is a selective AT III antagonist [5.83]. [Sar^1, Ile^7] AT III is a selective vascular AT III antagonist. On the other hand, [Ile^7]AT III inhibits both AT III as well as AT II [5.24].

Applications. [Asn^1, Val^5] AT II (angiotensin amide, Hypertensin CIBA), an AT II agonist, is administered in states of shock and collapse in which a normal blood pressure should be restored as quickly as possible.

The AT II antagonist Saralasin (Sarenin) is used in the diagnosis of AT II dependent forms of hypertonia and in the prior treatment of donor kidneys before a kidney transplantation (to minimize loss of functions of the transplanted kidney caused by ischemia).

5.1.2. Kinins [5.84]

5.1.2.1. Occurrence

The kallikrein/kininogen/kinin system is widely distributed, e.g., in the liver, in acinar cells of the rat submandibular gland, in distal tubular and upper collecting duct cells of the human kidney, in platelets, endothelial cells, rat vascular smooth muscle cells (important for the regulation of the local vascular tone) [5.85], in the brain [5.86], Leydig cells, ovaries, and in milk [5.87]. Apart from bradikinin (BK) and kallidin (KD), large amounts of [Hyp^3]Lys-BK and [Hyp^3]BK were found in human urine [5.88], plasma [5.89] and in ascites of patients suffering from stomach cancer [5.90].

In addition, T-kininogens (TK) have also been found in the rat [5.91], [5.92]. These precursors are cleaved by T-kininogenase [5.93] (endopeptidase K [5.94]) or high concentrations of trypsin to give T-kinin (Ile-Ser-BK).

Structural similarities have been observed between the kinins and the adipokinetic hormones of *Locusta, Schistocera,* and *Manduca* (AKH I and II), the red pigment concentrating hormone (RPCH) from crustaceans, the periplanetin CC-1 and CC-2 from the cockroach (*Periplaneta americana*) which have tachycardiac and hyperglycemic activity, the cardioactive corazonin from the American cockroach, the hypertrehalosemic hormones from Heliothis (he-HTH) [5.95], the German cockroach (*Blatella germanica*) (bl-HTH) [5.96], *Tenebrio molitor* and *Zophobas rugipes* (te-HTH) [5.97], *Carausius* (car-HTH)[5.98], and from the Colorado potato beetle (*Leptinotarsa decemlineata)* (identical with periplantin CC-1 and CC-2 [5.99], Ro I and Ro II from *Romealea* [5.95], and the APGW-amides (with muscle contracting activity) from the mollusc *Fusinus ferrugineus* [5.100].

The C-terminal region of leucopyrokinin is also found in locustamyotropin (loc-MTO) from *Locusta migratoria* and in the pheromone biosynthesis activating neuropeptide (PBAN) from *Heliothis zea* [5.101].

AKH I and AKH II are formed from similar prepropeptides which contain 63 and 61 amino acids respectively [5.98].

Kinins from different species	Amino acid sequence																								
a) Mammals:																									
Bradykinin (BK)											R	P	P	G	F	S	P	F	R						
[Hyp³]BK											R	P	Ph	G	F	S	P	F	R						
Kallidin (KD = Lys-BK)										K	R	P	P	G	F	S	P	F	R						
[Hyp³] Lys-BK										K	R	P	Ph	G	F	S	P	F	R						
Met-Lys-BK									M	K	R	P	P	G	F	S	P	F	R						
T-Kinin									I	S	R	P	P	G	F	S	P	F	R						
T-Kinin-Leu									I	S	R	P	P	G	F	S	P	F	R	L					
b) Chicken:																									
[Thr⁶]BK											R	P	P	G	F	T	P	F	R						
Ornitho-Kinin											R	P	P	G	F	T	P	L	R						
c) Amphibians:																									
[Val¹, Thr⁶]BK											V	P	P	G	F	T	P	F	R						
Phyllokinin											R	P	P	G	F	S	P	F	R	I	Ys				
Rana rugosa: Rr I-3											R	P	P	G	F	S	P	F	R	I	A	P	E	I	V
Rr I-2											R	P	P	G	F	S	P	F	R	I	A	P	E		
Rr 3											R	P	P	G	F	S	P	F	R	I	A	P	E	I	
Rana nigro maculata											R	P	P	G	F	S	P	F	R	V	A	P	A	S	
Bombina orientalis											R	P	P	G	F	S	P	F	R	G	K	F	H		
Heleophryne purcelli											R	P	Ph	G	F	S	P	F	R						
d) insects:																									
Gly-BK										G	R	P	P	G	F	S	P	F	R						
Vespakinin M										G	R	P	Ph	G	F	S	P	F	R	I	D				
Vespakinin X										A	R	P	P	G	F	S	P	F	R	I	V				
Policystekinin	*Q	T	N	K	K	K	L	R	M	G	R	P	P	G	F	S	P	F	R						
Leucopyrokinin	*Q	T	S	/	/	/	/	/	/	/	/	/	/	/	F	R	P	/	T	La					
Periplanetin CC-2	*Q	L	T	/	/	/	/	/	/	/	/	/	/	/	F	T	P	/	/	N	Wa				
Periplanetin CC-1	*Q	V	N	/	/	/	/	/	/	/	/	/	/	/	F	S	P	/	/	N	Wa				
sch-, lo-AKH-I	*Q	L	N	/	/	/	/	/	/	/	/	/	/	/	F	T	P	/	/	N	W	G	Ta		
ma-AKH	*Q	L	T	/	/	/	/	/	/	/	/	/	/	/	F	T	S	/	/	S	W	Ga			
sch-AKH-II	*Q	L	N	/	/	/	/	/	/	/	/	/	/	/	F	S	T	/	/	G	Wa				
loc-AKH-II	*Q	L	N	/	/	/	/	/	/	/	/	/	/	/	F	S	T	/	/	A	Wa				
Ro-I	*Q	V	N	/	/	/	/	/	/	/	/	/	/	/	F	T	P	/	/	N	W	G	Ta		
Ro-II	*Q	V	N	/	/	/	/	/	/	/	/	/	/	/	F	S	T	/	/	G	Wa				
RPCH	*Q	L	N	/	/	/	/	/	/	/	/	/	/	/	F	S	P	/	/	G	Wa				
Corazonin	*Q	T	/	/	/	/	/	/	/	/	/	/	/	/	F	Q	Y	S	R	G	W	/	T	Na	
he-HTH	*Q	L	T	/	/	/	/	/	/	/	/	/	/	/	F	S	S	/	/	G	W	G	/	Na	
bl-HTH	*Q	V	N	/	/	/	/	/	/	/	/	/	/	/	F	S	P	/	/	G	W	G	Ta		
car-HTH	*Q	L	T	/	/	/	/	/	/	/	/	/	/	/	F	T	P	/	/	N	W	G	Ta		
te-HTH	*Q	L	N	/	/	/	/	/	/	/	/	/	/	/	F	S	P	/	/	N	Wa				
APGW-amide																A	P	/	/	G	Wa				

Kinins from different species	Amino acid sequence
	1 21
he-PBAN	L S DD M P A T P A D Q E M Y R Q D P E Q-
	22 33
	I D S R T K Y F S P R La
loc-MTP	G A V P A A Q F S P R La
Leucopyrokinin	*Q T S F T P R La

5.1.2.2. Release

The cascade of biochemical reactions initiated by tissue injury → lowering of tissue respiration → falling pH [5.102] which results in the release of BK is caused by the activation of the Hagemann factor (blood clotting factor XII) [5.103].

While plasma kallikrein produces BK, tissue kallikrein cleaves kininogen to form KD (Lys-BK), which can in turn be converted to BK by aminopeptidases. After cleavage of kininogen, an assymetric disulfide is formed which consists of the heavy chain (N-terminal of kinin) and the light chain (C-terminal of kinin) of kininogen.

Kinins are released during muscle work [5.104] (not in the case of Type II diabetics), during PMSG/h-CG induced ovulation [5.105] and during inflammation. Acrosin stimulates the formation of kinin from kininogen in the Leydig cells.

Prostaglandins (PGs) inhibit the formation of kinins. Therefore, inhibitors of prostaglandin synthesis potentiate the formation of bradykinin [5.106]. Aspirin, for instance, potentiates the antihypertensive effect of p.o. captopril in spontaneously hypertensive rats [5.107]. Hence, aspirin exerts a stimulating effect of bradykinin through the inhibition of PGs. Is this the cardioprotective effect of aspirin?

Inhibitors of kallikrein retard the release of kinins. Endogenous kallikrein inhibitors are, e.g. α-macroglobulin or aprotinin (Trasylol orAntagosan), while synthetic inhibitors of kallikrein are substrate analogs of the kininogens. For instance, D-cyclohexylglycinyl-L-cyclohexylglycinyl-L-arginine-4-nitroanilide (IC_{50} = 2.1 μM) imitates the C-terminal cleavage sequence, Pro-Phe-Arg-Ser [5.108]. The bradykinin B_2 receptor antagonists also have an inhibitory effect on kallikrein [5.109], [5.110].

BK is deactivated primarily in the lungs by the dipeptidase kininase II, which is identical to ACE. Inhibitors of ACE (Captopril, Enalapril, Ramipril) potentiate BK [5.111]-[5.113]. A series of other endopeptidases, carboxy- (kininase I activity) and aminopeptidases also degrade BK [5.114]-[5.116].

In patients suffering from essential hypertension, the concentration of kallikrein, pre-kallikrein and kinin in the urine is significantly lower than in normotensive persons [5.117]. Captopril causes a drop in blood pressure in hypertensive patients with a normal urine kallikrein content, but not in hypertensives who have a greatly reduced urine kallikrein content [5.118]. The kallikrein activity decreases in old age.

An increased level of kinin or kallikrein is found in diabetics with orthostatic hypotension [5.119], patients suffering from inflammatory stomach diseases, ulcerative colitis, Dumping syndrome [5.120], pancreatitis, rheumatic inflammations or myocardial infarction, and in allergic persons (in nasal fluid) [5.121].

In the rat, pre-T-kininogen mRNA [5.122] and T-kininogen [5.123], [5.124] appear to be formed primarily in response to inflammatory stimulants. The level of T-kininogen is raised on administration of estrogen [5.125], in lactating female rats, in newborn rats, and in thyroidectomized rats [5.126]. In rats, the expression of the T-kininogen gene increases with age [5.127]. Indomethacin, dexamethasone and testosterone [5.128] reduce the formation of T-kininogen.

5.1.2.3. Receptors

Until now, the B_1 and B_2 receptors have been studied intensively. The B_1 receptors bind more strongly to de-Arg^9-BK than to BK, while the B_2 receptors have a greater affinity for BK. Studies with various agonists and antagonists on different tissues have led to the assumption that there are multiple B_2 receptors for kinins [5.129]. Two other BK receptors (B_3 and B_4 receptors) which react with BK as well as de-Arg^9-BK have been found in the longitudinal muscles of the oppossum esophagus [5.130].

The vasorelaxing effects of BK exerted via B_2 receptors in isolated large vessels require intact vascular endothelial cells. Vasoconstricting effects of BK (no matter whether via the activation of B_1 or B_2 receptors) require no endothelium.

The BK receptor complex is internalized, causing a temporary down regulation of the BK receptor [5.131].

5.1.2.4. Biological Effects [5.132], [5.133]

B_1-Receptor Mediated Effects [5.134]:

Contraction of rat duodenum, rat renal vessels [5.135] and rabbit aorta (via PLC/IP_3)
Relaxation of rabbit coeliac artery [5.136], rabbit mesenteric artery (via PGs [5.137], dog saphenous vein [5.138], dog renal artery (without endothelium, via PLA_2 and PGs)
Capsaicin-induced inflammation in mice [5.139]
Release of IL-1 and TNF-α in macrophages
Cell proliferation und collagen synthesis [5.84].

B_2-Receptor Mediated Effects [5.134]:

Release of EDRF and calcium from intracellular stores, leading to activation of PLA_2 and increase in PGE_2 production and the PGE_2-dependent activation of adenylate cyclase in arterial smooth muscle cells [5.140]
Activation of phosphatidylinositol-specific phospholipase C (PLC) and formation of inositol triphosphate (IP_3), diacylglycerol (DAG) and lysophosphatidyl-inositol [5.141]
Contraction of rat vas deferens (via PLC/IP_3), rabbit jugular vein (via PLC/IP_3) and hamster urinary bladder (via PLA_2)
Relaxation of rat duodenum (via cAMP), rat renal vessels [5.135], guinea pig ileum, dog carotid artery (via PLC/IP_3 and EDRF) and dog renal artery (via PLC/IP_3 and EDRF)
Hypotensive activity: reduces AT II [5.142], adrenaline [5.143] and vasopressin [5.84]

induced blood pressure
Increase of vascular permeability [5.144] and pain [5.145] in rat skin
Improvement of heart function [5.146]
Facilitation of norepinephrine release from sympathetic nerves
Activation of Cl^- transport in intestinal epithelia
Stimulation of airway epithelial Cl^- secretion via PGE_2, cAMP or IP_3 and calcium release [5.147]
Stimulation of the glomerular rate of filtration [5.48]
Diuretic and natriuretic activity [5.149] (by inhibition of vasopressin via PGs [55.150]).
Increase in human sperm motility
Nociception in the rat spinal cord [5.151]-[5.153]
Neurotransmitter activity in the guinea pig brain [5.154]
Increase in arterial pressure by the i.c.v. application to rats [5.155].

Other Effects of the Kinins:
Stimulation of ovulation [5.105], [5.156]
Causing rhinitis [5.84]
Release of prolactin [5.157] and β-endorphin [5.158] from the pituitary
Release of SP- and CGRP-like material in neurons sensitive to capsaicin [5.159]
Enhanced prostaglandin formation in mouse osteoblasts and subsequent bone resorption [5.160]
Stimulation of miosis (abnormal contraction of the pupil) by the release of SP [5.161]
Glucose uptake in working skeletal muscle via PGs and phospho-fructokinase [5.162]
Formation of edema in the brain [5.163]
Antinociceptive activity (i.t. application) [5.164].

5.1.2.5. Structure-Activity Relationships

The ligands for B_1 receptors are de-Arg^9-BK and de-Arg^{10}-Met-kallidin (stronger), while the C-terminal arginine is very important for the B_2 receptors. De-Arg^9-[D-Phe^8]BK has four times and Sar-[D-Phe^8]de-Arg^9BK ten times the effect of de-Arg^9-BK on B_1 receptors [5.165].

Strong B_1 receptor antagonists are de-Arg^9-[Leu^8]BK, the corresponding KD derivative, de-Arg^9-[Gly^7]BK, and de-Arg^9-[D-Ala^7]BK [5.166].

The action of KD on B_2 receptors is greater than that of BK. The guanido groups of the two arginines and the two prolines in positions 2 and 3 are important for the biological action on B_2 receptors. Stronger agonists are obtained by the substitution of β-(2-thienyl)-alanine (Thi) or dehydro (Δ)Phe for Phe^5 and Phe^8, Hyp for Pro^3, α-aminoisobutyric acid for Pro^7, Tyr(Me) for Phe^8 or by reduction of the peptide bonds between positions 8 and 9 [5.167] and 6 and 7 [5.168]. The effect of [Hyp^3, Tyr$(OMe)^8$]BK on B_2 receptors is about three times that of BK [5.165]. [Phe^8 Ψ (CH_2-NH)Arg^9]BK is a selective B_2 receptor agonist, which is five times as active as BK [5.167]. The effect of [Δ-Phe^5]BK on the uterus and ileum is twice that of BK and the effect on blood pressure is 23 times that of BK. Phyllokinin is about 2.5-3.5 times as active as BK, the O-sulfate residue being responsible for the high activity.

The replacement of Pro^7 in BK by D-Phe, D-Nal, D-Pal or, still better, by D-tetrahydroisoquinolinecarboxylic acid (D-Tic) [5.169], and the substitution of octahydroindolecarboxylic acid (Oic) [5.169] for Phe^8 gives rise to strong B2 receptor antagonists. Depending on the test, D-Arg-[Hyp^3,Thi^5,D-Tic^7,Oic^8]BK (Hoe 140) [5.169] is an approx. 100-1000 times stronger BK antagonist than the standard peptide, D-Arg-[Hyp^3,$Thi^{5,8}$,D-Phe^7]BK [5.170]. The alkylation of D-Arg-[Hyp^3,$Thi^{5,8}$,D-Phe^7]BK with adamantane acetic acid gives rise to an antagonist that is about ten times as active as the starting compound [5.171].

If the BK antagonists are modified in position 1, 2, 3, 8 and 9, the tissue selectivity can be altered. Arg° improves the affinity and inhibits enzymatic degradation [5.172].

Specific BK antagonists are, e.g., [5.172]:

[D-Phe^2, D-Phe^7]BK	for the rat uterus
[D-Phe^7, Hyp^8]BK	for the guinea pig ileum
[D-Phe^1, $Thi^{5,8}$, D-Phe^7]BK	for rat blood pressure

The BK receptor antagonists with Phe or Thi in position 8 also act as B_1 receptor blockers [5.173] because carboxy-peptidases convert these peptides to biologically active B_1 receptor antagonists by cleaving the C-terminal arginine [5.174]. Antagonists with Oic^8 or a reduced peptide bond between Phe^8 and Arg^9, such as Hoe 140 or D-Arg-[Hyp^3,D-Phe^7,$Phe^8\psi(CH_2NH)Arg^9$]BK should be selective B_2 antagonists.

The histamine releasing effect of BK antagonists can be reduced by the acetylation of the N-terminal amino group [5.175].

The contraction of the smooth muscles of the rat uterus by BK and kallidin is selectively and competitively inhibited by extracts of *Mandevilla velutina* [55.176].

5.1.2.6. Applications

BK antagonists may find application for treating inflammations, pain, rheumatic arthritis, osteoarthritis, inflammatory stomach diseases, pancreatitis, rhinitis, asthma and gout [5.84].

5.2. Substance P, Neurokinins and Tachykinins [5.177], [5.178]

5.2.1. Occurrence

The neurokinins include Substance P (SP), neurokinin A (NKA), and neurokinin B (NKB). Similar compounds which occur in cold blooded animals are called tachykinins.

SP was the first of a large number of peptides that was found in the gastrointestinal tract as well as in the brain. SP and the neurokinins are produced in the form of several prepro-neurokinins. In fact, three prepro-neurokinins have been characterized in the cow [5.179], [5.180] as well as in the rat [5.180], [5.181]. b-β-prepro-neurokinin and the shortened b-γ-prepro-neurokinin [5.180] contain both SP as well as NKA (substance K, neuromedin L [5.182]), while b-α-prepro-neurokinin [5.179] processes only SP. There is a very great similarity between h-β-prepro-neurokinin and the corresponding b-β-prepro-neurokinin [5.183]. Neuropeptide K (NPK) is NKA, that is extended by 26 amino acids at the N-terminus [5.184], [5.185]. It is found with NKA [5.186], NKA-(3-10), and NKA-(4-10) [5.187] primarily in the plasma and tumor tissue of carcinoid patients and in the bronchial tubes. NPK is processed only from β-prepro-neurokinin. The neuropeptide γ, which corresponds to the amino acid sequence of r-γ-prepro-neurokinin-(72-92) [5.188], has been isolated from the small intestine of the rabbit [5.188] and the rat [5.189]. g-SP and NKA corresponding to β-prepro-neurokinin have been isolated from the small intestine of chicken [5.190]. Apart from NKA, NKB (neuromedin K) has also been isolated from the spinal cord of the pig.

Substance P is found centrally in the spinal cord and in the sensorial nuclei of the brain stem and peripherally in the ends of the sensorial fibers of the gastrointestinal tract, urogenital tract, bile duct, bronchial tubes, skin [5.181] and thymus [5.182]. Neurokinin A has been isolated primarily from the rabbit iris [5.193].

Tachykinins, such as eledoisin, are found in the salivary glands of the octopus. Physalaemin, uperolein, phyllomedusin, kassinin, enterokassinin, hylambatin and enterohylambatin have been isolated from the skin of amphibians. Indeed, five tachykinins have been isolated from the skin of the Australian frog, *Pseudophyrne güntheri*, two of which are similar to SP (pg-SP I and pg-SP II) and three to kassinin (pg-kassinin I, II and III) [5.194]. Scyliorhinin I and II are the tachykinins of the mud fish, *Scyliorhinus caniculus* [5.195]. Two tachykinin- and kinin-like peptides have been isolated from the insect, *Locusta migratoria*, locustatachykinin I and II [5.196].

There are signs that point to the fact that in addition to the neurokinins found until now, tachykinin like substances also occur in mammals [5.185], [5.187]

There is a structural similarity between the tachykinins and the neurotoxic β-amyloid protein-(25-35) [5.197].

5.2.2. Release

Substance P is released in response to stimulation of the sensorial neurons. It is assumed that SP is also a pain transmitter, especially since the level of SP in the spinal cord is increased by mechanical pressure or hot water on the extremities [5.198], and morphine and related analgesics inhibit the release of SP from sensorial nerve ends [5.198]. In patients with reduced sense of pain, less SP is found in the substance gelatinosa of the spinal cord than in healthy persons. On the other hand, rats suffering from chronic pain have increased levels of SP in the nervus ischiadicus.

Substance P and NKA are released by capsaicin, bombesin [5.199], CCK-8 [5.200], and by cholinergic [5.201], serotoninergic [5.202], and dopaminergic [5.203] mechanisms.

h, b-Substance P (SP)	R P K P Q Q F F G L Ma
g-Substance P	R P R P Q Q F F G L Ma
pg-SP I	•Q P N P D E F F G L Ma
pg-SP II	•Q P N P N E F F G L Ma
h, g, p-Neurokinin A (NKA)	H K T D S F V G L Ma
h-Neuropeptide K (NPK)	D A D S S I E K- Q V A L L K A L Y G H G Q I S H K R H K T D S F V G L Ma
r, cu-Neuropeptide-γ	D A G H G Q I S H K R H K T D S F V G L Ma
p-Neurokinin B (NKB)	D M H D F F V G L Ma
Elodoisin (ELE)	•Q P S K D A F I G L Ma
Physalaemin (PHY)	•Q A D P N K F Y G L Ma
[Lys^5,Thr^6] Physalaemin	•Q A D P K T F Y G L Ma
Uperolein	•Q P D P N A F Y G L Ma
Phyllomedusin	•Q / N P N R F I G L Ma
Kassinin (KAS)	D V P K S D Q F V G L Ma
[Glu^2, Pro^5] Kassinin	D E P K P D Q F V G L Ma
Enterokassinin	D E P N S D Q F I G L Ma
pg-Kassinin I	•Q P H P D E F V G L Ma
pg-Kassinin II	•O P N P D E F V G L Ma
pg-Kassinin III	•Q P H P N E F V G L Ma
Hylambatin	D P P D P D R F Y G M Ma
Enterohylambatin	D P P N P D R F Y G M Ma
Scyliorhinin I (SCY I)	A K F D K F Y G L Ma
Scyliorhinin II (SCY II)	S P S N S K C P D G P D C F V G L Ma
Locustatachykinin I	G P S G F Y G V Ra
Locustatachykinin II	A P L S G F Y G V Ra
Amyloid β protein-(25-35)	G S N K G A I I G L M

A higher SP content has been found in the anterior hypophyseal lobe of male rats than in that of female rats. Here, estradiol lowers the SP content and dihydrotestosterone raises it [5.204].

A high NKA and SP like immunoreactivity is found in tissues of carcinoid tumors. Indeed, plasma NKA immunoreactivity can be used as a tumor marker. Thus, the flushing episodes and diarrhea correlate positively with plasma NKA [5.205]. The neurokinin level is also raised in the joint fluids of patients with rheumatic inflammatory diseases [5.206] and in rats suffering from adjuvant induced arthritis [5.207].

5.2.3. Metabolism of SP

The most important enzymes that cleave SP in the brain are metalloendopeptidase 3.4.24.11, which is also called enkephalinase or neutral endopeptidase, and the post-proline cleaving enzyme. The N-terminal heptapeptide, SP-(1-7) is formed on cleavage with enkephalinase and the C-terminal heptapeptide SP-(5-11) with post-proline-cleaving enzyme [5.208].

Normally, cleavage in the brain occurs preferentially through enkephalinase. On the other hand, digestion in the plasma occurs preferentially through a post-prolin-cleaving enzyme to give dipeptides [5.209]. SP is also a substrate for the angiotensin converting enzyme (ACE). SP is potentiated by smoking (inhibition of the neutral endopeptidase) [5.210] and by ACE inhibitors [5.211].

5.2.4. Receptors

Until now, three different receptors have been found for the neurokinins: NK1 for SP, NK2 for NKA and NK3 for NKB [5.212], [5.213]. The r-NK1 receptor [5.214], [5.215], the r-NK2 receptor [5.214] and the b-NK2 receptor [5.216] have been characterized via the cDNA.

Types of receptors in various tissues [5.212]

Isolated tissue	NK1	NK2	NK3
Guinea pig ileum	±	+	+
Guinea pig ileum (atropin treated)	+	+	
Guinea pig urinary bladder	+	+	+
Guinea pig trachea	+	+	
Guinea pig gall bladder [5.217]		+	
Rat submandibular salivari gland [5.218]	+		
Rat duodenum	+	±	+
Rat vas deferens		+	
Rat urinary bladder	+	+	
Rat portal vein (RPV)			+
Rat cerebral cortex [5.219]			+
Rat cerebral cortex synaptosomes [5.220]			+
Rat brain synaptosomes	+		
Rat thymus and spleen	+		
Rat urether [5.221]		+	+
Rat stomach [5.222]		+	
Rat skin (plasma extravasation [5.223]	+		
Mouse urinary bladder [5.225]		+	
Hamster urinary bladder		±	+
Hamster trachea (HT) [5.226]		+	
Rabbit pulmonary artery (RPA)		+	
Rabbit mesenteric vein	+	+	
Rabbit vas deferens		+	
Opossum oesophagus [5.227]	+		
Dog carotid artery	+		
Dog urinary bladder	+	+	
Human urinary bladder		+	
Human gastrointestinal tract [5.228]	+	+	

The NK2 receptors in the hamster trachea are different from those in the rabbit pulmonary artery. Although they do not differ from natural or synthetic agonists, the selective binding to NK2 antagonists varies greatly [5.226].

Selective Ligands of Neurokinin Receptors [5.212]

NK1 receptor-agonists:

SP-methylester,
[Sar^{9}, Met(O_2)^{11}]SP,
[β-Ala^{4}, Sar^{9}, Met(O_2)^{11}]SP-(4-11),
Ac-Arg-[Sar^{9}, Met(O_2)^{11}]SP-(7-11),
[Pro^{9}]SP, [Pro^{10}]SP [5.213],
N-1.6-(β-D-Glucopyranosyl)-[Glu^{6}, Pro^{9}]SP-(6-11) [5.229],
pGlu-[Pro^{9}]SP-(7-11) (Septide),
δ-Aminovaleryl-[Pro^{9}, MeLeu^{10}]SP-(7-11) (GR 73632) [5.230],
[Cys$^{3.6}$, Tyr^{8}, Pro^{9}]SP [5.213],
[Cys$^{3.6}$, Tyr^{8}, Gly^{9}]SP [5.231],
[MeVal^{7}] NKA-(4-10) [5.231].

NK2-receptor-agonists:

NKA-(4-10),
[Nle^{10}]NKA-(4-10),
[β-Ala^{8}]NKA-(4-10) (10 times stronger than NKA-(4-10) [5.232])
[Lys^{3}, Gly^{8}-R-γ-lactam-Leu^{9}]NKA-(3-10) (GR 64349) [5.230]
[Leu^{9} ψ (CH_2NH)Leu^{10}]NKA-(4-10) [5.233]
[pGlu^{6}, Phe^{8} ψ (NHCO)Gly^{9}]SP-(6-11) [5.234]

NK3-receptor-agonists:

[MePhe^{7}]NKB,
[β-Asp^{4}, MePhe^{7}]NKB-(4-10),
Succinyl-[Asp^{5}, MePhe^{7}]-(5-10) (Senktide),
[Pro^{7}]NKB [5.213]
[Cys$^{2.5}$]NKB
Scyliorhinin II-(7-18) [5.219]

NK1-receptor-antagonists:

[Leu^{10} ψ (CH_2NH) Leu^{11}]SP [5.236]
[D-Arg^{1}, D-Pro$^{2.4}$, D-Phe^{7}, D-His^{9}]SP (3 times stronger than Spantide) [5.237],
[D-Phe^{7}, D-His^{9}]SP-(6-11) (2 times stronger than Spantide) [5.238],
(GR 71251) [5.2388]

RPKPQQFF—N—C(S)—(C=O)—N—(S)CH—CO—Wa (pK_b = 7.7)

NK2-receptor-antagonists:

[D-Tyr^{4}, D-Trp$^{7.9}$, Nle^{11}]SP-(4-11) [5.239],
[D-Pro^{4}, Ala^{6}, D-Trp$^{7.9.10}$, Phe^{11}]SP-(4-11) [5.240],
[D-Pro^{4}, Lys^{6}, D-Trp$^{7.9.10}$, Phe^{11}]SP-(4-11) [5.239], [5.240]

[D-Pro^4, D-Trp^{7,9}]SP-(4-11) [5.218]
[Tyr^5, D-Trp^{6,8,9}, Arg^{10}]NKA-(4-10) (very selective at RPA ($pA_2=7.9$)) [5.241]

NK3 receptor-antagonists:
[D-Pro^2, D-Trp^{6,8}, Nle^{10}]NKB [5.198]
[Trp^7, β-Ala^8]NKA-(4-10) (very selective at RPV ($pA_2=7.46$)) [5.242],

5.2.5. Biological Effects

NK1-Receptor Mediated Effects [5.212]:
Hyperalgesia
Hypotension
Salivary secretion
Increase of capillary permeability
Plasma extravasation in rat skin [5.224]
Release of histamin in rat mast cells
Proliferation of human fibroblasts [5.243]
Stimulation of 6-keto-$PGF_1\,\alpha$ from human umbilical vein endothelial cells [5.244]

NK2-Receptor Mediated Effects [5.212]:
Bronchoconstriction [5.185], [5.245], [5.246]
Activation (superoxid anion production) of guinea pig alveolar macrophages [5.247]
Protection of gastric lesions in rats [5.248]
Inhibition of water intake in neonatal rats [5.249]
Contraction of rat bladder [5.245]
Tachycardia
Inhibition of gastric SRIF-release [5.241]

NK3-Receptor Mediated Effects [5.212]:
Algesia (The NKB-antagonist [D-Pro^2,D-Trp^{6,8},Nle^{10}]NKB shows analgetic activity [5.226]
Bradycardia
Capillary permeability

Other Effects of SP:
Activation of phospholipase C and D [5.250]:
Stimulation of prolactin release in the pituitary
Inhibition of somatotropin release by central mechanism [5.251]
Stimulation of calcitonin release [5.252]
Stimulates the proliferation of synoviocytes and the release of PGE_2 and collagenase (pathogenesis of rheumatic arthritis?) [5.253]
Contraction of lower esophageal sphincter [5.254]
Modulation of gastrointestinal motility (relaxation in low doses and stimulation in high doses) [5.255], [5.256]
Stimulation of gastrointestinal transit [5.257]

Inhibition of vagus stimulated gastric acid secretion [5.258]
Species specific modulation of exocrine [5.259], [5.260] and endocrine [5.261], [5.262] pancreatic secretion
Modulation of diuresis and natriuresis [5.263]
Immunostimulating activity [5.264], [5.265]
Stimulation of locomotor activity [5.266] and LH-release [5.267] by central SP
Increase of blood pressure [5.268] and hypoventilation [5.269] by central SP.

5.2.6. Structure-Activity Relationships

(see also: Selective ligands of neurokinin receptors)

In the brain, SP-(5-11) is the neurotransmitter for pain. It causes aggressive and stress induced behavior, reduces memory and increases blood pressure when applied centrally. On the other hand, SP-(1-7) exerts an analgetic and relieving effect, increases the learning capacity, improves memory and lowers blood pressure [5.208], [5.270]-[5.274]. Based on the opposing effects of the N- and C-terminal peptides of SP, Stewart [5.208] postulated SP-N and SP-C receptors. The regeneration of nerve fibers is stimulated by N-terminal fragments of SP (and also by Arg-Pro) [5.275].

SP Agonists

The shortest active SP analogs are acylated SP-(7-11) compounds [5.276], which are in part more active than SP. The most effective compound in the GPI test is 4-OH-C_6H_4-CH_2-CO-SP-(7-11).

The C-terminal amide group, a neutral lipophilic amino acid in position 11 [5.277], Leu^{10} and Gly^9 are important for the biological activity. Phe^8 can be replaced by Tyr(OMe), cyclohexylalanine or Ile without loss of activity, while the substitution of these residues for Phe^7 produces only slightly active compounds. The remaining N-terminal amino acids are not essential for the biological activity.

An SP analog with long lasting effect is pGlu-Gln-Phe-MePhe-Sar-Leu-Met-amide. It has only 10% of the spasmogenic activity, but all of the CNS activity and is resistant to the enzymes of the hypothalamus.

The derivative, that is dimerized with succinic acid via the N-terminal amino groups of SP-(3-11), has 2.4 times the receptor affinity and 75 times the salivatory effect of SP [5.278].

It has been found that NKA-(4-10) is just as active as NKA. The amino acids Asp^4, Phe^6 and Val^7 are important for the biological activity of NKA and NKB [5.279].

SP Antagonists [5.177]

The standard antagonist is spantide I. It binds primarily to the NK1 and NK2 receptors. The activity of SP antagonists is increased by the replacement of Gln^6 by Asn^6 [5.280]. As a result of the substitution of D-Lys(Nic) for D-Arg^1, spantide II, in contrast to spantide I, has no neurotoxic properties [5.281] and a reduced histamine releasing activity [5.282]. While spantide I also acts as a bombesin antagonist, spantide II has no effect on contractions caused by bombesin. Spantide II evidently binds to all three neurokinin receptors [5.282] and exerts a short antinociceptive effect when applied intrathecally [5.283].

	pA_2 (guinea pig Taenia coli)
Spantide I: [D-Arg^1, D-Trp$^{7.9}$, Leu^{11}]SP	7.1
[D-Arg^1, D-Cl_2Phe^5, Asn^6, D-Trp$^{7.9}$, Nle^{11}]SP	8.0
[D-Arg^1, Pal^3, D-Cl_2Phe^5, Asn^6, D-Trp$^{7.9}$, Nle^{11}]SP	8.0
Spantide II: [D-Lys(Nic)^1, Pal^3, D-Phe(Cl_2)^5 Asn^6, D-Trp$^{7.9}$, Nle^{11}]SP	7.7

D-Phe(Cl2) = 3.4-Dichloro-D-phenylalanin, D-Lys(Nic) = NεNicotinoyl-D-lysin, Pal = 3-(3-Pyridyl)-alanin.

The reduction of individual peptide bonds also gives rise to SP antagonists, which were evaluated by the displacement of SP bound to SP receptors of guinea pig acinar cells [5.284]. It was found that [Leu$^{10}\psi$(CH_2NH)Nle^{11}]spantide is equipotent to spantide (K_i = 2.8 μM).

While the SP antagonists with basic N-terminal amino acids release histamine, shortened antagonists (e.g., [D-Pro^4,D-Trp^{7,9,10}]SP-(4-11) inhibit the SP induced release of histamine *in vitro*.

[D-Arg^1,D-Phe^5,D-Trp^{7,9},Leu^{11}]SP was the most suitable SP-antagonist for the inhibition of the bombesin and vasopressin stimulated mitogenic effect on small cell lung cancer cells. It has the same affinity for the bombesin and the SP receptor and is ten times as active as spantide on the bombesin receptor [5.285], [5.286]. Under comparable conditions (50 μM), spantide and other N-terminal shortened SP antagonists showed no inhibitory effect on the bombesin or vasopressin induced growth of Swiss 3T3 cells.

5.2.7. Amyloid A4 Protein [5.287]

Occurrence. In Alzheimer's disease and in trisomy 21 (Down's syndrome), fibrillary amyloid is found within the neurons of the cortex as neurofibrillary tangles (NFT) and extracellularly in meningeal and intracortical blood vessels as amyloid plaques. These deposits are also found in the skin, subcutaneous tissue and in the intestine of Alzheimer patients and in old healthy persons [5.288].

The plaques mainly consist of an insoluble peptide (amyloid A4 protein (A4) [5.289] or amyloid β protein [5.290]) containing 42 amino acids [5.291]. The gene that codes for A4 produces at least three mRNAs, which form the following three A4-precursor proteins: APP_{695}, APP_{751} and APP_{770} [5.292].

APP_{695} consists of 695 amino acids and contains A4 in positions 597-638. Apart from an insert of 56 amino acids (HL 124i) C-terminal from Arg^{288} and the substitution of Ile for Val^{289}, APP_{751} is identical to APP_{695}. Apart from an insert of 19 amino acids C-terminal from HL 124i and the substitution of Leu for Val^{289}, APP_{770} is identical with APP_{751}. The insert HL 124i with 56 amino acids has the structure of a Kunitz inhibitor, which specifically inhibits serine proteases, such as trypsin, chymotrypsin, elastase, plasmin and cathepsin G. In Alzheimer's disease, the total increase in the mRNA for APP in neurons from the locus ceruleus and nucleus basalis is based solely on an increase in APP_{695}, which is lacking the inhibitor domain. Therefore, it is assumed that A4 is formed especially easily from this precursor by the action of cerebral proteases [5.292]. These precursors of A4 are transmembrane proteins and receptors for neuronal adherons [5.293]. A4 is found, in part, in the membrane bridging region [5.289], [5.294]. $APP_{751}$is identical to the protease inhibitor, protease nexin II [5.295].

The gene for APP is expressed in certain neurons, some glia cells and in brain macrophages. The A4 formed in the glia cells and macrophages is responsible for the extracellular formation of the amyloid plaques [5.291].

The A4 of patients suffering from Alzheimer's and Down's syndrome is identical to the intraneural amyloid of the Parkinson dementia of Guam and the vascular amyloid of sporadic cerebral amyloid angiopathy (SCAA). The amyloid protein of patients with leptomeningiosis haemorrhagica interna is shorter by three amino acids at the C-terminus [5.296].

No i.r.-A4, but i.r.-APP [5.297], has been found in serum. The mRNA for APP has also been found in muscles and other tissues [5.298], senile plaques, brain extracts and the cerebrospinal fluid [5.299] from adults. Human platelets contain APP_{751} in high concentrations [5.300].

```
APP                      ..S G  L  T  N  I  K  T  E  E  I  S  E  V  K  M596-
            1                                                           21
h-A4        D  A  E  F  R  H  D  S  G  Y  E  V  H  H  Q  K  L  V  F  F  A-
m-A4        D  A  E  F  G  H  D  S  G  F  E  V  R  H  Q  K  L  V  F  F  A-

            22                                                          42
h-A4        E  D  V  G  S  N  K  G  A  I  I  G  L  M  V  G  G  V  V  I  A-
m-A4        E  D  V  G  S  N  K  G  A  I  I  G  L  M  V  G  G  V  V  I  A-
Eledoisin            E  P  S  K  D  A  F  I  G  L  Ma

                  <------------- transmembrane -------------------

            639
APP         T  V  I  T  L  V  H  L...
            sequence ------------->
```

Release of APP. Proteolysis of APP in membranes C-terminal of A4 did not change the aggregation properties. However, the enzymatic cleavage at the N-terminus of A4 is of great importance for plaque formation. Proteinase K cleaves a 10-18 kDa pro-A4 fragment to peptides of the size of A4 [5.301].

The APP gene localized on chromosome 21 is not amplified in Alzheimer's disease. However, the formation of this gene is greatly enhanced in Down's syndrome. It is assumed that the A4 protein is formed by increased proteolytic activity in Alzheimer's disease [5.302].

Both NGF and EGF stimulate the release of two C-terminal shortened APPs (125 kDa and 120 kDa), from PC12 cells [5.303]. Platelets release APP during degranulation [5.304].

Biological Effects. APP is widely distributed in the neurons of the rat brain and appears to play a role in the formation of cell-cell contacts, which is required for the memory capacity of the animal [5.289], [5.294].

APP released from cells has an autocrine effect on growth regulation [5.305], [5.306]. A form of APP_{751} that is shortened at the C-terminus is an inhibitor of the coagulation factor XIa and trypsin [5.307].

Low concentrations of A4 have a neurotrophic and higher concentrations a neurotoxic effect on mature neurons. In fact, A4-(25-35), which is structurally similar to the tachykinins, possesses this activity. This neurotoxic activity of A4 and A4-(25-35), which is also exhibited by some tachykinin antagonists, can be abolished by tachykinin agonists [5.308].

5. 3. Vasopressin, Oxytocin and Vasotocin [5.309]

5. 3.1. Occurrence

Oxytocin (OT) and vasopressin (VP) are formed as prohormones. OT and VP are localized at the N-terminal end of these prohormones, while the C-terminal residue contains the specific neurophysins (neurophysin I with OT and neurophysin II with VP). Together with the neurophysins as transport molecules, OT and VP are transported in neurosecretory vesicles down the axon to the neurohypophysis. Other sites of synthesis of OT and VP are the pineal body, the retina of mammals, the thymus [5.310], corpus luteum [5.311], testes [5.312], pheochromocytoma [5.313] and the cerebrospinal fluid.

Vasotocin (VT) has been found in the pineal body of mammals and in the cerebrospinal fluid of humans. Precursors of VT (hydrin 1 and 2) have been isolated from the pars intermedia of *Rana esculenta* [5.314]. Nα-Acetyl-OT has been found in the pineal body of the cow [5.315] and the rat [5.316], and, along with N-acetyl-VP, also in the pars intermedia and the brain of the rat [5.316].

5.3.2. Release

Pneumadin, a decapeptide amide from the lungs that has been characterized recently, releases VP from the hypophysis and has antidiuretic activity [5.319a]:

h-Pneumadin:	A	G	E	P	K	L	D	A	G	Va
r-Pneumadin:	Y	G	E	P	K	L	D	A	G	Va.

Naturally occuring peptides with OT- and VP-activity [5.317]		Occurrence
Vasotocin (VT)	C Y I Q N C P R Ga	birds, reptiles, amphibians fish, mammals (pineal gland)
Hydrin 1	C Y I Q N C P R G G K R	*Rana esculenta*
Hydrin 2	C Y I Q N C P R G G	*Rana esculenta*
Vasopressin (VP)	C Y F Q N C P R Ga	mammals except the pig
[Lys[8]]VP	C Y F Q N C P K Ga	pig, hippopotamus
Phenypressin	C F F Q N C P R Ga	marsupials
Lys-Conopressin-G	C F I R N C P K Ga	*Conus geographus* [5.318]
Arg-Conopressin-S	C I I R N C P R Ga	*Conus striatus* [5.318]
	C L I T N C P R Ga	*Locusta migratoria* [5.319]
Oxytocin (OT)	C Y I Q N C P L Ga	mammals
Mesotocin	C Y I Q N C P I Ga	*Bufo japonicus*, ostrich
Ichthyotocin (= Isotocin)	C Y I S N C P I Ga	fish
Glumitocin	C Y I S N C P Q Ga	cartilaginous fish
Valitocin	C Y I Q N C P V Ga	cartilaginous fish
Aspartocin	C Y I N N C P L Ga	cartilaginous fish

The plasma VP level increases in response to potassium uptake [5.320], dehydration and thirst [5.321], blood loss, i.c.v. infusion of hypertonic synthetic CSF, i.v. infusion of a hypertonic NaCl solution or hydrochloric acid [5.322], smoking [5.323], hypoxia (lack of oxygen) and hypercapnia (increase in the CO_2 concentration in arterial blood) and to sexual arousal in men [5.324]. In rats, estrogen potentiates the increased release of VP caused by osmotic stimulation [5.325]. The level of OT in the Plasma also increases on infusion with NaCl [5.326]. The release of VP is stimulated by acetyl choline (mediated by the hypothalamus) [5.327], AT II, AT II-(1-7) [5.328], histamine [5.329] and by dopamine (i.c.v.) [5.330]. The activation of central α_1-adrenoreceptors increases the release of VP [5.331], whereas the stimulation of the α_2-adrenoreceptors decreases the release of VP [5.332]. The plasma VP level is raised in many different high pressure animal models [5.333] and in human essential hypertension of the low renin type [5.334].

A low NaCl diet [5.335], water uptake [5.336], chronic alcohol consumption [5.337], and GABA [5.338] reduce the release of VP

The vasoactive intestinal peptide (VIP) [5.339] angiotensin II (AT II), corticoliberin (CRH) [5.340], fever [5.341] and endotoxin [5.342] release both VP and OT. β-Endorphin (β-EP) and other endogenous opiates inhibit the release of OT and VP.

The plasma level of OT is increased by manual stimulation before milking [5.343], during the third stage of labor [5.344], by CCK, after the intake of food [5.345], [5.346], during orgasm in the woman and ejaculation in the man [5.347].

In rats, it has been found that the VT content in the pineal body increases almost 1000 fold in August and falls in old age.

5. 3. 3. Receptors

There are two VP receptors, the V_1 receptor for the pressor activity and the V_2 receptor for the antidiuretic activity.

The VP receptors are down-regulated by VP agonists [5.348], [5.349].

The density of the OT receptors in various regions of the brain is increased by estrogen or testosterone [5.350]. Progesterone increases the binding of OT in the posterior part of the ventromedial nuclei of the hypothalamus. This is the site at which feminine mating behavior is induced by infusion with OT [5.351].

5. 3.4. Biological Effects of VP

V_1-Receptor Mediated Effects [5.352]
Stimulation of phosphatidylinositol turnover and aldosteron synthesis in rat adrenal glomerulosa cells [5.353]
Increase of cytosolic free calcium [5.354]
Vasoconstrictor effect and hypertension [5.333]
Stimulation of PGE_2 and $PGF_2\alpha$ in isolated perfused rabbit kidneys [5.355]
Indomethacin induced antipyresis [5.356]
Inhibition of mucosal acid secretion in rats [5.357]
Central inhibition of corticoliberin [5.358] and proopiomelanocortin release [5.359]
Periferal stimulation of proopiomelanocortin, ACTH and β-Endorphin [5.359]
Longlasting hypothermia by injection into the preoptic area of rats [5.360], [5.361]
Central analgesia
Acceleration of glycogenolysis in heptocytes
Promotion of platelet aggregation via serotonin release from platelets [5.362]
Stimulation of the clotting factors vWF and Factor VIII [5.363]

V_2-Receptor Mediated Effects [5.352]
Stimulation of adenylate cyclase
Antidiuretic effect
Formation of hypoosmolarity and cerebral edema in rats [5.364], [5.365]

Other Effects of VP:
Direct effect of VP on the release of ACTH [5.366]-[5.368]
Potentiation (by VP and OT) of CRH stimulated ACTH release [5.366]-[5.368]
Improvement of the memory and learning capacity of older persons [5.369]
Inhibition of the diuretic [5.370] and potentiation of the natriuretic effect of ANF [5.371]
Reduction of the pressure in the pulmonary arteries of hypoxic rats by the release of ANF [5.372]
Reduction of renin activity and ACE
Stimulation of kallikrein

5. 3.5. Biological Effects of OT

Stimulation of adenylate cyclase and the release of arachidonic acid and $PGF_2\alpha$ from human decidual cells [5.373]
Potent constrictor of the umbilical placental vasculature [5.374]
Contraction of the uterus and stimulation of lactation
Hypotensive effect [5.374] (release of VIP [5.346], [5.375], [5.376])
Stimulation of diuresis and natriuresis [5.338]
Prevents the fall in blood pressure associated with hemorrhage, possibly by increasing the release of renin [5.377]
Stimulation of sexual maturity in male mice [5.378]
Enhanced sexual receptivity and maternal behavior in female rats when administered centrally [5.379]
Enhanced penis erection in male rats when applied centrally [5.380] and increased sexual bahavior in old rats after i.p. application
Release of relaxin
Stimulation of the release of insulin and glucagon [5.381] and increase in plasma glucose levels [5.382]
Release of POMC peptides α-MSH, β-endorphin and ACTH from human placenta [5.383]
Reduction of addiction to narcotic analgetics [5.384]
Reduction of food intake after i.c.v. or i.p. application [5.385]

5. 3.6. Biological Effects of VT [5.386]

Extremely high frog bladder or hydroosmotic activity
High hen oxytocic activity
Inhibition of the stimulating action of pregnant mare's serum on mice uteri and ovaries
Extremely specific ratio of frog bladder to rat uterus activities
Extremely specific ratio of frog bladder to rat antidiuretic activities
Isometric contraction of eel ventral aortic strips
Lowers emotional-arousal in rats [5.387]
Increases REM-phases during sleep in children via serotonin release

5. 3.7. Structure-Activity Relationships [5.388]

The ring structure is of special importance for the activity of OT and VP agonists. However, the disulfide bridge is not essential. Indeed, 1-carba-OT is more potent than OT. An N-terminal extension, e.g., by means of triglycine, causes a depot effect. For instance, triglycyl-[Lys^8]VP (glypressin and terlipressin) has clear advantages over VP in the treatment of bleeding of esophageal varices. De-9-glycinamide-VP (Org 5667) has no cardiovascular or antidiuretic activity. However, this substance can still improve the learning and memory capacity of rats in behavior tests and also has the VP stimulating effect on penis erection [5.389]. The memory promoting effect of VP is possibly caused by a metabolite. Thus, the compound,

```
*Q N C P R Ga
     |
     C
```

isolated from the synaptic membrane of the rat brain and the corresponding deglycinamide derivative as well as its cysteine methylester [5.390] no longer have pressor activity, but possess a greater memory promoting effect than VP.

While the substitution of Phe for Ile^3 increases the vasopressor activity at the expense of the OT activity, the replacement of Arg^8 or Lys^8 by Leu has the opposite effect. Other examples for the alteration of activity as a result of amino acid substitution are given in Table 14.

Highly specific analogs can be produced by combining various characteristics. Compounds in which the OT activity predominates are deamino-di-carba-[Gly^7]OT (cargutocin), deamino-1-monocarba-[$Tyr(Me)^2$]OT (carbetocin is weaker but acts longer than OT) [5.400], [Thr^4,Gly^7]OT and [Thr^4,Sar^7]OT [5.401]. Deamino-[p-Et-Phe^2]6-carba-OT (narcartocin) has a higher natriuretic and diuretic activity, but is less effective in uterine contraction than OT [5.402]. Deamino-[D-p-Et-Phe^2]6-carba-OT is an OT antagonist, which inhibits the uterogenic and galactogenic activity of OT [5.403].

A strong antidiuretic preparation with only slight pressor activity but with vasodilating activity [5.404] is 1-deamino-[D-Arg^8]VP (dDAVP, desmopressin). The ratio of antidiuresis (A) to pressor activity (P) in 1-deamino-[Val^4,D-Arg^8]VP (dVDAVP) is practically infinitely large because of the lacking pressor activity. Like desmopressin, 1-deamino-[$Tyr(Me)^2$, Val^4,D-Arg^8]VP is a strong antidiuretic compound, but also a strong OT antagonist (pA_2 = 8.02) [5.405]. Another antidiuretic with only slight vasopressor activity is 1-deamino-[$MeAla^7$,D-Arg^8]VP (A/P ratio = 5260) [5.392].

The pressor activity is especially pronounced in the preparations, [Phe^2,Lys^8]VP (felypressin or octapressin) and [Phe^2,Ile^3,Orn^8]VP.

5.3.8. Antagonists [5.406]

Strong inhibitors with antioxytocin and antivasopressor activity and slight anti-antidiuretic activity are structural variants of β, β-dialkylated or β,β cyclopentamethylene-3-mercapto-propionic acid (Pmp) in position 1, O-alkylated tyrosine in position 2, and ornithine in position 8, e.g., the OT antagonists, [β-mercapto-β, β-diethylpropionic $acid^1$, $Tyr(Me)^2$,Orn^8]VT or deamino-[$Tyr(Me)^2$,Orn^8]VT. Relative to [Pen^1]OT and [Pen^1,Phe^2,Thr^4]OT, the prototypes for OT antagonists, additional Orn^8 and alkylated Tyr^2 do increase the duration of action but not the potency, while D-amino acids in position 2 increase the activity but not the duration of action. Thus, [Pen^1,D-$Phe(Me)^2$,Thr^4,Orn^8]VT ($pA_2 \approx 7.5$) is somewhat more potent than [Pen^1,$Phe(Me)^2$,Thr^4,Orn^8]VT ($pA_2 \approx 7.3$), but has

Table 14. Alteration of the activity spectrum as a result of the substitution of individual amino acids in OT and VP

	Increased quotient
Substitution in [Arg^8]VP:	
Desamino	antidiuresis: pressor effect
Tyr(Me)^2, Phe^2	antidiuresis: pressor effect
Increased lipophilia at position 4 (Val^4)	antidiuresis: pressor effect
Pro^4 [5.390a], Arg^4 [5.391]	antidiuresis: pressor effect
Δ^3 Pro^7, 4-hydroxy-Pro^7	antidiuresis: pressor effect
Sar^7, MeAla^7 [5.392]	antidiuresis: pressor effect
D-Arg^8	antidiuresis : pressor effect
Ala^9, D-Ala^9 [5.393]	antidiuresis: pressor effect
1-Aminocyclopropan-1-carbonic acid^9 (Acc) [5.394]	antidiuresis: pressor effect
4-Hydroxy-Pro^7 [5.395]	uterine effect: pressor effect
Ala° [5.396]	pressor effect: antidiuresis
Substitution in [Lys^8]VP:	
Phe^2	pressor effect: antidiuresis
Ile^3	pressor effect: antidiuresis
Orn^8	pressor effect: antidiuresis
Substitution in OT:	
p-F-Phe^2	uterine effect: antidiuresis
Thr^4	uterine effect: antidiuresis
Gly^7, thiazolidine carbonic acid^7, Δ^3Pro^7[5.397],Sar^7	uterine effect: antidiuresis
Desamino	antidiuresis: uterine effect
Carba-1	antidiuresis: uterine effect
Tyr (Me)^2	antidiuresis: uterine effect
Carba-6	antidiuresis: uterine effect
Gln^8	antidiuresis: uterine effect
Asn^4, Val^4, Glu(OMe)^4	milk ejection: uterine effect
β-Acetylamino-L-α-aminopropionic acid^4	milk ejection: uterine effect
Carba-1	milk ejection uterine effect
MeAla^7, Gly^7, Acc^7 [5.397], Aze^7 [5.398]	milk ejection: uterine effect
tert.-Leu^8	milk ejection uterine effect
Desamino^1, Hmp^1, Carba-1 [5.353]	uterine effect: milk ejection
Thr^4 [5.399]	uterine effect: milk ejection
Leu^4	diuretic effect: VP antagonist
3-Iod-Tyr^2	oxytocin inhibitor

Hmp = L-2-hydroxy-3-mercaptopropionic acid

a shorter duration of action [5.407]. Very strong and selective OT antagonists are [Pmp^1,Tyr(Me)^2,Thr^4,Orn^8,Leu^9]VT with a low antivasopressor and antidiuretic activity and [Pmp^1,Tyr(Me)^2,Thr^4,Orn^8]VT-(1-8) with a weak antivasopressor and anti-antidiuretic activity [5.408].

Other strong OT antagonists are deamino-[D-Tyr(Et)^2,Thr^4,Orn^8]VT (ORF 22164) [5.409] and [MePmp^1]OT [5.410]. Above all, ORF 22164 strongly inhibits contraction of the human myometrium, and has a very slight vasopressor and antidiuretic activity [5.411].

Deamino-[Pen^1,Tyr(Me)^2]VP and [Pmp^1,Tyr(Me)^2]VP (SK&F-100273) have high antivasopressor activity (V_1 antagonists) [5.412]. Highly selective V_1 antagonists with low antidiuretic activity are produced by introducing a 4-methyl or 4-phenyl group in Pmp^1: [Me-Pmp^1]VP [5.413] and [phenyl-Pmp^1,Tyr(Me)^2]VP [5.414].

In these analogs, the anti-antidiuretic activity (aquaretic) can be increased by substituting D-Phe, D-Tyr, O-alkylated D-Tyr or D-Ile for Tyr^2. The anti-antidiuretic/antivasopressor ratio is further improved by Ile^4, Val^4, Abu^4 or Ala^4: [Pmp^1, D-Ile^2,Abu^4]VP [5.415], [5.416]. The introduction of a cis-4-methyl group in Pmp^1 increase the antagonistic activity and greatly reduce the agonistic activity of antagonists of the type [Pmp^1,D-Tyr(Et)^2,Val^4]VP. The strongest anti-antidiuretic compound of this type is [Me-Pmp^1,D-Tyr(Et)^2,Val^4]VP [5.417], [5.418]. The deletion of Gly-amide and Pro^7 is of no great significance for the anti-antidiuretic activity of the antagonists [5.419], [5.420]. The replacement of the C-terminal tripeptide by, e.g., Arg-amide, Arg-D-Arg-amide [5.421] or diaminoalkylamides [5.422], [5.423] give rise to poten V_2 antagonists with aquaretic activity. One of the strongst aquaretic VP antagonists is dicarba-[Pmp^1,D-Tyr(Et)^2,Val^4,Arg^7,D-Arg^8]VP-(1-8)-amide (SK&F-105494) [5.424]-[5.426]. A very selective V_2 antagonist with an ED ratio of 440 is [Pmp^1,D-Ile^2,Ile^4]VP-(1-8) [5.427]. Numerous non-peptide structures are also tolerated at the C-terminal end [5.428].

The cyclic structure is, however, not required for the antagonistic activity. Thus, a reduced antagonist with free SH groups of the type [Pmp^1,D-Tyr(Et)^2,Val^4]VP still has a high anti-antidiuretic activity. Adamantyl-acetyl-[D-Tyr(Et)^2,Val^4,Abu^6, Arg^9]VP-(2-9)-amide has a higher anti-antidiuretic activity than [Pmp^1,D-Tyr(Et)^2,Val^4] VP [5.429].

In VP antagonists, substitution of Sar or N-methylalanine for Pro^7 give rise to OT antagonists. Thus, [Pmp^1,D-Phe^2,Sar^7]VP or [Pmp^1,D-Phe^2,MeAla^7]VP, like OT, binds to OT receptors and, like VP, to V_1 receptors, but has only slight antidiuretic activity. The OT and V_1 antagonist, [Pmp^1,D-Phe^2,MeAla^7]VP is a very strong antiglycogenolytic compound with antigalactogenic activity [5.430].

Completely new OT antagonists are based on a natural product of *Streptomyces silvestris* (L-156,373: cyclo-(Pro-D-Phe-N-OH-Ile-D-piperazyl-L-piperazyl-N-Me-D-Phe)). L-156,373 binds to OT receptors with a K_i = 150 nM and an approximately 20 fold selectivity compared with V_1 and V_2 receptors. Dehydroxylation of N-hydroxy-Ile and oxidation of piperazic acid (dehydrogenation in the ring) leads to L-365,209, which with a K_i = 1.7 nM, is a competitive OT antagonist having only slight affinity for V_1 and V_2 receptors [5.431]. Further derivatization led to the selective OT receptor ligands, cyclo-(Pro-D-Nal-Ile-D-Pip-Pip-D-His) (K_i = 1.6 nM) [5.432] and cyclo-(Pro-D-Pheψ(CSNH)Ile-DΔPiz-ΔPiz-NMe-D-Phe) [5.433] (Pip = pipecolic acid, ΔPiz = dehydro-piperazic acid).

5.3.9. Applications

Oxytocin (OT) (Orasthin, Orasthin "stark", Oxytocin 3 "Horm", Oxytocin forte "horm", Partocaon, Pitocin, Syntocinon is used to induce birth in primary and secondary uterine inertia, in placenta retention, postpartum hemorrhage, uterine atonia, lactation disturbances, Cesarean section and in mastitis prophylaxis.

Vasopressin (VP) (Pitressin, Pitressin Tannat) is used in the bleeding of esophageal varices, in intestinal atonia, to stop bleeding in gynecological operations and in diabetes insipidus. [Lys^8]VP (lypressin) is available as Postacton or Vasopressin-Sandoz for hypophyseal function tests, paralytic ileus and for diabetes insipidus. [Orn^8]VP (ornipressin) is available as Por 8 Sandoz and triglycyl-[Lys^8]VP (glypressin, terlipressin) as Glycylpressin. These compounds are used primarily in the bleeding of esophageal varices, acute uterine bleeding, and in operations. Deamino-[D-Arg^8]VP (desmopressin) is applied as Minirin in diabetes insipidus, after hypophysectomy, for increasing Factor VIII activity in hemophiliacs and in von Willebrand's disease [5.434]. Since desmopressin reduces blood loss during operations, it is recommended by Jehovah's Witnesses for this purpose [5.435].

In clinical studies, the V_1 antagonist SK&F-100273 ([Pmp^1,$Tyr(Me)^2$]VP) inhibits the [Lys^8]VP induced rise in blood pressure [5.436].

The OT antagonist deamino-[$Tyr(Et)^2$,Thr^4,Orn^8]VT can be administered in premature labor [5.437].

5.3.10. Melanostatin [5.438]

Occurrence. Melanostatin (MIF) has been isolated from extracts of the median eminence and has been found to have the structure: H-Pro-Leu-Gly-amide. It is identical to the C-terminal tripeptide of oxytocin (OT).

Biological Effects. MIF inhibits the release of α-MSH from the hypophysis. As a weak μ receptor antagonist, it impedes the antinociceptive activity of morphine, β-endorphin or morphiceptin [5.439]. It reduces aggression [5.440], impedes morphine [5.441] or alcohol [5.442] induced tolerance, and has an antiamnesic [5.443] and inhibitory effect on tremor induced by oxotremorin in mice. MIF increases the affinity of dopamine agonists for dopamine receptors and reduces the dopamine D_2 receptor supersensitivity [5.444]. Some of the CNS effects also appear to proceed via dopamine receptors [5.445].

Structure-Activity Relationships. Structure-activity studies using the oxotremorin tremor test have shown that Pro can be replaced by a pGlu residue and the amide group can be alkylated. For instance, pGlu-Leu-Gly-$N(CH_3)_2$ has 32 times the potency of MIF. The compound, [N-Me-D-Leu^2]MIF (pareptide) potentiates L-dopa induced behaviour patterns and has anti-parkinsonian activity and anti-depressive activity. [D-Pip^1]MIF and D-Pip-amide have a stronger antiamnesic activity than MIF [5.443]. The compounds, [3-(+)-thiazolidine-2-carboxamide]MIF, [3-(−)-

thiazolidine-2-carboxamide]MIF and [3-L-3,4-dehydroprolinamide]MIF [5.446] cause a two to three fold stimulation of the affinity of dopamine agonists for dopamine receptors and 3(R)-(N-L-prolylamino)-2-oxo-1-pyrrolidine acetamide is 10 000 times as potent as MIF [5.447].

Applications [5.448]. MIF and its analogs are supposed to exert an antidepressive effect, potentiate the learning capacity and serve in the treatment of Parkinson's disease.

5.4. Endothelin [5.449]

5.4.1. Occurrence

Endothelin (ET) is formed in the endothelial cells. ET is found, e.g., in the kidneys [5.450], hypothalamus, hypophysis [5.451], and in human milk [5.452].

Three different endothelins have been found in human beings (h-ET 1, h-ET 2, h-ET 3) [5.453]. p-ET [5.454] is identical to h-ET 1 and r-ET is identical to h-ET 3. ET-1 and a vasoactive intestinal contractor (VIC), that occurs only in the gastrointestinal tract, have been found in mice [5.455]. VIC is identical to h-ET 2 except for position 4. There is a great structural similarity between the endothelins and the sarafotoxins (SRT-A, SRT-B, SRT-C, SRT-D [5.456]) from the venom of the snake, *Atractaspis engaddensis* [5.457]. The structure of the prepro-compounds of p-ET 1, h-ET 2 [5.458], [5.459] and h-ET 3 [5.460], [5.461] have been elucidated via cDNA. Small amounts of a p-big-ET 1-(1-39) and the cleavage products p-big-ET 1-(22-39) and p-big-ET 1-(23-39) can also be isolated [5.462]. p-big-ET 1 is an intermediate product in the synthesis of pET 1.

h-,p-, m-ET 1	C	S	C	S	S	L	M	D	K	E	C	V	Y	F	C	H	L	D	I	I	W.
h-ET 2	C	S	C	S	S	W	L	D	K	E	C	V	Y	F	C	H	L	D	I	I	W.
m-ET 2 (m-VIC)	C	S	C	N	S	W	L	D	K	E	C	V	Y	F	C	H	L	D	I	I	W.
h, r-ET 3	C	T	C	F	T	Y	K	D	K	E	C	V	Y	Y	C	H	L	D	I	I	W.
SRT-A	C	S	C	K	D	M	S	D	K	E	C	L	N	F	C	H	Q	D	V	I	W.
SRT-B (SRT-6b)	C	S	C	K	D	M	T	D	K	E	C	L	Y	F	C	H	Q	D	V	I	W.
SRT-C	C	T	C	N	D	M	T	D	E	E	C	L	N	F	C	H	Q	D	V	I	W.
SRT-D	C	T	C	K	D	M	T	D	K	E	C	L	Y	F	C	H	Q	D	I	I	W.

5.4.2. Release

Endothelin is released in response to shear stress [5.463], thrombin, angiotensin II, vasopressin [5.464], the endogenous digitalis like factor (EDLF) [5.465], protein kinase C activating phorbol ester, the calcium ionophore ionomycin [5.464], TGF-β, TNF-α and IL-1β [5.466].

On the other hand, the release of ET is reduced by EDRF [5.467], dehydration

[5.451] and lack of oxygen (hypoxia) [5.468]. The plasma level of ET is lowered in genetically hypertensive rats [5.469].

The plasma ET levels are raised in patients suffering from essential hypertension [5.470], cardiogenic shock, pulmonary hypertension and in dialysis patients, but not in persons with congestive heart failure [5.471].

5.4.3. Receptors

Two different ET receptors are found in the lung of the rat: ET 1 and ET 2 bind tightly to the 44 kDa receptor and ET 3 binds preferentially to the 32 kDa receptor [5.472]. The receptors in the guinea pig bronchus which bind ET-(16-21) are called ET B receptors, while the receptors on the rat aorta which do not bind ET-(16-21) are known as the ET A receptors [5.473].

Treatment with ET [5.474] or AT II [5.475] leads to a reduction of the ET receptors.

5.4.4. Biological Effects

Increase in the intracellular calcium level [5.476]

Increase in the turnover of phosphatidyl inositol and in the concentration of 1,2-diacylglycerol 5.477]

Stimulation of phospholipase C [5.478]

Stimulation of 6-keto-$PGF_{1}\alpha$ [5.479], TXA_2, PGI_2, PGE_2 and prostacyclin [5.480]

Calcium dependent contraction of the vascular smooth muscles (stronger than AT II) [5.481] (important for the regulation of blood pressure and the circulation [5.454])

Contraction of the non-vascular smooth muscles (duodenum, stomach, bronchial tubes, trachea, uterus, urinary bladder [5.481], but not the guinea pig ileum [5.482]

Endothelium dependent [5.483] hypotensive effect via EDRF [5.484]

Strong pressor effect after a quick drop in blood pressure [5.485]

Positive inotropic effect on the left atria of the rat and the spontaneously active right atria [5.486]

Hemorrhagic and necrotic damage to the gastric mucosa [5.486], [5.487]

Reduction of the glomerular filtration rate which can be abolished by α-ANP [5.488]

Calcium dependent inhibition of the renin secretion [5.489]

Stimulation of aldosterone and corticosterone in dispersed zona glomerulosa cells [5.490]

Modulation of the synthesis of noradrenaline (reduced by low doses and increased by high doses)

Strong stimulation of the release of α-ANP in the isolated rat heart [5.491] and cultivated rat atrial cardiocytes [5.492]

Secretion of gonadotropins from anterior hypophyseal cells independent of gonadoliberin receptors [5.493]. (While ET 3 has no effect on the release of lutropin, it inhibits the secretion of prolactin (PRL). On the other hand, ET 1 has no effect on the release of PRL [5.494])

Inhibition of the *ex vivo* aggregation of platelets by i.v. application of ET 1 or ET 3 through the prostaglandins and/or EDRF [5.495], [5.496]

Cell proliferation of smooth muscle cells (potentiated by PDGF) [5.497]

Hyperalgesia (rat paw) and long lasting itching with local erythema and heat (intradermal application)

Possible pathological conditions arising from an increased formation of endothelin: coronary and cerebral vasospasm, cardiovascular diseases, bronchospasm, atherosclerosis and hypertension, stomach and duodenal ulcers, and acute renal failure [5.498].

5.4.5. Structure-Activity Relationships

ET 1 has about 100 times the contracting activity of big ET [5.499]. ET 1 and ET 2 have a higher contracting activity than ET 3 [5.500]. SRT B has a higher vasodilative activity than the endothelins (SRT B > ET 1 = ET 2 > ET 3). The vasoconstrictive effect of the SRTs on the rabbit aorta is: SRT B > SRT A > SRT C. SRT C exhibits vasodilative activity when applied in high doses [5.501]. SRT D is less active than SRT A and SRT B [5.456].

The C-terminal extended procompounds of ET 1 (p-ET-1-39 and h-ET-1-38) are 100 times less active *in vitro*. However, the activity of these compounds *in vivo* is practically identical to that of ET [5.502].

The free N-terminal amino group, the C-terminal carboxyl group, the carboxyl groups of Asp^8 and Glu^{10}, the aromatic ring in position 14, and Trp^{21} are important for activity [5.503]. Thus, the loss of Trp^{21} results in a thousand fold reduction of activity. Further shortening of the molecule at the C-terminus reduces the activity still further. The N-terminal 1-16 fragment has no vascular activity. In the guinea pig isolated bronchus, ET-(16-21) still exerts a contracting effect (ED_{50} = 0.228 μM). The C-terminal pentapeptide still acts at about 25 μM and the C-terminal tetrapeptide is devoid of activity. In ET-(16-21)-Trp^{21}, Asp^{18}, His^{16}, and evidently also Leu^{17} may be important for activity because the corresponding SRT-(16-21)-hexapeptides have no activity [5.504]. However, ET-(16-21) has no contracting effect on the rat aorta [5.473].

Both disulfide bridges are important, but not necessary, for activity. Thus, a cyclic derivative with a disulfide bridge between Cys^1 and Cys^{15}, Acm protection in positions 3 and 11 and a C-terminal amide has no effect on the arterial blood pressure [5.485]. Incorrectly formed disulfide bridges also lead to less active derivatives. However, [$Ala^{1,15}$]ET 1 and [$Ala^{3,11}$]ET 1 (weaker) still have vasoconstrictive activity, but [$Ala^{1,3,11,15}$]ET 1 is devoid of activity [5.483]

5.5. Atrial Natriuretic Factor [5.505]-[5.507]

5.5.1. Occurrence

A group of peptides has been isolated from the atrium of the mammalian heart which exerts natriuretic, diuretic and vasodilating effects. These peptides are synthesized as prepro-atrial natriuretic factor (prepro-ANF). A large number of peptides that are shortened at the N- and C-terminus are formed from prepro-ANF by enzymatic degradation. Depending on the species, 23-25 N-terminal amino acids of prepro-ANF belong to the pre-sequence. Thus, Asn^1 represents the start of ANF with 126 amino acids. It could be isolated from atrial tissue and was also called γ-ANP or cardionatrin IV. In the granulocytes of the atrium, r-ANF is stored and not metabolized further [5.508]. Apart from α-ANP-(3-28), α-ANP-(ANF (99-126)) represents the circulating peptide [5.509]. In human plasma, h-β-ANP [5.510], an antiparallel dimer of h-α-ANP, was found along with α-ANP. Indeed, h-β-ANP can be converted to h-α-ANP in plasma [5.511]. In the rat, atriopeptin III has also been found as a circulating peptide [5.512]. The N-terminal fragment ANF-(1-98) and α-ANP are secreted from the cardiocytes of the atrium into the circulation, where ANF-(1-98) accumulates to a greater extent than α-ANP because of its longer half life [5.513], [5.514]. In addition, two N-terminal sequences of ANF are also found in the serum of human beings, ANF-(1-30) and ANF-(31-67), which also show ANF activity [5.515].

Immunoreactive ANF was not only found in the atrium of the heart and in the plasma, but also in the rat heart ventricle [5.516], in the brain [5.517], hypothalamus [5.518], cerebrospinal fluid [5.519], posteria lobe of the hypophysis [5.520], [5.521], ovaries [5.522], intestine [5.523], gastrointestinal tract [5.524], thymus [5.525] peripheral autonomic nervous system [5.526], eye tissue [5.527], lung [5.528] and in the adrenal medulla. ANP like peptides have also been isolated from the hearts of chicken [5.529], eels [5.530], and frogs [5.531].

An ANF like peptide, the brain natriuretic peptide (p-BNP) [5.532], has been found in the bovine brain. The precursors of the brain natriuretic peptides have been characterized through the cDNAs: p- [5.533], h- [5.534], ca- [5.534] and the r-BNP-prepro peptide [5.535]. The BNP structure is located in the C-terminal part. The b-prepro-BNP is processed by the cleavage of the signal peptide to b-γ-BNP with 106 amino acids and to the smaller forms, such as b-BNP-26 and b-BNP-32 [5.533]. The highest concentrations of i.r. BNP are found in the atrium. However, this is only 1/50 of the amount of ANF. Compared with ANF, only 1/20 of the amount of i.r. BNP is found in the plasma [5.536]. In the rat, i.r. BNP is found primarily as r-BNP-45 (iso-ANP [5.537]) and γ-BNP in the atrium, but not in the brain [5.538]. BNP like peptides have also been found in the brain of eels and chicken [5.539].

Another ANF like peptide, which is called the C type natriuretic peptide (CNP), has been found in the porcine brain. The pharmacological spectrum of CNP is similar to that of ANP and BNP [5.540] see Table 15 and 16.

Table 15. ANF of various species [5.541]

Species	Amino acid sequence
	+1 … 16 … 21 … 32
h	N P M Y N A V S N A D L M D F K N L L D H L E E K M P L E D E V-
ca	N P V Y G S V S N A D L L D F K N L L D R L E D K M P L E D E A-
cu	N P V Y N A M S N A D L M D F K N L L D H L E D R M P F E D E A-
r	N P V Y S A V S N T D L M D F K N L L D H L E E K M P V E D E V-
m	N P V Y S A V S N T D L M D F K N L L D H L E E K M P V E D E V-
b	N P V Y G S V S N A D L M D F K N L L D R L E D K M P L E D E A-
	33 … 49 … 54 … 64
h	V P P Q V L S E P N E E A G A A L S P L P E V P P W T G E V S P-
ca	E S P Q A L S E Q N A E A G A A L S P L P E V P P W T G E V S P-
cu	V P P Q A L S E Q S D E A G A A L S P L P E V P P W T G E V S P-
r	M P P Q A L S E Q T D E A G A A L S S L S E V P P W T G E V N P-
m	M P P Q A L S E Q T E E A G A A L S S L P E V P P W T G E V N P-
b	V P S Q V L S E Q N E E A G A P L S P L S E M P P W M G E V T P-
	65 … 79 … 94 … 96
h	A Q R D G G A L G R G P W D S S D R S A L L K S K L R A L L T A-
ca	A Q R D G G A L G R S P W D S S D R S A L L K S K L R A L L A A-
cu	A Q R D G E A L G R S T W E A S E R S A L L K S K L R A L L T A-
r	S Q R D G G A L G R G P W D P S D R S A L L K S K L R A L L A G-
m	P L R D G S A S R R S P W D P S D R S A L L K S K L R A L L A G-
b	A Q R E G G V L G R G P W E S S D R S A L L K S K L R A L L T A-
	97 … 105 … 126
h	P R S L R R S S C F G G R M D R I G A Q S G L G C N S F R Y.
ca	P R S L R R S S C F G G R M D R I G A Q S G L G C N S F R Y.
cu	P R S L R R S S C F G G R I D R I G A Q S G L G C N S F R Y R R.
r	P R S L R R S S C F G G R I D R I G A Q S G L G C N S F R Y R R.
m	P R S L R R S S C F G G R I D R I G A Q S G L G C N S F R Y R R.
b	P R S L R R S S C F G G R M D R I G A Q S G L G C N S F R Y R R.

——— α-ANP ———

——— Atriopeptin III ———

——— Atriopeptin II ———

——— Atriopeptin I ———

Species	Amino acid sequence																															
g-α-ANP				M	M	R	D	S	G	C	F	G	R	R	I	D	R	I	G	S	L	S	G	M	G	C	N	G	S	R	K	N.
anq-ANP				S	K	S	S	S	P	C	F	G	G	K	L	D	R	I	G	S	Y	S	G	L	G	C	N	S	/	R	K.	
ra-ANP-24							S	S	D	C	F	G	S	R	I	D	R	I	G	A	Q	S	G	M	G	C	/	G	/	R	R	F.
p-BNP-32	S	P	K	T	M	R	D	S	G	C	F	G	R	R	L	D	R	I	G	S	L	S	G	L	G	C	N	V	L	R	R	Y.
p-BNP-26							D	S	G	C	F	G	R	R	L	D	R	I	G	S	L	S	G	L	G	C	N	V	L	R	R	Y.
h	S	P	K	M	V	Q	G	S	G	C	F	G	R	K	M	D	R	I	S	S	S	S	G	L	G	C	K	V	L	R	R	H.
ca	S	P	K	M	M	H	K	S	G	C	F	G	R	R	L	D	R	I	G	S	L	S	G	L	G	C	N	V	L	R	K	Y.
r	N	S	K	M	A	H	S	S	S	C	F	G	Q	K	I	D	R	I	G	S	L	S	R	L	G	C	D	G	L	R	L	F.
g-BNP				M	M	R	D	S	G	C	F	G	R	R	I	D	R	I	G	S	L	S	G	M	G	C	N	G	S	R	K	N.
anq-BNP					G	W	N	R	G	C	F	G	L	K	L	D	R	I	G	S	L	S	G	L	G	C.						
r-Iso-ANP (= r-BNP-45)																		S	Q	D	S	A	F	R	I	Q	E	R	L	R.		
	N	S	K	M	A	H	S	S	S	C	F	G	Q	K	I	D	R	I	G	A	V	S	R	L	G	C	D	G	L	R	Q	F.
p-CNP					G	L	S	K	G	C	F	G	L	K	L	D	R	I	G	S	M	S	G	L	G	C.						

Table 16. ANF peptides isolated from the atrium of the heart, serum and the brain [5.505]

γ-ANP = Cardionatrin IV	= ANF-(1-126)
h-ANF-(1-30) [5.515]	
h-ANF-(31-67) [5.515]	
r-Atriopeptigen	= r-ANF-(16-126)
r-ANF-H2 = r-Pronatriodilatin	= r-ANF-(21-126)
r-Cardionatrin III	= r-ANF-(49-126)
r-ANF-H1	= r-ANF-(54-126)
r-β-ANP	= r-ANF-(79-126)
r-β-ANP-(14-48) = ANF-I	= r-ANF-(92-126)
r-β-ANP-(16-48)	= r-ANF-(94-126)
r-β-ANP-(17-48)	= r-ANF-(95-126)
h-β-ANP-(17-48) = h-Urodilatin [5.542]	= h-ANF-(95-126)
r-β-ANP-(18-48) = ANF-II	= r-ANF-(96-126)
r-β-ANP-(18-47) = ANF-III	= r-ANF-(96-125)
h-α-ANP = b-α-ANP [5.543] = Cardilatin	= h-ANF-(99-126)
r-α-ANP = Cardionatrin I	= r-ANF-(99-126)
r-α-ANP-(3-28)	= r-ANF-(101-126)
r-α-ANP-(4-28) = ANF-IV = Auriculin B	= r-ANF-(102-126)
p-α-ANP-(4-28) [5.544]	= p-ANF-(102-126)
h-α-ANP-(4-28) = Anaritide [5.545]	= h-ANF-(102-126)
p-α-ANP-(5-28) = p-Atriopeptin III [5.544]	= p-ANF-(103-126)
r-α-ANP-(4-27) = Auriculin A	= r-ANF-(102-125)
r-Atriopeptin III	= r-ANF-(103-126)
r-Atriopeptin II	= r-ANF-(103-125)
r-Atriopeptin I	= r-ANF-(103-123)
r-Atriopeptin I-(2-21)	= r-ANF-(104-123)
r-Atriopeptin I-(2-20)	= r-ANF-(104-122)
r-Atriopeptin I-(3-21)	= r-ANF-(105-123)

5.5.2. Release

The calcium dependent [5.546] formation of α-ANP appears to proceed via the phosphoinositide and cAMP system [5.547]. The plasma levels of ANF correlate positively with the pressure of the pulmonary artery, with the pressure in the right ventricle and with the heart index [5.548].

The release of α-ANP is stimulated after an increase in volume of the blood, in hypertension, by the activation of protein kinase C [5.549], by PGE_2 and $PGF_{2}\alpha$ (greater) [5.550], vasopressin [5.551], oxytocin, angiotensin II [5.552], phenylephrine [5.553], adrenaline [5.554], by the stimulation of α_1-adrenergic [5.555] and α_2-adrenergic receptors [5.556], β-receptor blockers (in patients with essential hypertension [5.557] or in healthy persons who are engaged in sports [5.558], but not in normal healthy persons), verapamil (in hypertensive patients [5.559]), Substance P [5.560], NPY [5.561], endothelin [5.562], thyreoliberin [5.563], thyroid hormones [5.564], by the stimulation of muscarinic and cholinergic receptors, stress [5.565], glucocorticoids [5.566], [5.567], mineralocorticoids [5.568], testosterone [5.567], NaCl infusion and a high NaCl diet [5.569], e.g., by eating salty potato chips (the presence of gluco- and mineralocorticoids is necessary), a bath [5.570] or by anesthesia produced by halothane [5.512], morphine [5.571], diethyl ether, chloral hydrate or ketamine chlorohydrate [5.565].

Plasma ANP levels are especially high in newborn children, particularly in premature babies [5.572]. The plasma level of α-ANP in old healthy persons is higher than in young persons [5.573].

The plasma α-ANP level in human beings is raised in congestive cardiac defects [5.574], attacks of supraventricular tachycardia and atrial bradycardia, supraventricular tachycardia, in burns (stress!) [5.575], Cushing's syndrom [5.576] primary aldosteronism [5.577], liver cirrhosis [5.578], SIADH (the syndrome of inappropriate secretion of antidiuretic hormone) [5.579], hyperthyroidism (falls in hypothyroidism) [5.580], in anorexia nervosa [5.581], during exercise [5.582], in pregnant women, and in chronic kidney failure [5.583]. In patients suffering from cardiac defects, an eight fold increase in the plasma α-ANP level is observed after exercise, while a two fold increase at the most is found in healthy persons [5.584].

The secretion of ANF is reduced by the activation of protein kinase A [44], by capsaicin [5.585], hypophysectomy [5.586], hypothyroidism [5.587], pithed rats, after operations [5.588], after a decrease in the plasma volume, e.g., after the application of furesemide [5.589] or on dehydration [5.528], by a low salt diet, neurotensin [5.560], VIP [5.560] and PYY [5.560], [5.561].

The endopeptidase 3.4.24.11 cleaves h-α-ANP first within the disulfide ring and then at the Ser-Phe bond. It is assumed that cleavage occurs at the following sites [5.590], [5.591]:

UK 69 578, an inhibitor of endopeptidase 3.4.24.11, increases the plasma level of ANF in healthy persons [5.592] and in patients with chronic cardiac defects, reduces the blood pressure and increases natriuresis and diuresis [5.593].

5.5.3. Receptors

Until now, three different receptors have been isolated by molecular cloning and their structures have been elucidated [5.594].

ANF A (120-140 kD) is a membrane bound guanylate cyclase. ANF A binds both ANF as well as BNP peptides. ANF B (120-140 kD) is a membrane bound guanylate cyclase. However, ANF B preferably binds BNP peptides. ANF C (60-70 kD) is not a guanylate cyclase and may play a role in the clearance of ANF. ANF C is coupled to the adenylate cyclase/cAMP system [5.595]. There is a high sequence similarity between the three receptors.

Only the completely biologically active ANF sequences bind to the cGMP bound receptor, whereas smaller ANF sequences, e.g., r-ANF-(108-114): G R I D R I G [5.596] or de-(18-22)-α-ANP-(4-23)-amide bind to the receptors that do not mediate the release of cGMP. The latter sequence stimulates diuresis and natriuresis and lowers blood pressure *in vivo* via an endogenous increase in α-ANP [5.597].

There are specific receptors on the smooth muscles also for the N-terminal fragments of ANF-(1-126), namely ANF-(1-30) and ANF-(31-67). These receptors differ from those for ANF-(99-126) [5.598].

Glucocorticoids [5.599] and dehydration [5.600] increase the binding sites for ANP. The receptor density on the blood platelets decreases with the intake of a high salt diet and increases with a low salt diet [5.601]. The ANF receptors in the rat aorta are reduced by pretreatment with α-ANP (receptor down regulation) [5.602].

5.5.4. Biological Effects [5.603]

The peripheral application of ANF peptides results in the following endothelium independent [5.604] effects: vasodilation, reduction of plasma volume [5.605], increase in hematocrit, hypotension, natriuresis, diuresis, and in the inhibition of the renin-angiotensin system and the stress hormone cascade.

The vasodilating activity of ANF peptides is based on the inhibition of the calcium effect and of the release of calcium [5.606].

In animals that are not dependent on renin, α-ANP reduces the cardiac output and raises the total peripheral resistance. In hypertensive models in which the renin-angiotensin system is stimulated, α-ANP lowers the blood pressure by reducing the total peripheral resistance [5.607]. The ANF dependent reduced cardiac output is explained by the lower amount of venous blood reaching the heart and the reduced volume of blood [5.608]. This also results in reduced blood pressure which is not very pronounced in healthy animals, but is clearly observed in spontaneously hypertensive animals or in animals treated with noradrenaline or AT II [5.609]. Endothelin induced contractions are also reduced by ANP peptides [5.610].

In human beings, the renal activity of α-ANP (Wy 47663) is not dose dependent, but corresponds to a bell shaped curve. The diuretic activity is highest on infusion (2 hours) with 0.075 μg/kg.min. In comparison, the natriuretic activity is increased at 0.015 μg/kg.min. A hypotensive effect has only been observed at higher doses. The renal effects disappeared at the highest doses (0.3 μg/kg.min) and the plasma levels of renin and noradrenaline were increased [5.611]. In healthy test persons, vasopressin inhibits the diuretic effect and potentiates the natriuretic effect of ANF [5.612]. In numerous animal models for kidney failure, ANF peptides improve kidney function [5.613]-[5.618].

Other Effects of ANF Peptides:

Relaxation of the pulmonary artery [5.619], [5.620] and of the vessels of the lung in patients with chronic obstructive pulmonary disease (COPD) [5.621].

Reduction of pulmonary edema [5.622]

Reduction of the inner pressure in the rabbit eye [5.623]

Stimulation of the $Na^+/K^+/Cl^-$ cotransport in the vascular smooth muscles [5.624]

Inhibition of EDRF release [5.625]

Inhibition of adenylate cyclase activity [5.626]

Stimulation of guanylate cyclase and release of cGMP [5.627], which can be abolished by AT II [5.628]

Stimulation of erythropoietin (via cGMP) [5.629]
Inhibition of the formation [5.630] and activity of noradrenaline [5.613]
Inhibition of the release of CRH and proopiomelanocortin [5.631]
Inhibition of the release and activity [5.632], [5.633] of vasopressin (the vasorelaxing [5.634] and the diuretic [5.635], but not the natriuretic activity of ANF are inhibited by vasopressin)
Inhibition of the AT II stimulated formation of aldosterone [5.636], water uptake [5.637], renal water resorption [5.638] and cellular calcium increase [5.639]
Inhibition of the synthesis of pregnenolone, deoxycorticosterone, progesterone, cortisol and corticosterone
Stimulation of the formation of testosterone in interstitial cells of the mouse [5.640]
Inhibition of the secretion of thyroid hormone [5.641]
Inhibition of the endothelin induced pressor activity [5.642] (endothelin inhibits the ANF induced natriuresis)
Inhibition of PDGF and TGF-β [5.643] stimulated proliferation of smooth muscle cells
Release of histamine from rat peritoneal mast cells [5.644]
Inhibition of intestinal water and ion absorption [5.645]
Stimulation of the exocrine pancreatic secretion [5.646], [5.647]
The i.c.v. application results in the inhibition of prolactin secretion via dopamine [5.648], stimulation of somatotropin secretion via the inhibition of somatostatin formation [5.649] and stimulation of gastic acid secretion [5.650].

5.5.5. Structure-Activity Relationships [5.651]

The α-ANP nomenclature will be used for small C-terminal molecules and the ANF-(1-126) for larger molecules.

```
            1           7 8 9       12                              23    26  28
h-α-ANP     S L R R S S C F G G R M D R I  G A Q S  G L G C N S  F R Y
(h-ANF-(99-126))        |_____________________________________|
```

α-ANP peptides that are cyclized via the disulfide bridge or by means of Asu [5.652] exhibit good activity. The C-terminal Tyr can be replaced by the amide function. An N-terminal extension of α-ANP leads to more active derivaties [5.653]. Indeed, urodilatin (ANF-(95-126)), exerts a stronger renal effect than α-ANP [5.654]. The aromatic ring in Phe^8 of α-ANP, but not that in Phe^{26} is essential [5.655]. The deletion of Gly^9 in h-α-ANP produces an inactive product, while the full biological activity is retained by the substitution of D-Ala for Gly^9 [5.652]. The position 12 of α-ANP must be lipophilic (Met or Ile) since the oxidation of Met to the polar Met sulfoxide [Met(O)] leads to a substantial loss of activity. Nle^{12} is not as favorable as Ile^{12} [5.656]. Asp^{13} is evidently not essential because $[Asp(OMe)^{13}]\alpha$-ANP-(5-27)-methylester is still quite active, whereas $[D\text{-}Asp^{13}]$h-α-ANP has practically no activity [5.652]. The activity of α-ANP appears to be increased by the phosphorylation of Ser^6: $[Ser(PO_3H)^6]\alpha$-ANP-(3-28) is about 100 times as active as the unphosphorylated peptide in stimulating the $Na^+/K^+/Cl^-$ cotransport in vascular cells of the smooth muscles [5.657].

The replacement of each of three amino acids by 8-amino-octanoic acid shows in r-α-ANP-(5-28)-amide that above all the sequence Arg^{11}-Gly^{16} is important for receptor binding and Phe^8-Gly^{16} for the vasodilating activity [5.658].

Superactive analogs are obtained by deleting the N-terminal peptide chain up to Cys and by replacing Cys by 3-mercapto-propionic acid (Mpr), 2-mercaptoacetic acid or 4-mercapto-butyric acid [5.659]. [Mpr^7, D-$Ala^{9,16}$] α-ANP-(7-27)-amide (PL 68) has about two to three times the biological activity (receptor binding, drop in blood pressure and natriuresis in rats) of h-α-ANP [5.660]. [Mpr^7,D-Ala^9]α-ANP-(7-28) has twice the hypotensive effect and five times the cGMP stimulating effect of α-ANP-(5-28) [5.661]. r-[D-Tyr^6]α-ANP-(6-28) (dextronatrin) has the same diuretic activity as α-ANP, but it exerts a lower hypotensive effect [5.662]. [$Aoc^{6,28}$, D-Ala^9] α-ANP-(6-28) [5.663] exhibits strong receptor binding (IC_{50} = 10^{-11}–10^{-12} M), results in a considerably longer diuresis and natriuresis and is more active than h-α-ANP in animal models for kidney failure.

A cyclopeptide from [Phe^6, Pro^7, Glu^{13}]r-α-ANP-(6-22),
cyclo-(F P F G G R I E R I G A Q S G L G),
still exerts the full biological effect of ANF, but has reduced activity. The activity of this type of compound could be increased by further substititions [5.664].

The binding to the membranes of the glomerulosa zone correlates with the inhibition of the PGE_1 stimulated secretion of aldosterone: r-α-ANP-(3-27) > r-α-ANP, r-α-ANP-(3-28) or r-α-ANP-(4-28). α-ANP-(7-28) peptides exert a greatly reduced inhibitory effect on the PGE_1 stimulated secretion of aldosterone [5.665].

The inhibition of the CRH stimulated release of ACTH requires an intact amino end of α-ANP-(1-28). Indeed, α-ANP-(5-28) no longer exhibits activity, whereas α-ANP-(1-11) still exerts a weak inhibitory effect on the CRH stimulated release of ACTH [5.666].

Surprisingly, N-terminal fragments of ANF (ANF-(1-30), ANF-(31-67), ANF-(79-98)), also have a vasodilating and a cGMP stimulating activity similar to that of ANF [5.667]. ANF-(1-30), -(31-67), -(79-98) and -(99-126) increase sodium excretion by 231%, 973%, 167% and 1405% respectively. Apart from ANF-(79-98), all these peptides have diuretic activity. ANF-(79-98) exhibits kaliuretic activity [5.668].

5.5.6. Inhibitors

h-α-ANP-(7-23)-amide and [$Cys(R)^{7,23}$]h-α-ANP (R = Acm, Me or carboxyamidomethyl) inhibit h-α-ANP-stimulated cGMP [5.669].

A β-ANP like carba analog shows antagonistic activity (pA_2 = 7.45) [5.670]:

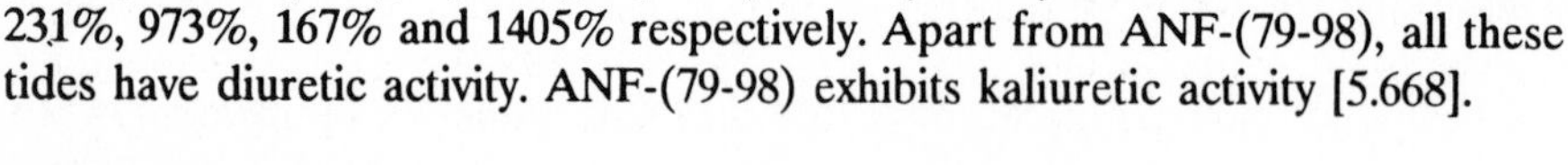

```
Y ←R←F←S ←N←C←G←L←G←S ←Q←A←G←I ←R←D←M←R←G←G←F ─┐
                /                                               │
              C  F  G  G  R  M  D  R  I   G  A  Q  S   G  L  G   Asu N  S  F  R  Y
```

In rats, cyclo-(Phe-Gly-Gly-Arg-Leu-Asp) and cyclo-(Phe-Gly-Gly-Arg-Met-Asp-

Arg-Ile-Gly) have hypertensive activity [5.655].

5.6. Pancreatic Peptide Family [5.671]

The three peptide amides containing 36 amino acids, pancreatic peptide (PP), neuropeptide Y (NPY), and the peptide with N- and C-terminal tyrosine (PYY), belong to the pancreatic peptide family.

There is a certain structural similarity between these peptide and neuromedin U (NMU) and the delta sleep inducing peptide (DSIP).

	1																	18
h-PP	A	P	L	E	P	V	Y	P	G	D	N	A	T	P	E	Q	M	A-
h-, r-NPY [5.672]	Y	P	S	K	P	D	N	P	G	E	D	A	P	A	E	D	M	A-
h-PYY [5.673]	Y	P	I	K	P	E	A	P	G	E	D	A	S	P	E	E	L	N-
cu-DSIP							W	A	G	G	D	A	S	G	E			

	19																	36
h-PP	Q	Y	A	A	D	L	R	R	Y	I	N	M	L	T	R	P	R	Ya.
h, r-NPY	R	Y	Y	S	A	L	R	H	Y	I	N	L	I	T	R	Q	R	Ya.
h-PYY	R	Y	Y	A	S	L	R	H	Y	L	N	L	V	T	R	Q	R	Ya.
p-NMU-B											Y	F	L	F	R	P	R	Na.

5.6.1. Pancreatic Peptide

Occurrence. Cells that produce pancreatic peptide (PP) are found primarily in the pancreas. In the pancreas of ground squirrels (*Gopherus polyphemus*), PP was found in the same cells that produce glucagon [5.674]. The precursor of h-PP has been characterized [5.675].

Release. The pulsatile release of PP in the pancreas is coupled to an insulin pulse [5.677]. The plasma level of PP is increased by the electrical stimulation of the vagus, food intake (particularly protein and oleate [5.678]), stimulation of β-adrenergic receptors, β-casomorphine, the μ receptor ligands morphine and morphiceptin, hypoglycemia induced by insulin or fasting, HB 699 [5.679], acetylcholine, CCK, caerulein, secretin, neurotensin (via cyclooxygenase products [5.680]), bombesin [5.681] [5.682], gastrin, GIP and by VIP. The PP level was found to oscillate in a 90-minute rhythm in test persons with an empty stomach. This PP secretion has been connected with the migratory motorial complex which also runs at a cycle of 90 minutes.

The plasma level of PP increases continuously with age [5.683]. In general, patients with a reduced exocrine secretion from the pancreas appear to have an excessively high plasma PP level. In 45% of the patients suffering from pancreatic endocrine tumors, plasma PP levels are raised and cannot be lowered by atropine [5.684].

PP Sequences of various species [5.676].

	1												13					18
h-PP	A	P	L	E	P	V	Y	P	G	D	N	A	T	P	E	Q	M	A-
p-PP	A	P	L	E	P	V	Y	P	G	D	D	A	T	P	E	Q	M	A-
b-PP	A	P	L	E	P	E	Y	P	G	D	N	A	T	P	E	Q	M	A-
o-PP	A	S	L	E	P	E	Y	P	G	D	N	A	T	P	E	Q	M	A-
ca-PP	A	P	L	E	P	V	Y	P	G	D	D	A	T	P	E	Q	M	A-
f-PP	A	P	L	E	P	V	Y	P	G	D	N	A	T	P	E	Q	M	A-
r-PP	A	P	L	E	P	M	Y	P	G	D	Y	A	T	H	E	Q	R	A-
m-PP	A	P	L	E	P	M	Y	P	G	D	Y	A	T	P	E	Q	M	A-
cav-PP	A	P	L	E	P	V	Y	P	G	D	D	A	T	P	Q	Q	M	A-
g-PP	G	P	S	Q	P	T	Y	P	G	D	D	A	P	V	E	D	L	I-
ans-PP	G	P	S	Q	P	T	Y	P	G	N	D	A	P	V	E	D	L	I-
st-PP	G	P	A	Q	P	T	Y	P	G	D	D	A	P	V	E	D	L	V-
ra_1-PP	A	P	S	E	P	H	H	P	G	D	Q	A	T	P	D	Q	L	A-
ra_2-PP	A	P	S	E	P	H	H	P	G	D	Q	A	T	Q	D	Q	L	A-
al-PP	T	P	L	Q	P	K	Y	P	G	D	G	A	P	V	E	D	L	I-

	19																	36
h-PP	Q	Y	A	A	D	L	R	R	Y	I	N	M	L	T	R	P	R	Ya.
p-PP	Q	Y	A	A	E	L	R	R	Y	I	N	M	L	T	R	P	R	Ya.
b-PP	Q	Y	A	A	E	L	R	R	Y	I	N	M	L	T	R	P	R	Ya.
o-PP	Q	Y	A	A	E	L	R	R	Y	I	N	M	L	T	R	P	R	Ya.
ca-PP	Q	Y	A	A	E	L	R	R	Y	I	N	M	L	T	R	P	R	Ya.
f-PP	Q	Y	A	A	E	L	R	R	Y	I	N	M	L	T	R	P	R	Ya.
r-PP	Q	Y	E	T	Q	L	R	R	Y	I	N	T	L	T	R	P	R	Ya.
m-PP	Q	Y	E	T	Q	L	R	R	Y	I	N	T	L	T	R	P	R	Ya.
cav-PP	Q	Y	E	T	E	M	R	R	Y	I	N	T	L	T	R	P	R	Ya.
g-PP	R	F	Y	D	N	L	Q	Q	Y	L	N	V	V	T	R	H	R	Ya.
ans-PP	R	F	Y	D	N	L	Q	Q	Y	R	L	V	V	F	R	H	R	Ya.
st-PP	R	F	Y	D	N	L	Q	Q	Y	L	N	V	V	T	R	H	R	Ya.
ra_1-PP	Q	Y	Y	S	D	L	Y	Q	Y	I	T	F	I	T	R	P	R	Fa.
ra_2-PP	Q	Y	Y	S	D	L	Y	Q	Y	I	T	F	V	T	R	P	R	Fa.
al-PP	Q	F	Y	D	D	L	Q	Q	Y	L	N	V	V	T	R	P	R	Fa.

Plasma PP level is lowered by stimulating the α-adrenergic receptors, by δ-opioid receptor ligands, [D-Ala^2,D-Leu^5]enkephalin and [D-Ala^2]enkephalin, by the intravenous application of glucose or fatty acids, by glucocorticoids, p-PYY, galanin [5.685], [5.686], atropine and somatostatin [5.681].

Biological Effects. PP inhibits the exocrine secretion from the pancreas which is stimulated e.g., by i.v. secretin, i.v. caerulein, i.v. cholecystokinin, i.v. bethanechol, i.d. HCI or i.d. sodium oleate [5.687], and the caerulein induced pancreatic hypertrophy [5.684]. The effect in experimental pancreatitis is being disputed [5.688].

Other effects of PP:

Antagonization of VIP induced vasodilatation of vascular smooth muscles [5.689]
Decrease in plasma insulin levels in the rat
Relaxation of Oddi's sphincter in small doses and contraction in high doses [5.690]
Contraction of the lower esophageal sphincter [5.691]

Inhibition of phase III of the migratory motorial complex and decreased motilin release [5.692]
Release of $NaHCO_3$ from the gastric and duodenal mucosa of dogs [5.693]
Lowering of plasma lutropin (LH) and somatotropin when applied centrally to ovariectomized rats [5.694], [5.695], but increase in plasma LH if the rats are pretreated with estradiol benzoate and progesterone
Reduction of food intake

Structure-Activity Relationships. The C-terminal Tyr-amide is very important for the biological activity. Since the C-terminal hexapeptide simulates the inhibitory effect of PP on the exocrine pancrease secretion, the active center appears to be located in the C-terminal region of PP. In contrast to PP, however, both the C-terminal hexapeptide and the C-terminal decapeptide were unable to lower the insulin level in rats. Even on i.c.v. application, the C-terminal hexapeptide had no influence on the plasma LH level in ovariectomized rats [5.694].

5.6.2. Neuropeptide Y [5.696]

Occurrence. Neuropeptide Y (NPY) is widely distributed in the central and peripheral nervous system, usually in the vicinity of arteries. In the brain, the highest concentrations are found in the hypothalamus. NPY is found associated with somatostatin in many i.r. NPY neurons of the cerebral cortex and in some NPY containing cells of the diencephalon. In neurons of the arcuate nucleus, NPY is found associated with somatoliberin [5.697]. In the adrenal medulla, NPY is found in the majority of norepinephrine releasing granula. Met- and Leu-enkephalin have also been observed in some of these cells. NPY has been found together with VIP and PHI in the non-adrenergic nerve fibers of the small intestine [5.698].

The peptides, h- [5.699] and r-NPY [5.700] are processed in the form of a preprocompound consisting of 97 amino acids. The 30 amino acid C-terminal peptide of prepro-NPY (C flanking peptide of neuropeptide Y = CPON) is found with NPY both in the brain as well as in the peripheral nervous system [5.701]. A NPY fragment, p-NPY-(12-36)-amide (p-NPY-25), has been isolated from the porcine brain [5.702].

Sequences of various species [5.676]

	1												13					18
h-, r-, cav-NPY [5.703]	Y	P	S	K	P	D	N	P	G	E	D	A	P	A	E	D	M	A-
p-NPY	Y	P	S	K	P	D	N	P	G	E	D	A	P	A	E	D	L	A-
o-NPY [5.704]	Y	P	S	K	P	D	N	P	G	D	D	A	P	A	E	D	L	A-
sa-NPY	Y	P	P	K	P	E	N	P	G	E	D	A	P	P	E	E	L	A-

	19																	36
h, r-, cav-NPY	R	Y	Y	S	A	L	R	H	Y	I	N	L	I	T	R	Q	R	Ya.
p-NPY	R	Y	Y	S	A	L	R	H	Y	I	N	L	I	T	R	Q	R	Ya.
o-NPY	R	Y	Y	S	A	L	R	H	Y	I	N	L	I	T	R	Q	R	Ya.
sa-NPY	K	Y	Y	T	A	L	R	H	Y	I	N	L	I	T	R	Q	R	Ya.

Release. NPY is released along with noradrenaline on activation of the sympathetic nerves. The release of NPY and noradrenaline from the perfused guinea pig heart depends on extracellular calcium, calcium influence via N type calcium channels and on the activation of protein kinase C [5.705]. The increase in the plasma NPY level parallels that of noradrenaline, and the heartbeat and blood pressure during exercise [5.706]. NPY is released by tyramine [5.707], insulin stimulated hypoglycemia [5.708], loss of blood [5.709], stress [5.710]. deprivation of food [5.711], dexamethasone [5.712], and testosterone [5.713], [5.714].

An increased level of NPY is found in platelets of spontaneously hypertensive rats [5.715]. It is assumed that NPY plays a role in the pathophysiology of congestive heart failure [5.716]. A high plasma NPY level is observed in patients with pheochromocytoma (tumor of the adrenal medulla) [5.717] and in septicemia [5.718].

The release of NPY is reduced by sympathectomy, reserpin or CCK-8 [5.719]. The NPY concentration in the hypothalamus decreases in old rats [5.720].

Receptors. There are two receptors. The Y_1 receptor is a 70 kDa glycoprotein which binds NPY, but not NPY-(13-36). The Y_2 receptor is a 50 kDa glycoprotein which binds both NPY and NPY-(13-36). Both receptors also bind PYY [5.721], [5.722].

The vascular effects of NPY are presumably mediated by the Y_1 receptors and the cardiac effects by the Y_2 receptors [5.723].

A selective Y_1 receptor agonist is [Leu^{31},Pro^{34}]NPY, a cross between PP and NPY [5.724].

The Y_1 receptors are found on the human neuroblastoma cell line MC-IXC and on the rat pheochromocytoma cell line PC-12 [5.721]. The Y_2 receptors are found on porcine hippocampal membranes, human neuroblastoma cell line SMS-MSN [5.721], proximal tubule cells in rabbit kidney [5.725] and on rat vas deferens [5.726].

Biological Effects. NPY increases intracellular calcium in cells of the vascular smooth muscles [5.727], inhibits the formation of cAMP which is stimulated by, e.g., forskolin [5.728] or isoproterenol [5.729], and the calmodulin stimulated phospodiesterase [5.730].

NPY exerts a strong, calcium dependent vasoconstricting effect on cerebral arteries *in vitro*. NPY inhibits the VIP induced relaxation of arteries [5.731] and causes a persisting increase in blood pressure when applied parenterally.

Other effects of NPY:

Inhibition of renin release in the kidneys [5.732] as well as vasopressin release [5.733] and vasopressin stimulated antidiuresis [5.734]

Stimulation of diuresis and natriuresis through the release of ANF [5.735]

Inhibition of the release of noradrenaline [5.736] and ACTH stimulated synthesis of aldosterone in the adrenal gland, but also stimulation of the secretion of 11-deoxycorticosterone and corticosterone [5.737]

Inhibition of the PGE_2 induced fluid and electrolyte secretion in the jejunum of the rat [5.738]

Inhibition of the secretin and cholecystokinin stimulated exocrine pancreatic secretion and reduction of the glucose or β-adrenergic stimulated release of insulin [5.739]
Inhibition of the pentagastrin stimulated secretion of gastric acid [5.740]
Increase in the motility of the gallbladder [5.741]
Bronchoconstriction in isolated guinea pig airways [5.742]
Increase in leukotriene LTD_4 in the rat lung [5.743] and inhibition of the release of tachykinins in the capsaicin sensitive airway sensory nerves [5.744]
Inhibition of lipolysis in adipocytes [5.745]
Inhibition of the adrenaline stimulated aggregation of platelets [5.746]
Potentiation of the TSH and VIP induced release of iodothyronine
Inhibition of the formation of α-MSH in the pars intermedia of the frog hypophysis [5.747]
Increase in the serum LH level in normal male rats [5.748] and in ovariectomized, estrogen treated rats via the release of gonadoliberin [5.749] and the potentiation of gonadoliberin activity [5.750]
Inhibition of the secretion of LH in ovariectomized rats but without estrogen substitution [5.751]

When applied centrally, NPY stimulates the release of ACTH and cortisol via the potentiation of CRH [5.752], inhibits CRH or stress induced cecal motor response [5.753] and the reflex contraction of the bladder [5.754], increases the intake of food (especially carbohydrate [5.755]) and water in the rat, delays the emptying of the stomach [5.756], reduces stress induced stomach erosions [5.757], cleaning behavior, and sexual behavior [5.758], and reduceds the blood pressure [5.759] (however, i.c.v. NPY-(13-36) increases blood pressure in spontaneously hypertensive rats via Y_2 receptors [5.760]) and the body temperature [5.761]. It is posible that NPY plays a role in bulimia nervosa.

Structure-Activity Relationships. Both the N-terminal and the C-terminal end of NPY are important for binding to the Y_1 receptors, while C-terminal fragments allow the binding to the Y_2 receptors.

The sequence 5-24 is not necessary for the binding to Y_1 receptors. Thus, shortened sequences, e.g., cyclo-[D-Cys^{7},Aoa$^{8\text{-}17}$,Cys^{20}]p-NPY (IC_{50} = 2.3 nM) and cyclo-[Cys^{5},Aoa$^{7\text{-}20}$,D-Cys^{24}]p-NPY (IC_{50} = 150nM), bind to mouse brain membranes [5.762] (Aoa = 8-amino octanoic acid residue). [Aca$^{5\text{-}24}$]NPY (Aca = ε-amino caproic acid) still has about 1/3 of the receptor binding capacity of NPY to rat renal membranes [5.763] and [Ahx$^{5\text{-}24}$]NPY (Ahx = amino hexanoic acid) has about 1/10 of the activity of NPY [5.764].

Agonists selective for Y_1 receptors are [Pro^{34}]p-NPY (binds 1000 times better) [5.765], [His^{34}]p-NPY (19 times better) and [Leu^{31},Pro^{34}]NPY (50-100 times better) [5.766]. [Leu^{31},Pro^{34}]NPY has a higher blood pressure increasing activity than NPY [5.766].

Arg^{33} and Arg^{35} are important for the biological activity, whereas Arg^{25} can be replaced, e.g., by Lys [5.764].

The 12 C-terminal amino acids and the amide function are required to activate the Y_2 receptor (electrically stimulated rat vas deferens). NPY-(26-36)-amide and PYY-(26-36)-amide have no activity. Among the fragments, NPY-(16-36)-amide is most active in this test [5.767]. The C-terminal fragments, NPY-(12-36)-amide, NPY-(22-36)-amide, and NPY-(22-33)-amide inhibit calmodulin stimulated phosphodiesterase more than NPY. On the other hand, N-terminal fragments are devoid of activity [5.730].

NPY-(18-36) is a NPY antagonist in membranes of the heart [5.716].

5.6.3. Peptide with N-terminal Tyrosine and C-terminal Tyrosinamide

Occurrence. Peptide with N-terminal tyrosine and C-terminal tyrosinamide (PYY occurs primarily in the small intestine. In human beings, PYY forming cells were found, above all, in the duodenum, while in the dog and in the pig [5.768], PYY is formed primarily in the mucous membranes of the ileum and the colon [5.769].

Sequences of various species [5.676]

	1												13					18
h-PYY [5.673]	Y	P	I	K	P	E	A	P	G	E	D	A	S	P	E	E	L	N-
p-PYY	Y	P	A	K	P	E	A	P	G	E	D	A	S	P	E	E	L	S-
r-PYY [5.770]	Y	P	A	K	P	E	A	P	G	E	D	A	S	P	E	E	L	S-
ca-PYY [5.771]	Y	P	A	K	P	E	A	P	G	E	D	A	S	P	E	E	L	S-
co-PYY	Y	P	P	Q	P	E	S	P	G	G	N	A	S	P	E	D	W	A-
lo-PYY	Y	P	P	K	P	E	T	P	G	S	N	A	S	P	E	D	W	A-

	19																	36
h-PYY	R	Y	Y	A	S	L	R	H	Y	L	N	L	V	T	R	Q	R	Ya.
p-PYY	R	Y	Y	A	S	L	R	H	Y	L	N	L	V	T	R	Q	R	Ya.
r-PYY	R	Y	Y	A	S	L	R	H	Y	L	N	L	V	T	R	Q	R	Ya.
ca-PYY	R	Y	Y	A	S	L	R	H	Y	L	N	L	V	T	R	Q	R	Ya.
co-PYY	K	Y	H	A	A	V	R	H	Y	V	N	L	I	T	R	Q	R	Ya.
lo-PYY	S	Y	Q	A	A	V	R	H	Y	V	N	L	I	T	R	Q	R	Ya.

Release. The plasma level of PYY increases postprandially (especially after oleic acids [5.769] and after the application of p-gastrin releasing peptide [5.772].

A raised basal and postprandial level of PYY is observed in steatorrhea (fat diarrhea), which is caused by atrophy of the small intestine (tropical sprue), and in chronic pancreatitis [5.773]. In patients suffering from Dumping syndrome (sweating, weakness and diarrhea caused by stomach operations), plasma levels of PYY increase 6-7 fold on oral glucose administration; this is not the case in healthy persons. Somatostatin inhibits this increase in plasma PYY [5.773].

Biological Effects. PYY reduces the plasma CCK level in dogs [5.774], the postprandial [5.775], or secretin or CCK-8 stimulated [5.776] exocrine secretion from the pancreas, VIP induced diarrhea [5.777] and the postprandial [5.775], pentagastrin [5.778], 2-deoxyglucose or bethanechol stimulated [5.779] secretion of gastric acid.

The vasoconstricting and the antisecretory effect on the pancreas is partially abolished by adrenalectomy [5.780].

Other Effects of PYY:
Increase in blood pressure [5.781]
Reduction of plasma PP and motilin levels [5.781]
Prolongation of the duration of the migratory motorial complexes [5.782] and the small bowel transit [5.777]
Inhibition of the CCK induced opening of Oddi's sphincter [5.783]
Strong emetic activity (PYY > NPY > PP) [5.784]
In dogs, PYY inhibits the 2-deoxy-D-glucose or β-adrenergic stimulated release of insulin [5.785]. It inhibits the release of ANF and reduces diuresis and natriuresis [5.786].

5.6.4. Neuromedin U

Occurrence. Neuromedin U (NMU) were first isolated from the spinal cord of the pig (p-NMU-25 and p-NMU-8). r- [5.787], ra$_2$- [5.788] and cav-NMU [5.789] were recovered from the gastrointestinal tract.

However, NMU like immunoreactivity is widely distributed. In the rat, it is found primarily in the hypophysis and in the gastrointestinal tract. It is also found in low concentrations in the brain, spinal cord and in both the male and female genital tract [5.790].

r-NMU	Y	K	V	N	E	/	Y	Q	G	P	/	V	A	P	S	G	G	F	F	L	F	R	P	R	Na
p-NMU-25	F	K	V	D	E	E	F	Q	G	P	I	V	S	Q	N	R	R	Y	F	L	F	R	P	R	Na
p-NMU-8																		Y	F	L	F	R	P	R	Na
cav-NMU																	G	Y	F	L	F	R	P	R	Na
ra$_2$-NMU	L	K	P	D	E	E	L	Q	G	P	G	G	V	L	S	R	G	Y	F	V	F	R	P	R	Na

Biological Effects. NMU causes an increase in blood pressure and the contraction of the rat uterus [5.787], the longitudinal muscle of the isolated human ileum [5.791] and of muscle strips from the human bladder [5.791].

Structure-Activity Relationships. r-NMU-25 (0.12 nM) is about twice as active as p-NMU-25 on the rat uterus. The C-terminal heptapeptide amide is essential for activity [5.787].

5.6.5. δ-Sleep Inducing Peptide [5.792]

Occurrence [5.793], [5.794]. δ-Sleep inducing peptide (DSIP) was first isolated from a dialysate of rabbit venous blood (cu-DSIP) while the animals were kept awake artificially by electrically stimulating the thalamic region. Although this peptide is widely distributed in the brain and other tissues, it is mot concentrated in the thalamus and pineal gland. DSIP has also been found in the hypophysis (together with TSH) [5.795], in urine, in CSF and in human milk. The i.r.-DSIP in plasma, urine and CSF is bound to proteins which protect it from proteolysis.

cu-DSIP: W A G G D A S G E

Release. The concentration of DSIP in the brain increases during hibernation. In rats addicted to alcohol, however, the DSIP concentration in the brain falls. Corticoliberin (CRH) and vasopressin inhibit the release of DSIP from cells of the anterior hypophysis [5.795].

Biological Effects [5.793], [5.794]. There are various studies that confirm that DSIP passes the blood brain barrier. The infusion of this peptide into the ventricular system of the rabbit brain stimulates the δ waves of the EEG, which are characteristic of natural sleep. For the sleep inducing activity in rabbits, DSIP exhibits a bell-shaped activity/dose relationship in which 30 nmol/kg (i.v.) produces the greatest effect. In mice, the i.c.v. injection of DSIP causes an analgetic effect (1.6 times as active as morphine) [5.796]. Although naloxone abolishes the sleep inducing [5.797] and analgetic effect [5.796] of DSIP, DSIP does not bind to opioid receptors [5.798]. DSIP increases the resistance to acute emotional stress in rats [5.799], inhibits the release of prolactin [5.800] and the CRH stimulated increase in ACTH [5.801] and glucocorticoids [5.802].

DSIP has a direct stimulating effect on the release of somatotropin (STH) and inhibits the release of somatostatin via dopaminergic mechanisms [5.803]. Since STH bursts have been observed in slow wave sleep, DSIP could be of physiological importance as a STH releaser.

Similar to NPY and PP, the i.c.v. application of DSIP increases the release of lutropin (LH) in estrogen and progesterone treated ovariectomized rats. On the other hand, DSIP has no LH releasing activity *in vitro* (hypophysis) [5.804].

Structure-Activity Relationships. Shortening of the peptide chain at the N-terminal and C-terminal ends or the rearrangement of α-Asp to β-Asp produces inactive peptides. The replacement of Trp^1 by D-Phe or p-nitrophenylalanine appears to be permitted, while [D-Trp^1]- and [Phe^1]DSIP have the opposite effect, namely a slight increase in the motorial activity. Phosphorylation of the hydroxyl group of Ser^7 leads to a more potent and longer acting product (DSIP-P). It is more stable to enzymatic degradation than DSIP.

Applications. In patients suffering from disturbed sleep, the time of sleep has been significantly prolonged by the administration of DSIP. The ability to concentrate and the tolerance to stress in the waking phase is increased. DSIP exerts an analgetic and antidepressive effect in patients with migraine, vasomotorial head aches, psychogenic attacks of pain, and tinnitus [5.805]. Treatment with DSIP produces a favorable effect in cases of opiate or alcohol withdrawal. In these patients, raised blood pressure, tachycardia, tremor and sweating could be abolished by DSIP [5.792].

6. Cholecystokinin and Gastrin

Cholecystokinin (CCK), which was known to stimulate contraction of the gallbladder, and pancreozymin, which was made responsible for enzyme release from the pancreas, denote the same peptide. The counterparts of CCK in amphibians are caerulein (CRL) and related peptides. The common C-terminal pentapeptide of these peptides is identical to the C-terminus end of gastrin. While the characteristic activity spectrum of the gastrins is dictated mainly by the C-terminus pentapeptide, the activity of CCK/CRL is based on the C-terminal heptapeptide, which should contain a sulfated tyrosine, especially for the peripheral activity. The shortest sequence which still appears to be of physiological importance is the common C-terminal tetrapeptide.

	1																			20
p-CCK-39	Y	I	Q	Q	A	R	K	A	P	S	G	R	V	S	M	I	K	N	L	Q-
						1														15
p-GT-34						*Q	L	G	L	Q	G	P	P	H	L	V	A	D	L	A-

	21																		39
p-CCK-39	S	L	D	P	S	H	R	I	S	D	R	D	Ys	M	G	W	M	D	Fa
	16																		34
p-GT-34	K	K	Q	G	P	W	M	E	E	E	E	E	A	Y	G	W	M	D	Fa
										1									10
CRL										*Q	Q	D	Ys	T	G	W	M	D	Fa
GT-4																W	M	D	Fa

A trypsin inhibitor with a structure similar to that of the epidermal growth factor (EGF) has been characterized as the CCK releasing peptide. The gastrin releasing peptide is structurally related to bombesin from amphibians.

Another peptide which can be included in this group is valosin.

6.1. Trypsin Sensitive Cholecystokinin Releasing Peptide

A peptide has been isolated from the pancreatic juice of rats which stimulates pancreatic enzyme secretion through the release of cholecystokinin (CCK). This CCK-releasing peptide (CCK-RP) is evidently released in response to a diet rich in protein [6.1].

CCK-RP is structurally similar to the pancreatic trypsin inhibitors of the Kazal type (PSTIs, pancreatic secretory trypsin inhibitors) and is classified as one of the trypsin inhibitors of the Kazal type [6.1]. Other structural similarities are also found between CCK-RP and the epidermal growth factor (EGF) [6.2] and PEC-60, a peptide from the entrails of the pig which inhibits the glucose induced release of insulin from the perfused pancreas, but not trypsin activity [6.3]. CCK-RP also binds to EGF receptors and exerts a growth promoting effect [6.2].

	1																								23
r-CCK-RP	G	N	P	P	A	E	V	N	G	K	T	P	N	C	P	/	/	K	Q	I	M	G	C	P	R-
p-PEC-60				E	K	Q	V	F	S	R	M	P	I	C	E	H	M	T	E	S	P	D	C	S	R-
PSTI						T	S	P	Q	R	E	A	T	C	T	/	/	S	E	V	S	G	C	P	K-
m-EGF-(1-28)																		N	S	Y	P	G	C	P	S-

	24																								48
r-CCK-RP	I	Y	D	P	V	C	G	T	N	G	I	T	Y	P	S	Z	C	S	L	C	F	E	N	R	K-
p-PEC-60	I	Y	D	P	V	C	G	T	D	G	V	T	Y	E	S	E	C	K	L	C	L	A	R	I	E-
PSTI	I	Y	N	P	V	C	G	T	D	G	I	T	Y	S	N	E	C	V	L	C	S	E	N	K	K-
m-EGF	S	Y	D	G	Y	C	L	N	G	G	V	C	M	H	I	E	/	S	L	D	S...				

	49												61
r-CCK-RP	F	G	T	S	I	H	I	Q	R	R	G	G	C.
p-PEC-60	N	K	Q	D	I	Q	I	V	K	D	G	E	C.
PSTI	R	Q	T	P	V	L	I	Q	K	S	G	P	C.

6.2. Cholecystokinin/Pancreozymin and Caerulein [6.4]

6.2.1. Occurrence

Cholecystokinin/Pancreozymin (CCK) is formed as prepro-CCK. The structures of h- [6.5], p- [6.6] and r-prepro-CCK [6.7] have been elucidated via the cDNA.

CCK has been isolated and characterized in the form of peptides containing 58 (CCK-58) [6.8], [6.9], 39 (CCK-39), 33 (CCK-33), 22 (CCK-22) [6.10], 12 (CCK-12), 8 (CCK-8) [6.10], 5 (CCK-5) [6.11] and 4 (CCK-4 and GT-4) C-terminal amino acids. They are synthesized in the brain, small intestine and in the pancreas.

The r-prepro-(24-32)-nonapeptide (V-9-M) is found in the rat brain. It exerts a persisting memory improving effect when injected into the lateral ventricle [6.12].

Caerulein (CRL) has been isolated from the skin of various amphibians. It was first obtained from the Australian Frog, *Litoria caerulea*, and received its name from it. Phyllo-CRL was isolated from *Phyllomedusa sauvagii* and [Asn^2,Leu^5]CRL from *Hylambates maculatus*. CCK-8, but not CRL, could be detected in the brain of amphibians. The prepro-CRL from the skin of *Xenopus laevis* codes for five heterogeneous pro-CRL peptides [6.13].

h-CCK-22	N	L	Q	N	L	D	P	S	H	R	I	S	D	R	D	Ys	M	G	W	M	D	Fa
p-CCK-22	N	L	Q	S	L	D	P	S	H	R	I	S	D	R	D	Ys	M	G	W	M	D	Fa
r-CCK-22	N	L	Q	G	L	D	P	S	H	R	I	S	D	R	D	Ys	M	G	W	M	D	Fa
h-CCK-8															D	Ys	M	G	W	M	D	Fa
CRL													*Q	Q	D	Ys	T	G	W	M	D	Fa
Phyllo-CRL													*Q	/	E	Ys	T	G	W	M	D	Fa
[Asn^2 Leu^5] CRL													*Q	N	D	Ys	L	G	W	M	D	Fa

6.2.2. Release

The oral intake of unsaturated fats and fatty acids causes a greater release of CCK than the intake of the corresponding saturated compounds. Corn oil (diunsaturated fatty acids) is twice as active as olive oil (monounsaturated fatty acids) in stimulating the release of CCK [6.14]. CCK is also released in response to the acidification of the duodenum [6.15], to bombesin [6.16], depolarizing K^+ concentrations, veratridin (a Na^+ channel activator) and to the δ-agonists DTLET (delta-kephalin) and DPDPE ([D-$Pen^{2,5}$]enkephalin) [6.17].

Bile acids [6.18], somatostatin, atropine [6.19] and the μ selective opioid receptor agonist, DAGOL [6.20] inhibit the release of CCK. The castration of male rats leads to a dramatic reduction of the i.r.-CCK content in the diomorphic cell groups in the forebrain of the rat [6.21]. Stress modulates the CCK content in certain regions of the rat hypothalamus [6.22].

The postprandial level of CCK in the plasma of patients with pancreatic insufficiency is lower than that of healthy persons [6.23]. On the other hand, the plasma CCK level in patients with non-insulin-dependent diabetes [6.24], hepatic cirrhosis [6.25] and chronic pancreatitis [6.26] is raised.

6.2.3. Receptors [6.27]

It is possible to differentiate between two CCK receptors: CCK-A (120 and 80 kDa) and CCK-B (55 kDa) [6.13]. The CCK-A receptors (alimentary) are found on pancreatic acinar cells, on the gallbladder, on anterior pituitary cells and on inhibitory neurons of the lower esophageal sphincter. Only sulfated CCK derivatives bind to these high affinity CCK receptors. Gastrin exhibits a 500 to 1000 times weaker binding. CCK-B receptors (brain) are observed in the cerebral cortex and in the central nervous system. They have only a ten times higher preference for sulfated CCK derivatives. Another class of receptors is found on the parietal cells and on the gastrointestinal smooth muscles. They have about the same affinities for sulfated and non-sulfated CCK derivatives and have been named the gastrin receptors. However, CCK-A receptors are also found here [6.28].

The binding of CCK on pancreatic receptors leads to an up regulation of the CCK receptors [6.29]. In newborn rats, the CCK receptors are increased by hydrocortisone [6.30]. The CCK receptors on rat pancreatic acini are increased four fold in diabetic rats [6.31].

Selective ligands for the CCK-A receptor [6.32]:

CCK antagonist devazeptide (= MK-329, = L-364 718): 3S(–)-N-(2,3-dihydro-1-methyl)-2-oxo-5-phenyl-1H-1,4-benzodiazepine-3-yl)-1H-indole-2-carboxamide

CCK antagonist L-365 022: indolincarbonyl-Glu-di-n-pentylamide

CCK antagonist lorglumide (CR-1409): 3,4-Dichlorobenzoyl-Glu-di-n-pentylamide

CCK antagonist loxiglumide (CR-1505): 3,4-dichlorobenzoyl-Glu-n-pentyl-3-methoxypropylamide

Selective ligands for the CCK-B receptor:

CCK antagonist L-365,260 [6.12]: (3R(+)-N-(2,3-dihydro-1-methyl-2-oxo-5-phenyl-1H-1,4-benzodiazepine-3-yl-N'-(3-methyl-phenyl urea), acts also as a gastrin antagonist [6.33]

CCK antagonists PD 134 308 and PD 135 158 with anxiolytic activity [6.34]

CCK agonist Boc-Tyr(SO_3Na)-gNle-mGly-Trp-MeNle-Asp-Phe-amide (K_i = 0.15 nM on guinea pig brain membranes, K_i = 78 nM on guinea pig pancreatic membranes), and propionyl-Tyr(SO_3H)-gNle-mGly-Trp-MeNle-Asp-Phe-amide (pBC 264) [6.35];

CCK agonist Boc-D-γ-Glu-Tyr(SO_3H)-Nle-D-Lys-Trp-Nle-Asp-Phe-amide (K_i = 0.56 nM on guinea pig brain membranes, K_i = 2500 nM on guinea pig pancreatic membranes [6.36];

CCK agonist Boc-[Nle^3,Phe^6]CCK-8-(2-8) [6.37].

6.2.4. Biological Effects

The activity spectrum of CCK was predominantly studied with synthetically produced CCK-8 (Sincalid) and CRL (Ceruletid).

Receptor activation leads to the breakdown of phosphoinositides, mobilization of cellular calcium, activation of protein kinase C and to the phosphorylation of numerous intracellular proteins.

The most important physiological effects of CCK are the CCK-A receptor mediated stimulating effect on the flow of bile [6.38] (contraction of the gallbladder, relaxation of the bile duct and of Oddi's sphincter) and on the exocrine pancreatic secretion [6.39] (relaxation of the pancreatic duct sphincter, secretion of amylase trypsin, chymotrypsin, lipase and $NaHCO_3$ [6.40], and potentiation of the secretin induced exocrine pancreatic secretion). The CCK-8 induced contraction of the gallbladder is increased in patients suffering from constipation and lowered in patients with diarrhea [6.41] and coeliac disease [6.42]. VIP and secretin potentiate the effect of CCK on the gallbladder and on the pancreas. Somatostatin [6.43], atropine [6.44] and loperamide [6.45] inhibit the CCK induced contraction of the gallbladder. The insulin releasing effect that has been described often is of no physiological importance in human beings [6.46].

cAMP is activated in CCK induced relaxation. In contractions, cGMP is activated and intracellular calcium and acetyl choline are released [6.47].

Other CCK-A Receptor Mediated Effects:

Decreased feeding in rats by peripherally injected CCK-8 [6.48]

Trophic effects on the pancreas [6.49] (p-PP inhibits the trophic activtiy of CCK [6.50]

Growth of human ductal pancreatic adenocarcinoma xenografts in nude mice [6.51]

Rise in plasma PP [6.52]

Increased insulin and glucagon secretion in mice [6.53]

Inhibition of pentagastrin-stimulated gastric acid secretion [6.54]

Miosis in monkeys [6.55]

Facilitated dopamine efflux in rat brain [6.56]

The postero-median nucleus accumbens mediated reduced exploratory behavior and increased emotional state in rats [6.57]
Release of acetyl choline from Auerbach's plexus in guinea pigs [6.58]

CCK-B Receptor Mediated Effects:
CCK/opiate interactions [6.59]
Anxiety [6.34]

Other Effects of CCK:
Relaxation of the Corpus ventriculi and the lower esophageal sphincter
Contraction of the pylorus and prolongation of the time required to empty the stomach (stimulation of the emptying of the stomach when applied centrally [6.60])
Increase in motility from the distal part of the duodenum up to the rectum [6.61]
Stimulation (also when applied centrally [6.62]) of gastric acid secretion, but inhibition of the gastrin stimulated secretion of gastric acid
Causes pancreatitis in mice [6.63] and rats [6.64]
Stimulation of SRIF release in the pancreas and stomach [6.65]
Stimulation of the release of PRL [6.66] and STH [6.67]
Stimulation of the release of ACTH [6.68] and β-endorphin [6.69] and increase in the plasma cotricosterone level [6.70]
Increase in the survival time in hemorrhagic shock [6.71]
Contraction of the guinea pig trachea [6.72]
Immunostimulating effect (increase in IgA, IgG antibodies) [6.73]
Analgetic effect through δ-opioid receptors via the release of endogenous opioids [6.74]
Antagonization of the activity of endogenous opioids (endogenous opioids block the CCK activity)
Reduction of the morphine or β-endorphin induced analgesia [6.75] via μ-receptors [6.76] (opiates inhibit the analgetic activity of CCK)
Hypothermic effect via 5-hydroxytryptamine [6.77]
Increase in the memory capacity (vagus mediated [6.78]) and reduction of amnesia [6.79]

6.2.5. Structure-Activity Relationships

(see also 6.2.3. Selective ligands for CCK-A and CCK-B receptors)

The C-terminal octapeptide of CCK (CCK-8) has the full activity spectrum of the longer CCK compounds.

The C-terminal heptapeptide of CCK (CCK-7) or CRL (CRL-(4-10)) [6.80] is required for the characteristic CCK activity. *In vivo,* deamino-CCK-7 still has the same protein secretory activity as CCK-7 and is more effective in contracting the gallbladder than CCK-7. CCK-6 has no activity.

The sulfated tyrosine is important for the characteristic activity spectrum of CCK and CRL (binding to type A receptors). The activity of desulfated CCK-8 is more like that of gastrin [6.81]. The substitution of Phe(CH_2SO_3H) for Tyr(SO_3H) leads to more stable and still highly active derivatives [6.82]. If the N-terminal Tyr(SO_3H) is replaced by the 3-(4-sulfo-oxyphenyl)-2-methyl-propanoyl group, the analgetic activity is about five times that of CCK-7 [6.83].

Met^3 is not essential and can be replaced by Leu or Thr. On the other hand, Met^6 can be replaced only by Nle without loss of activity. Thus, the CCK derivatives Arg-[Thr^3,Nle^6]-CCK-8 [6.84], Boc-[$Nle^{3,6}$]CCK-8-(2-8) [6.85], [$Nle^{3,6}$]CCK-8 [6.86] and Ac-[$Nle^{3,6}$]CCK-8 [6.87] are stable to oxidation and at least fully active.

Gly^4 can be replaced by D-amino acids: [D-Ala^4]CCK-8 and [D-Trp^4]CCK-8 are almost as active as CCK-8, however, they act longer and are stable to brain proteases [6.88].

The effect of Boc-[$Nle^3\psi(COCH_2)Gly^4$]CCK-8-(2-8) on the release of amylase from disperse rat pancereatic acini is about the same as that of CCK-8. However, this compound exerts a central antagonistic effect [6.89]. Similar derivatives with modified peptide bonds exhibit agonistic activity [6.90], [6.91].

As a result of the rearrangement of β-Asp^7, CCK loses its cholecystokinin properties but retains its pancreozymin activity. The substitution of Ser(SO_3H), Thr(SO_3H) or Hyp(SO_3H) for Asp^7 in Ac-CCK-8-(2-8) gives rise to preparations that have 2- 3 times the activity of CCK-8 on isolated gallbladder strips [6.92].

Phe^8 can be replaced by 3-(2-naphthyl)-alanine or by 3-cyclohexylalanine without loss of acitvity [6.93].

The C-terminal amide function is very important for the activity of CCK peptides [6.94].

6.2.6. Antagonists [6.95]-[6.98]

CCK receptor antagonists are obtained by deleting C-terminal amino acids. The shortest still active fragment was CCK-8-(1-5) [6.99]. Stronger antagonists are the succinyl-CCK-8-(2-7)-2-phenethylamide (GE-410, pA_2 = 7.6-8.9) [6.100] and Boc-[D-Trp^5,Nle^6]CCK-8-(2-7)-2-phenethylester [6.101].

Surprisingly strong antagonists are also obtained by substituting D-Orn(Z) for Met^6 of CCK-8: [Nle^3,$Orn(Z)^6$]CCK-8-(2-8) [6.37].

Relatively simple derivatives of glutamic acid and tryptophan act as CCK antagonists. For instance, the inhibitor of acid secretion proglumide (N-benzoyl-L-glutamic acid-α-di-n-propylamide) and especially its derivatives lorglumide (CR-1409 = 3,4-dichlorobenzoyl-Glu-di-n-pentylamide), loxiglumide (CR-1505 = 3,4-dichlorobenzoyl-Glu-n-pentyl-3-methoxypropylamide) and L-365,022 (indolincarbonyl-Glu-di-n-pentylamide) as well as carbobenzoxy-Trp and p-chlorobenzoyl-Trp (benzotript) [6.102].

Strong CCK antagonists are asperlicin, a non-peptide component of *Aspergillus alliaceus* and its benzodiazepine variants, devazeptides (MK-329 = L-365,718), L-365,260, trifluadom, and L-156,440. Apart from L-365,260, which binds preferably to CCK-B receptors, these antagonists bind, above all, to CCK-A receptors.

6.2.7. Applications

CCK-8 (sincalid) and CRL (Ceruletid, Takus [6.36]) are used in X- ray diagnostics and in the diagnosis of pancreatic function. A 60 minute infusion is preferred to a bolus injection [6.103]. Since the time required to pass the small intestine is reduced by half, they permit a quick examination of the small intestine after an investigation of the stomach by using an oral contrast medium. Ceruletid is used therapeutically in postoperative intestinal atonia and paralytic ileus because of its stimulating effect on the small intestine.

Ceruletid can also be employed for the expulsion of gallstones [6.104] and in biliary colic [6.105]. Sincalid and Ceruletid (nasal application 3 times daily, 100 μg) produce a favorable effect in chronic pancreatitis [6.106]. Clinical reports confirm the analgetic properties of CCK-8 [6.108].

6. 3. Gastrin Releasing Peptide and Bombesin [6.108]

6. 3.1. Occurrence

Gastrin releasing peptide (GRP), a peptide amide consisting of 27 amino acids, has been isolated from the gastric tissue of pigs, chicken and dogs and from human lung tumors [6.109]. The C-terminal heptapeptides of GRP and bombesin (BB), which is a peptide from the skin of *Bombina bombina*, are identical. Other peptides with related structure and activity that are isolated from amphibians are alytesin, ranatensin, ranatensin R, litorin, [Glu(OMe)^2]litorin, phyllolitorin, [Leu^8]phyllolitorin and rohdei litorin [6.110].

The neuromdins B, B-32, B-30 and C have been isolated from the brain and spinal cord of the pig. p-Neuromedin C corresponds to the C-terminal decapeptide of p-GRP.

Very little sequence similarity is observed with the leucokinins, ion transport stimulators from *Leucophaea maderai* [6.111].

Immunoreactive GRP has been found in the central nervous system and in peripheral tissue, e.g., in the entire gastrointestinal tract. The highest concentrations were detected in the antrum, duodenum and in the lung. Material similar to GRP-(14-27) has been found in a large number of endocrine tumors [6.112] and in cow's milk [6.113].

6. 3.2. Release

Electrical stimulation of the vagus leads to an enhanced formation of GRP in the pancreas [6.113a]. The β-adrenergic stimulation of gastrin and somatostatin appears to be mediated here by GRP [6.113b].

	1													12				
h-GRP	V	P	L	P	/	/	A	G	G	G	T	V	L	T-				
p-GRP	A	P	V	S	/	/	V	G	G	G	T	V	L	A-				
ca-GRP	A	P	V	P	/	/	G	G	Q	G	T	V	L	D-				
g-GRP	A	P	L	Q	/	/	P	G	G	S	P	A	L	T-				
p-Neuromedin B-32	A	P	L	S	W	D	L	P	E	P	R	S	R	A-				

				13														27
h-GRP	/	/	/	K	M	Y	P	R	G	N	H	W	A	V	G	H	L	Ma
p-GRP	/	/	/	K	M	Y	P	R	G	N	H	W	A	V	G	H	L	Ma
ca-GRP	/	/	/	K	M	Y	P	R	G	N	H	W	A	V	G	H	L	Ma
g-GRP	/	/	/	K	I	Y	P	R	G	S	H	W	A	V	G	H	L	Ma
p-Neuromedin B-32	G	K	I	R	V	H	P	R	G	N	L	W	A	T	G	H	F	Ma

									1				5					10
p-Neuromedin B									G	N	L	W	A	T	G	H	F	Ma
p-Neuromedin C									G	N	H	W	A	V	G	H	L	Ma

					1				5					10				14
Bombesin					*Q	Q	R	L	G	N	Q	W	A	V	G	H	L	Ma
Alytesin					*Q	G	R	L	G	T	Q	W	A	V	G	H	L	Ma
Ranatensin								*Q	V	P	Q	W	A	V	G	H	F	Ma
Ranatensin R		S	N	T	A	L	R	R	Y	N	Q	W	A	T	G	H	F	Ma
Litorin										*Q	Q	W	A	V	G	H	F	Ma
Phyllolitorin										*Q	L	W	A	V	G	S	F	Ma
[Leu^{8}]Phyllolitorin										*Q	L	W	A	V	G	S	L	Ma
Rohdei-litorin										*Q	L	W	A	T	G	H	F	Ma

Leucokinin I						D	P	A	F	N	S	W	Ga	
Leucokinin II						D	P	G	F	S	S	W	Ga	
Leucokinin III						D	Q	G	F	N	S	W	Ga	
Leucokinin IV						D	A	S	F	H	S	W	Ga	
Leucokinin V						G	S	G	F	S	S	W	Ga	
Leucokinin VI						*Q	S	S	F	H	S	W	Ga	
Leucokinin VII						D	P	A	F	S	S	W	Ga	
Leucokinin VIII						G	A	D	F	Y	S	W	Ga	

6.3.3. Receptors

It was assumed that there are at least two different receptors for GRP and neuromedins. Bombesin or GRP bind to pancreatic acinar cells and murine 3T3 cells with high affinity, while neuromedin B binds with only low affinity. On the other hand, the esophageal muscularis mucosa possesses receptors which have a high affinity for neuromedin B, but bind GRP with only low affinity. The GRP antagonist, [D-Phe^{6}]BB(6-13)-OEt binds preferably on the pancreas while [Tyr^{4},D-Phe^{12}]BB and the SP antagonists, [D-Pro^{1},D-Trp^{7,9,10}]SP(4-11) and [D-Arg^{1},D-Trp^{7,9},Leu^{11}]SP bind better to the esophagus than to the pancreas [6.114].

6.3.4. Biological Effects

In healthy persons, an infusion of 2.4 pmol/kg. min of BB increases the plasma levels of gastrin (GT), cholecystokinin (CCK), motilin, pancreatic peptide (PP), vasoactive intestinal peptide (VIP), gastrin inhibiting peptide (GIP), glucagon, insulin, and trypsin [6.115]. as well as the plasma level of gonadoliberin stimulated lutropin and follitropin [6.116]. In this process, the glucose level in the blood is lowered and plasms Ca^{2+} levels are raised. The increased plasma levels of Ca^{2+} result in a decrease in parathyroid hormone and an increase in calcitonin. Somatostatin (SRIF) inhibits this BB induced hormone secretion [6.117]. In healthy test persons, the food induced release of insulin [6.118] and the thyroliberin (TRH) stimulated levels of thyreotropin (TSH) and prolactin (PRL) in the plasma [6.116] are lowered by an infusion with BB.

Other Biological Effects of GRP/BB:

Stimulation of the turnover of phosphatidylinositol in the CNS [6.119]

Stimulation of gastric acid secretion and exocrine pancreatic secretion (however, high doses of BB and GRP inhibit gastric acid secretion [6.120], they act as if applied centrally)

Reduction of alcohol induced stomach lesions in the rat via the formation of prostacyclins [6.121]

Reduction of food intake [6.122], reduced antral motility and delay of gastric emptying [6.123]

Contraction of muscles (e.g., gallbladder, canine antral muscle [6.124], feline esophageal sphincter [6.125])

Increase in blood pressure

Increase in the bile flow [6.126] (the pressure of Oddi's sphincter falls and the pressure in the bile duct rises [6.127])

Contraction of the mucosa muscularis via the release of Substance P [6.128]

Stimulation of the release of ACTH from rat hypophyseal cells and potentiation of the corticoliberin induced release of ACTH [6.129]

Increase in the survival rate in hypovolemic shock [6.130]

Increase in plasma SRIF. decrease in plasma somatotropin (STH) [6.131] and PRL [6.132]

Inhibition of the release of TRH from the rat hypothalamus [6.133]

Trophic action on the gastrin cells of the stomach [6.134] and on the pancreas [6.135]

Autocrine growth in cancer cells [6.136]

In vivo stimulation of the cytotoxic activity of natural killer cells [6.137]

Increase in the memory of mice [6.138]

The central application of BB results in a strong hypothermic and adrenaline releasing effect, which, in turn, leads to the inhibition of insulin release and an increase of blood glucose and plasma glucagon level. Other central effects are the reduction of the plasma STH level [6.139] (by the release of SRIF [6.140]), stimulation of dopamine synthesis in the hypothalamus [6.141], suppression of gastric acid secretion and pancreatic secretion, ulcus reduction in the small intestine, stimulation of the motility of the small intestine [6.142], inhibition of the gastrointestinal and colon transit in mice [6.143], and inhibition of the food and liquid uptake [6.139].

6.3.5. Structure-Activity Relationships

A parallel bioassay of peptides similar to GRP on different muscle preparations shows the existence of various GRP/BB receptors [6.144].

GRP-(14-27) acts like BB [6.145]. The C-terminal nonapeptide of BB and Ac-GRP-(20-27) has the minimum length with the maximum effect. The smallest, still active compound is GRP-(23-27). Shortening the molecule at the C-terminus gave rise to preparations with low affinity and low intrinsic activity. Acetylated peptides shortened at the N-terminus are biologically more active than the corresponding compound with a free amino group.

Trp^8 and His^{12} appear to be very important for the GRP activity.

6.3.6. Antagonists

A competitive BB antagonist is [D-Phe^{12},Leu^{14}]BB. It inhibits *in vitro* O.2 nM BB stimulated amylase secretion from dispersed guinea pig acini cells (IC_{50} = 3 μM). [D-Phe^{12}]BB and [Tyr^4,D-Phe^{12}]BB have similar activity [6.146]. A further introduction of D-Phe^6 enhances the antagonistic activity, while D-Phe^5 has no effect [6.147].

	[D-$Phe^{6,12}$,Leu^{14}]BB >	[D-$Phe^{5,12}$,Leu^{14}]BB =	[D-Phe^{12},Leu^{14}]BB
IC_{50}:	0.3 μM	1.3 μM	2 μM

BB antagonosts are also formed by deleting the C-terminal methionine [6.148]. $(CH_3)_3$-C-CO-His-Trp-Ala-Val-D-Ala-His-Leu-OMe exhibited good receptor binding (IC_{50} = 1.6 nM) and inhibition of the mitogenesis of mice Swiss 3T3 cells (IC_{50} = approx. 2 nM).

However, the best (100%) *in vivo* inhibition of BB stimulated amylase secretion was produced by the 2mg/kg s.c. application of $(CH_3)_2$-CH-CO-His-Trp-Ala-Val-D-Ala-His-Leu-NH-CH_3 (ICI 216140) [6.148].

Other good de-Met-antagonists in the receptor binding assay, are [D-Phe^6]BB-(6-13)-ethylamide (IC_{50} = 7 nM), [D-Phe^6]BB-(6-13)-propylamide (IC_{50} = 1.6 nM) [6.149], Ac-GRP-(20-26)-OEt (IC_{50} = 3.9 nM) [6.150] and [D-Phe^6]BB-(6-13)-OEt [6.151].

If the peptide bond between positions 13 and 14 is reduced, very potent BB antagonists are obtained: [Leu^{13}-ψ-(CH_2NH)-Leu^{14}]BB inhibits the BB stimulated growth of mice Swiss 3T3 cells with an IC_{50} value of 18 nM [6.152]. N-terminal shortening of this antagonist reduced the antagonistic activity. The introduction of D-Phe in position 6 did not significantly change the biological activity. However, a combination of D-Phe^6 and N-terminal shortening produced a compound with ten times the activity [6.153]: [D-Phe^6,Leu^{13}-ψ-(CH_2NH)-Leu^{14}]BB-(6-14) (IC_{50} for inhibition of amylase secretion is 5 nM). In comparison, [D-Phe^{6},Leu^{13}-ψ-(CH_2NH)-Phe^{14}]BB-(6-14) exerted a stronger inhibitory effect on the growth of mice Swiss 3T3 cells (IC_{50} = 0.72 nM) [6.153]. Ac-[Leu^{26}-ψ-(CH_2O)-Leu^{27}]GRP-(20-27)-amide is a more potent inhibitor than [Leu^{13}-ψ-(CH_2NH)-Leu^{14}]BB [6.154].

6.3.7. Applications

BB is used as a diagnostic aid in the gastrin stimulation test. For instance, patients suffering from antral gastritis have an abnormally low gastrin level after BB stimulation.

BB increases the plasma trypsin level in healthy persons, but not in patients with pancreatic insufficiency [6.155]. In severe chronic pancreatitis [6.156] and in patients with duodenal ulcers [6.157], BB increases the gastrin level to a greater extent than in healthy persons.

6.4. Gastrin [6.158]

6.4.1. Occurrence

In 1963, gastrin (GT-17) was the first hormone of the gastrointestinal tract to be structurally characterized (Kenner).

GT is synthesized primarily in the G cells of the antrum and duodenum as prepro-GT. Immunoreactive GT has also been found in the hypophysis, pancreas and in certain nerves, e.g., the vagus. In humans, the sulfation of the tyrosine of GTs depends on the site of synthesis and on the degree of development [6.159].

There is great structural similarity between GT, which occurs in several molecular sizes [GT-34 (big gastrin), GT-17 (little gastrin), GT-14 (mini gastrin), GT-6, and GT-4], and the transformation protein of polyoma virus (TPPV) and the chicken antral peptide (g-AP) which stimulates gastric acid in chickens [6.160]. The main physiologicallly active component is GT-17. Although GT-14 still has full biological activity, it is of no great significance because it occurs in only very small amounts in the plasma and tissues.

A sulfated myotropic neuropeptide, leucosulfakinin, has been isolated from extracts of the head of the cockroach, *Leucophaea maderae* [6.161]. The intestinal myotropic activities of this peptide are similar to those of GT. Perisulfakinin and leucosulfakinin II have been isolated from the corpora cardiaca of the American cockroach [6.162]. Leucomyosuppressin is structurally similar to leucosulfakinin [6.163]. A prepro-hormone from *Drosophila* has been cloned (drosulfakinin, DSK) which may process two GT-like peptides (DSK I and DSK II) [6.164].

TPPV-(838-869)				P	R	T	P	P	E	L	L	Y	P	E	S	D	Q	D	/-
	1																17		
h-GT-34	*Q	L	G	P	Q	G	P	P	H	L	V	A	D	P	S	K	K	/	/-
p-GT-34	*Q	L	G	L	Q	G	P	P	H	L	V	A	D	L	A	K	K	/	/-
ca-GT-34 [6.165]	*Q	L	G	L	Q	G	P	P	Q	L	V	A	D	L	S	K	K	/	/-
cap-GT-34 [6.165]	*Q	L	G	L	Q	D	P	P	H	M	V	A	D	L	S	K	K	/	/-
cav-GT-33 [6.166]	*Q	L	G	P	Q	V	P	A	H	L	R	T	D	L	S	K	K	/	/-
ch-GT-33 [6.167]	*Q	L	E	P	Q	G	P	P	H	L	G	T	D	L	S	K	K	/	/-
g-AP	F	L	P	H	V	F	A	E	L	S	D	R	K	G	F	V	Q	G	N-

TPPV	Q	L	E	P	L	E	E	E	E	E	E	Y	M	P	M	E	D
	18																34
h-GT-34	Q	G	P	W	L	E	E	E	E	E	A	Y	G	W	M	D	Fa.
p-GT-34	Q	G	P	W	M	E	E	E	E	E	A	Y	G	W	M	D	Fa.
ca-GT-34	Q	G	P	W	M	E	E	E	E	A	A	Y	G	W	M	D	Fa.
cap-GT-34	Q	G	P	W	V	E	E	E	E	A	A	Y	G	W	M	D	Fa.
cav-GT-33	Q	G	P	W	A	E	E	E	/	A	A	Y	G	W	M	D	Fa.
ch-GT-33	Q	G	P	W	A	E	E	E	/	A	A	Y	G	W	M	D	Fa.
g-AP	G	A	V	E	A	L	H	D	H	F	Y	P	D	W	M	D	Fa.

	1																17
h-GT-17	*Q	G	P	W	L	E	E	E	E	E	A	Y	G	W	M	D	Fa.
ca-GT-17	*Q	G	P	W	M	E	E	E	E	A	A	Y	G	W	M	D	Fa.
b,o-GT-17	*Q	G	P	W	V	E	E	E	E	A	A	Y	G	W	M	D	Fa.
f-GT-17	*Q	G	P	W	M	E	E	E	E	A	A	Y	G	W	M	D	Fa.
cu-GT-17 [6.168]	*Q	G	P	W	L	Q	E	E	E	E	A	Y	G	W	M	D	Fa.
cav-GT-16 [6.169]	*Q	G	P	W	A	E	E	E	/	A	A	Y	G	W	M	D	Fa.
ch-GT-16 [6.167]	*Q	G	P	W	A	E	E	E	/	A	A	Y	G	W	M	D	Fa.
h-GT-14				W	L	E	E	E	E	E	A	Y	G	W	M	D	Fa.
p-GT-6												Y	G	W	M	D	Fa.
GT-4														W	M	D	Fa.
DSK I								F	D	D	/	Y	G	H	M	R	Fa.
DSK II			G	G	D	D	Q	F	D	D	/	Y	G	H	M	R	Fa.
Leucosulfakinin						E	Q	F	E	D	/	Ys	G	H	M	R	Fa.
Leucosulfakinin II							*Q	S	D	D	/	Ys	G	H	M	R	Fa.
Perisulfakinin						E	Q	F	D	D	/	Ys	G	H	M	R	Fa.
Leucomyosuppressin							*Q	D	V	D	/	H	/	F	L	R	Fa.

6.4.2. Release

The atropine resistant secretion of GT-17 and GT-34 occurs postprandially, especially in response to proteins or on stimulation of the vagus. In fact, infusion with atropine increases the plasma gastrin level considerably. Furthermore, GT is released by the gastrin releasing peptide (GRP), by bombesin, carbachol [6.170], Ca^{2+}, vitamin D_3 [6.171], omeprazol which reduces gastric acid [6.172], tolbutamide [6.173], PGE_2 [6.174], hypoglycemia via β-adrenergic mechanisms [6.175], by amino acids (can be inhibited by somatostatin (SRIF)) and stronger by the corresponding amines (cannot be inhibited by SRIF) [6.176], inhibitors of monoamine oxidase [6.177] and by testosterone [6.178].

The release of GT is reduced by Mg^{2+}, acidification of the food in the antrum to pH 1.2-1.4 [6.179], SRIF through the inhibition of adenylate cyclase [6.180], indomethacin [6.174], inhibitors of amino acid decarboxylases, dermorphin [6.181], and by estrogen [6.178].

In human beings, the GT level rises during the day and reaches a maximum at about 18^{oo} [6.182]. The plasma level of gastrin is increased in the Zollinger-Ellison syndrome, a GT producing tumor, in pernicious anemia (some patients with pernicious anemia develop autoantibodies against gastrin receptors [6.183]), in gastric, but not in duodenal ulcers [6.182] and in hyperthyroidism [6.184].

6.4.3. Biological Effects

GT increases the formation of cAMP and of inositol phosphates [6.185] and decreases the plasma level of calcium, which is prevented by gastrectomy [6.186].

In the stomach, GT increases the formation of gastric acid and pepsin and stimulates the blood circulation in the mucosa [6.187]. GT-17 stimulates the secretion of gastric acid in patients with duodenal ulcers more than in healthy persons [6.188]. The gastrin inhibitory peptide (GIP), SRIF, secretin, neurotensin, glucagon, glicetin, thyroliberin, the anorexigenic peptide, caerulein, and cholecystokinin (CCK) all inhibit the GT stimulated secretion of gastric acid.

The synthesis and marketing of Boc-β-alanyl-GT-4 (pentagastrin), a GT analog that is fully active in humans, stimulated an intensive investigation of the activity of GT. An infusion with pentagastrin, like a protein meal, increases endogenous GT, insulin, somatotropin, and calcitonin, but decreases the plasma level of SRIF. The primary event here appears to be the inhibition of SRIF by pentagastrin. Once SRIF is inhibited, gastrin, insulin, STH and CT can be released.

Other Biological Effects of Gastrin:

Increased HCO_3^- secretion in the mucosa of the duodenum [6.189] and HCO_3^- and enzyme secretion in the perfused porcine pancreas (without stimulating PGE_2, pentagastrin has a cytoprotective effect on indomethacin or alcohol induced stomach lesions [6.190] and ulcers in the small intestine of the rat [6.191]).

Trophic effect on the oxyntic mucosa, enterochromaffin like cells, A-cells [6.192], pancreas (potentiated by secretin [6.193]), on human tumors of the stomach and colon [6.194], [6.195] and rat pancreatic adenocarcinoma [6.196]

Stimulation of histamine secretion

Increased pressure of the lower esophageal sphincter (can be inhibited by atropine, secretin, glucagon, VIP and calcitonin) and of the pylorus (inhibition of gastric emptying [6.197], [6.198]), but inhibition of CCK induced contraction of the pylorus.

Like CCK, stimulation of insulin release

Thymus dependent immunostimulating activity [6.199]

Biological Effects of GT-4 (Trymafan, Tetrin). The common C-terminal tetrapeptide amide of gastrin and CCK (GT-4) stimulates the secretion of insulin, glucagon, somatostatin and pancreatic peptide. On the other hand, the exocrine secretion from the pancreas is stimulated to a considerably greater extent by the higher molecular froms of gastrin and CCK than by GT-4. Infusion with GT-4 reduces experimental carcinogenesis in the rat colon [6.200].

In contrast to CCK-8, GT-4 stimulates locomotion and rearing in rats, which is enhanced still further by the CCK antagonist proglumide [6.201]. In healthy test persons, i.v. GT-4 (20-100 μg) causes short attacks of panic or symptoms of fear; this has not been observed with CCK-8. Lorazepam inhibits this effect [6.202]. In rats, i.c.v. application of GT-4 causes a deterioration of the memory capacity and can lead to complete amnesia [6.203].

6.4.4. Structure-Activity Relationships

GT-34 is more potent and acts longer than GT-17 when tested in humans [6.204]. The accumulation of glutamic acid in GT-17 is of importance for the biological activity. Thus, the biological activity falls with the stepwise shortening of [Nle^{15}]GT-17 to pGlu-[Nle^{15}]GT-17-(11-17) [6.205].

GT-4 already has the structure required for gastrin activity. The C-terminal amide is important for the biological activity. Esters and free carboxylic groups are devoid of activity. The methyl amide and the hydrazide are just tolerated. The substitution of other amino acids for Asp results in substantial loss of activity. The replacement of the amino acids of GT-4 by the corresponding D-amino acids produces a decrease in activity. Agonists and antagonists can be obtained by reducing the peptide bonds. Boc-[Trp^1-$\psi-(CH_2NH)$-Leu^2]GT-4 is an agonist which binds a little more tightly to gastric mucosa cells than Boc-GT-4 [6.206]. Trp^1 is not required for the insulin releasing activity of GT-4. In fact, this amino acid can be replaced by L- or D-Orn without loss of activity [6.207]. [Pro^2]GT-4 is more effective than GT-4 in releasing insulin from isolated islet cells of the rat pancreas [6.208]. [Pro^2,$Phe(NO_2)^4$]GT-4 (1 nM) is still more effective and almost as potent as CCK-8 in releasing insulin *in vitro*. However, high doses of this compound (100 nM) reduce the release of insulin [6.209].

Very potent analogs are obtained by the N-terminal extension of GT-4. Pentagastrin in which GT-4 is extended by Boc-β-alanine (Boc-β-Ala-Trp-Met-Asp-Phe-amide) is the best known of these analogs. Glutaroyl-[Leu^5]GT-7 (glutaroyl-Ala-Tyr-Gly-Trp-Leu-Asp-Phe-amide, deglugastrin) is more active by a factor of about two. Peptides having a GT-4 or GT-6 structure which contain a glycosylated N-terminal amino group exhibit enhanced water solubility as well as excellent biological activity. For instance, glucosyl-deoxyfructosyl-GT-4 and the corresponding [Thr^1]GT-6 derivative (CRL-6 derivative) are about four times as active as pentagastrin.

Oxidation of Met causes loss of gastrin activity. However, both Met residues of GT-7 can be replaced, e.g. by Leu, without loss of activity.

6.4.5. Antagonists

C-terminal GT sequeneces that lack the C-terminal Phe-amide inhibit the GT activity [6.210], [6.211] e.g., Boc-GT-8-(1-7)-amide has an ED_{50} = 1.5 mg/kg in rats. The smallest sequence that is still antagonistic is Boc-Trp-Met-Asp-amide (ED_{50} = 7.5 mg/kg) [6.210]. Somewhat stronger antagonists are obtained by substituting β-homo-aspartic acid for Asp, e.g., Boc-Trp-Leu-β-homo-Asp-Phe-amide and Boc-Trp-Leu- β-homo-Asp-D-Phe-amide with an ED_{50} = 0.1 mg/kg [6.212].

In human beings, GT-17-(1-13) also inhibits the postprandial or GT stimulated secretion of gastric acid [6.213].

Antagonists are obtained by reduction of the peptide bond between Leu and Asp (e.g., Boc-[$Leu^2\psi(CH_2NH)Asp^3$]-GT-4) [6.206].

Modified retro-inverso-pseudopeptide derivatives of GT-4 exhibit antagonistic activity. The most effective derivatives are Boc-Trp-gLeu-D-Asp-mPhe-amide (ED_{50} = 0.15 μM), Boc-Trp-Leu-gAsp-mPhe-amide (ED_{50} = 0.3 μM) and Boc-Trp-Leu-gAsp-CO-CH_2-CH_2-C_6H_5 (ED_{50} = 0.15 μM) [6.214].

6.4.6. Applications

Pentagastrin (Gastrodiagnost) is being used and deglugastrin has been used as a diagnostic for maximal stimulation of the secretion of gastric acid in the analysis of gastric juice.

6.5. Valosin

p-Valosin (p-VQY), a peptide consisting of 25 amino acids, was isolated from the small intestine of the pig [6.215]. A prepro-valosin was characterized via the cDNA [6.216]. The processing to valosin is not quite clear and valosin may be an artificial processing product [6.216].

Valosin exhibits a cholecystokinin or bombesin like stimulating effect on the exocrine pancreatic protein secretion [6.215] and on the formation of gastrin [6.217] and gastric acid [6.218]. The migratory myoelectric complex of the stomach and the small intestine is inhibited in the fasting state [6.217].

p-Valosin: V Q Y P V E H P D K F L K F G M T P S K G V L F Y

7. Secretin Family [7.1]

The first peptide hormone of the gastrointestinal tract to be discovered was secretin (SEC). This was achieved in 1902 by Bayliss and Starling. The structure of secretin was established in the sixties by V. Mutt and coworkers. In the meantime, this class of substances has grown into a large family. There is considerable structural similarity between SEC and the vasoactive intestinal peptide (VIP), p-PHI-27 (peptide with N-terminal histidine and C-terminal isoleucine amide), h-PHM-27 (peptide with N-terminal histidine and C-terminal methionine amide), pituitary adenylate cyclase activating polypeptide (PACAP), helodermin (HD) from *Heloderma suspectum*, helospectin I and II (HS I and HS II) from *Heloderma horridum*, glucagon (GG), two glucagon like peptides, GLP-1 and GLP-2, glucose dependent insulinotropic and gastrin inhibiting peptide (GIP) and somatoliberin (SRF).

There is some similarity between secretin and the dentinal fluid transport stimulating peptide (DFT), allatostatin from the cockroach *Diploptera punctata*, and the bombilitins from the poison of the bumble-bee *Megabombus pennsylvanicus*.

An important activity cascade of the growth hormones triggers somatoliberin: it releases somatotropin in the hypophysis which fairly ubiquitously stimulates the synthesis of the somatomedins. Hormones that are structurally related to somatotropin are placental lactogen, prolactin, proliferin, and prolactin like and prolactin related proteins. The somatomedins are structurally related to insulin and relaxin. On the basis of their activity, somatostatin, galanin and pancreastatin also belong to this group. Galanine starts the growth hormone cascade through somatoliberin and, like pancreastatin, inhibits the glucose induced release of insulin. Somatostatin inhibits not only the release and activity of somatotropin, insulin and glucagon, but also numerous other peptide hormones.

7.1. Secretin [7.1]

Occurrence. Secretin (SEC) is secreted primarily by the mucosa of the duodenum and upper jejunum. It is also found in the antrum, brain [7.2], hypothalamus and in the hypophysis. Pro-peptides of p-secretin have been isolated and characterized [7.3]. The structures of the prepro-hormones of r- and p-SEC have been established via the cDNA [7.4].

There is some similarity between secretin and the dentinal fluid transport stimulating peptide (DFT), also called the parotid hormone (PH). The latter has been isolated from the parotid gland and characterized [7.8]. The synthesis of DFT is stimulated by the intake of food or by a hypothalamus/thalamus extract [7.9]. DFT increases the transport of fluid in the teeth [7.9].

	1																					22
h-SEC	H	S	D	G	T	F	T	S	E	L	S	R	L	R	E	G	A	R	L	Q	R	L-
h-PHM	H	A	D	G	V	F	T	S	D	F	S	K	L	L	G	Q	L	S	A	K	K	Y-
h, p, b, r-VIP	H	S	D	A	V	F	T	D	N	Y	T	R	L	R	K	Q	M	A	V	K	K	Y-
o-PACAP-38	H	S	D	G	I	F	T	D	S	Y	S	R	Y	R	K	Q	M	A	V	K	K	Y-
HD	H	S	D	A	I	F	T	E	E	Y	S	K	L	L	A	K	L	A	L	Q	K	Y-
HS I	H	S	D	A	T	F	T	A	E	Y	S	K	L	L	A	K	L	A	L	Q	K	Y-
HS II	H	S	D	A	T	F	T	A	E	Y	S	K	L	L	A	K	L	A	L	Q	K	Y-
p, b, cr, h-GG	H	S	Q	G	T	F	T	S	D	Y	S	K	Y	L	D	S	R	R	A	Q	D	F-
h, cr-GLP-1	H	A	E	G	T	F	T	S	D	V	S	S	Y	L	E	G	Q	A	A	K	E	F-
h-GLP-2	H	A	D	G	S	F	S	D	E	M	N	T	I	L	D	N	L	A	A	R	D	F-
h-GIP	Y	A	E	G	T	F	I	S	D	Y	S	I	A	M	D	K	I	H	Q	Q	D	F-
h-SRF	Y	A	D	A	I	F	T	N	S	Y	R	K	V	L	G	Q	L	S	A	R	K	L-

	23				27																	
h-SEC	L	Q	G	L	Va.																	
h-PHM	L	E	S	L	Ma																	
h, p, b, r-VIP	L	N	S	I	L	Na																
o-PACAP-38	L	A	A	V	L	G	K	R	Y	K	Q	R	V	K	N	Ka						
HD	L	A	S	I	L	G	S	R	T	S	P	P	Pa.									
HS I	L	E	S	I	L	G	S	S	T	S	P	R	P	P	S	S.						
HS II	L	E	S	I	L	G	S	S	T	S	P	R	P	P	S.							
p, b, cr, h-GG	V	Q	W	L	M	N	T.															
h, cr-GLP-1	I	A	W	L	V	K	G	R	G.													
h-GLP-2	I	N	N	L	I	Q	T	K	I	T	D.											
h-GIP	V	N	W	L	L	A	Q	K	G	K	K	N	D	W	K	H	N	I	T	Q.		
h-SRF	L	Q	D	I	M	S	R	Q	Q	G	E	S	N	Q	E	R	G	A	R	A	R	La.

	1	10	20	27
h-SEC	HSDGTFT S ELSRL REGARL	Q RLL QGL		Va.
p,b-SEC	HSDGTFT S ELSRL RDSARL	Q RLL QGL		Va.
r-SEC [7.5]	HSDGTFT S ELSRL QDSARL	Q RLL QGL		Va.
ca-SEC [7.6]	HSDGTFT S ELSRL RESARL	Q RLL QGL		Va.
cu-SEC [7.7]	HSDGTLT S ELSRL RDRARL	Q RLL QGL		La.
g-SEC	HSDGLFT S EYSKM RGNAQV	Q KFI QNL		Ma.

r-DFT GVI AWE LQH NEPGR / KDSTAG.

Allatostatin	APSGAQ RLYGFG La
xe-PYL	YVRGMASKAGAI AGKI AKVALKA La
xe-PGL	GMASKAGAI AGKI AKVALKA La
Bombolitin I	IKI TTMLAKL GKVLAH/ Va.
Bombolitin II	SKI TDI LAKL GKVLAH/ Va.
Bombolitin III	IKI MDI LAKL GKVLAH/ Va.
Bombolitin IV	INI KDI LAKL VKVLGH/ Va.
Bombolitin V	INVLGI LGLL GKALSH/ La

Structural similarities with C-ternimal SEC are found in allatostatin, a peptide from the brain of the cockroach, *Diploptera punctata,* which inhibits the systhesis of the juvenile hormone [7.10], in bombilitins, mast cell degranulating peptides from the poison of the bumble-bee, *Megabombus pennsylvanicus* [7.11], and in xe-PGL [7.12], a fragment in the hypothetical xe-PYL [7.13].

Release. In the small intestine, hydrochloric acid (pH<4) stimulates the secretion of SEC into the blood stream. the acid stimulated release of SEC is mediated by a SEC releasing peptide which is sensitive to attack by trypsin [7.14]. Bombesin stimulates the secretion of SEC [7.15]. Stress produces a 3-6 fold increase in the plasma levels of secretin. High extracellular concentrations of K^+ have a stimulating effect and the absence of extracellular Na^+ and Ca^{2+} an inhibitory effect on the secretion of SEC from the mucosa of the small intestine [7.16]. Similar to tetraprenylacetone [7.17], camostate (trypsin inhibitor) [7.18], and plaunotol [7.19], extracts of the licorice root *(Glycyrrhiza glabra)* [7.20] cause intraduodenal stimulation of the secretion of SEC.

SRIF inhibits the release of SEC and the exocrine pancreatic secretion.

Patients with chronic kidney damage and Zollinger Ellison syndrome (ZES) have a significantly raised level of SEC.

Receptors. SEC binds to the perfused pancreas and to the gastric mucosa of rats. In contrast to SEC bound to the pancreas, SEC bound to the gastric mucosa can be displaced by tetragastrin [7.21].

Biological Effects. As a rule, the activity of SEC is expressed in clinical units (CU) which are determined in the dog. 1 mg of SEC has about 4000 to 5000 CU. The physiological SEC level of about 20 pg/ml, which is observed in the human jejunum after H^+ stimulus, should, in fact, be achieved after an infuion of 0.03 CU/kg.h.

In a process independent of the vagus [7.22], SEC stimulates the release of a pancreatic, $NaHCO_3$ containing aqueous secretion, which, in turn, neutralizes the hydrochloric acid that has found its way into the small intestine. CCK, neurotensin [7.23], and adenosine [7.24] potentiate the SEC stimulated pancreatic secretion. The SEC induced protein secretion from the pancreas that is described frequently is based on impurities, e.g., cholecystokinin (CCK) [7.25].

Other Effects of SEC. Activation of adenylate cyclase and formation of cAMP
Proliferative effect in the pancreas which is potentiated by CCK [7.26]
Antitrophic effect in organ culture of the rabbit small intestine [7.27]
Reduction of protein secretion in CCK induced acute pancreatitis in rats
Reduction of the plasma gastrin level and reduced gastric acid secretion (via SRIF and prostaglandins [7.28], potentiated by neurotensin [7.23])
Reduction of stress induced hemorrhage from the stomach [7.29]
Stimulation of the basal and potentiation of the pentagastrin stimulated secretion of pepsin in the chief cells of the stomach
Secretion of a highly viscous glycoprotein aggregate in the stomach of the cat (cytoprotective effect?)
Vasodilative and motility inhibiting effect (dilatation of the pancreatic ducts [7.30], reduction of the lower esophageal sphincter pressure [7.31], increased circulation in pancreas and duodenum)
Contraction of the pylorus (delayed emptying of the stomach [7.32])
Potentiation of the CCK stimulated bile flow [7.15]

Glycolytic and lipolytic activity in adipose tissue
Positive chronotropic, positive intropic and vasodilative effects [7.33] (improves left ventricular function in the normal circulation [7.34] and in acute ischemic left ventricular failure [7.35]
In the lung, inhibition of leukotriene release which is stimulated by the platelet activating factor (like VIP and PHI-27) [7.36]
Reduced plasma prolactin level in women [7.37]
Release of insulin *in vitro*

Structure-Activity Relationships. Until now, analogs of secretin that are more active than the native peptide have not been found. Shortening the molecule at either the N-terminal or the C- terminal end produces less active products. For instance, secretin-(2-27)-amide has only 1% of the activity. At present, the shortest active sequence is secretin-(1-13). In fact, 0.1 mg/kg of this substance still exerts a significant effect on the pancreas. However, 1 mg/kg of secretin-(1-11)-amide shows no activity in the rat. His^1, Asp^3, Phe^6, and Arg^{12} are important for secretin activity.

A compound extended at the C-terminus by Gly-Lys-Arg (natural pro-SEC) still has about 80% of the SEC activity [7.38].

The C-terminal hexapeptide is important for the insulin releasing activity of SEC. The glucose dependent release of insulin from the islet cells of the mouse is elicited by this hexapeptide [7.39].

Stability. The Asp-Gly bond in position 3-4 of secretin is the main cause for the frequently reported instability. The aspartoyl peptide formed in an acidic medium is then converted to the β-aspartic acid containing peptide, especially at a neutral or basic pH. HPLC studies have also shown that the Asp-Gly-(3-4) and the Asp-Ser (15-16) bonds in p-SEC are cleaved in aqueous solution at pH 5 [7.40]. The combination of the N-terminal His and Asp^3 is important for autolysis [7.40]. The main decomposition product in neutral phosphate buffer is [β-Asp^3]SEC [7.41]. In acidic acetate buffer (pH 4), SEC-(16-27), SEC-(4-27), [β-Asp^3] SEC, [3-aspartoyl]SEC and [β-Asp^{15}]SEC are formed [7.41]. These decomposition products have little or no biological activity.

Applications. Secretin is generally administered parenterally, but the intranasal application of a dry powder has also been described as effective. Secretin (sekretolin) is applied parenterally in doses of 75 to 100 CU to test pancreatic functions and to diagnose Zollinger-Ellison syndrome. In contrast to patients without chronic pancreatitis, the diameter of the pancreatic duct of patients with chronic pancreatitis does not increase after SEC [7.42]. Infusion of sekretolin (0.25-0.5 KE/kg.h) stops bleeding of the upper gastrointestinal tract [7.43]. SEC produces an improvement in patients suffering from peptic ulcers [7.44]. The relaxing effect on the lower esophageal sphincter can be exploited in patients with achalasia [7.31].

The insulin releasing effect of 2 CU SEC/kg can be used to differentiate between single insulinomas or nesidioblastosis and multiple B cell adenomas or hyperplasia. The insulin releasing effect is lacking in patients with insulinomas and nesidioblastosis and is increased in patients suffering from multiple B cell adenomas and hyperplasia [7.45].

In contrast to healthy persons, patients with Zollinger-Ellison syndrome (gastrin producing tumor) react to a diagnostic dose of secretin (75 KE i.v.) with an increased plasma level of gastrin and GIP and not with the inhibition of gastrin [7.40].

7.2. Vasoactive Intestinal Peptide and PHI-27/PHM-27 [7.46], [7.47]

Occurrence. Vasoactive intestinals peptide (VIP) was first isolated along with secretin from the intestine of the pig. It is distributed over the entire body, primarily in the nervous system. Thus, it is not only an intestinal peptide, but, in keeping with its abbreviation, also a very important peptide.

VIP, a peptide amide containing 28 amino acids, and PHI-27 or PHM-27 (human), peptide amides containing 27 amino acids, are formed in the same prepro-peptide. In addition to VIP, an extended version of VIP with C-terminal Gly-Lys-Arg has also been found [7.48]. A peptide consisting of 42 amino acids (h-PHV-42) has been isolated from a human adrenal pheochromocytoma. This peptide represents a C-terminally extended h-PHM-27 and is identical to h-prepro-VIP-(81-122) [7.49].

p-, b-, r-, ca-, cu-, cap-, o- and h-VIP are identical [7.50]. On the other hand, chicken VIP (g-VIP) and guinea pig VIP (cav-VIP) [7.51] differs in four positions. However, there is little difference in the sequences of h-, p-, b-[7.52] and r-PHI-27 [7.53]. cu-PHI-27 is identical to p-PHI-27 [7.54].

	Sequence (1 … 10 … 20 … 28)
h, b, o, cap, p, ca, cu, r-VIP	H S D A V F T D N Y T R L R K Q M A V K K Y L N S I L Na.
g-VIP	H S D A V F T D N Y S R F R K Q M A V K K Y L N S V L Ta.
cav-VIP	H S D A L F T D T Y T R L R K Q M A M K K Y L N S V L Na.

	Sequence (1 … 27)
h-PHM-27	H A D G V F T S D F S K L L G Q L S A K K Y L E S L Ma
p, cu-PHI-27	H A D G V F T S D F S R L L G Q L S A K K Y L E S L Ia
b-PHI-27	H A D G V F T S D Y S R L L G Q L S A K K Y L E S L Ia
r-PHI-27	H A D G V F T S D Y S R L L G Q I S A K K Y L E S L Ia
h-PHV-42 (1 … 27)	H A D G V F T S D F S K L L G Q L S A K K Y L E S L M-
h-PHV-42 (28 … 42)	G K R V S S N I S E D P V P V.

VIP and PHI-27 are found in the neurons of the central nervous system, in the digestive tract, lungs, urogenital tract, wall of the gallbladder [7.55], exocrine glands, thyroid gland, anterior lobe of the hypophysis [7.56] and in the adrenal gland. VIP is found associated with acetylcholine and galanine [7.57].

Release. VIP is released in response to the electrical stimulation (87-12 Hz) of nerve fibers, stimulation of the vagus (in the gallbladder wall [7.55]), prostaglandin E_1 (from the hypothalamus), bombesin, oxytocin, sexual stimulation in women, the sucking stimulus [7.58], physical stress [7.59] and operation stress [7.60] (in the plasma), corticosterone [7.61], Met-enkephalin and β-endorphin [7.62] (in the hypothalamus), insulin (in the intestine [7.63], estrogen (in the anterior hypophysis [7.64] and to sexual maturation [7.65].

In the stress induced release of prolactin via VIP, the following sequence of events appears to be plausible: stress —> β-endorphin (central) —> VIP —> prolactin.

The release of VIP is inhibited by glucose infusions (7.59), periphereal application of endogenous opioides [7.59], pancreatic peptide and by somatostatin.

Patients with multiple sclerosis have a significantly lower level of VIP in their cerebrospinal fluid than healthy persons. Impotent men have a low plasma VIP level. The VIP-level is reduced in the urinary bladder wall in patients suffering from bladder instability (increased bladder pressure during the filling phase). In patients suffering from cystic fibrosis, fewer VIP i.r. nerves are found in both the acini and ducts of the eccrine sweat glands [7.66]. Intravenous glucagon or secretin raises the plasma levels of gastrin and VIP in patients with Zollinger-Ellison syndrome [7.67]. The treatment of patients with congestive heart failure with ACE inhibitors increases plasma VIP levels [7.68].

Receptors. [7.69] There are indications of the existence of at least two high affinity VIP receptors and one low affinity VIP receptor. The VIP antagonist, [D-pCl-Phe^6,Leu^{17}] VIP differentiates between two high affinity VIP receptors in guinea pig acinar cells [7.70]. In dispersed acinar cells of the guinea pig, carbachol binds to low affinity VIP receptors whereas the high affinity VIP receptors whereas are reduced in the process [7.71]. The VIP receptors on human mononuclear leucocytes increase drastically in response to physical activity and an energy deficit [7.72].

Biological Effects. In the cell, VIP stimulates cAMP and the synthesis of catecholamines by activating tyrosine hydroxylase [7.73]. It also potentiates the muscarinic stimulation of the phosphoinositide turnover [7.74]. VIP inhibits the activation of the calmondulin sensitive phosphodiesterase [7.75]. VIP and acetylcholine act synergistically in secretion and circulation.

As a rule, PHI exhibits VIP like activity.

Other Effects [7.47]

Cardiovascular system: Vasodilation; hypotension; positive chronotropic and inotropic effects [7.76];

Respiratory system: Bronchodilation; augmented ventilation; pulmonary vasodilation; stimulation of bronchial sercetion (VIP and asthma [7.77]); protection against HCl-induced pulmonary edema [7.78] and injury triggered by xanthine/xanthinoxydase [7.79];

Lacrimal gland and eye: stimulation of fluid and protein secretion [7.80]; reduction of intraocular pressure [7.81].

Submandibular gland: Stimulation of salvia and EGF secretion [7.82];
Parotid gland: Stimulation of salivary secretion (potentiated by substance P) [7.83];
Esophagus: Relaxation of lower esophageal sphincter;
Stomach: Relaxation of fundic smooth muscles (causes gastric dilation and relaxes the pylorus), suppression of acid secretion;
Pancreas: Stimulation of water and HCO_3^- secretion;
Liver: increased bile flow;
Gallbladder: Relaxation of isolated smooth muscle, inhibition of contractile effect of CCK; In low doses mobility is increased and in high doses it causes relaxation [7.84];
Small and large intestine: Inhibition of absorption; stimulation of water and Cl^--secretion; relaxation of colonic circular muscles and contraction of longitudinal muscles [7.85], [7.86]; Inhibits motilin: increased circulation in the colon during defecation.
Uro-genital tract: relaxation of uterine muscle (potentiated by progesterone and estrogen) [7.87]; Stimulation of vaginal circulation and erection of the penis; antidiuretic activity; increases renal vascular resistance and natriuresis [7.88].
Biochemical/Metabolic: Inhibition of insulin-stimulated glucose transport in rat adipocytes [7.89]; glycogenolysis and lipolysis, bone resorption, hyperglycemia; hair growth [7.90]; regulates mitosis, differentiation and survival of cultered sympathetic neuroblasts [7.91]; stimulates meiosis of follicle-enclosed rat oocytes *in vitro* [7.92]; influences neuronal survival [7.93]; inhibits hamster ductal pancreatic cancer, but did not inhibit human pancreatic cancer growth [7.94].

Endocrine functions:
Stomach: Release of GRP [7.95] and somatostatin [7.96];
Pancreas: Release of insulin, glucagon or somatostatin (depending on experimental conditions); inhibition of GIP and glucose stimulated insulin release [7.97];
Pituitary: Stimulation of release of ACTH, prolactin (e.g., PRL release stimulated by sucking [7.98]), somatotropin and lutropin; suppression of somatostatin release;
Hypothalamus: Release of gonadoliberin (Gn-RH).
Thyroid gland: Release of iodine containing thyroid hormones.
Adrenal: ACTH-like action (stimulation of aldosteron and corticosteron synthesis [7.99]; increase of serum testosterone and dihydortestosterone in women [7.100]; elevates the level of enkephalin containing peptides in cultured bovine adrenal medullary chromaffin cells [7.101].
Kidney: Stimulation of renin release by juxtaglomerular cells.
Ovaries: Enhances aromatase activity in the neonatal rat ovary [7.102]; increase of serum estradiol in women [7.100]; stimulation of estradiol and progesterone release from rat ovaries [7.103].
Testis: Stimulation of testosterone, progesterone and pregnenolone in cultured neonatal testicular cells [7.104].
Immune System: Inhibition of mitogen-induced transformation of T-lymphocytes; stimulation of Ig synthesis; inhibition of mast cell histamine release; inhibition of platelet aggregation and secretion; inhibition of respiratory bursts in monocytes [7.105];
Central nervous system: Excitation of cerebral cortical and spinal cord neurons; hyperthermia; stimulation of glycogenolysis and of glucose utilization; hypnogenic

and antidipsogenic effects; generation and maintenance of rapid-eye movement (REM) sleep [7.106]; stimulation of gastric acid secretion [7.107]; generation of hypermobility in rats [7.108]; amnesia in mice [7.109].

Structure-Activity Relationships. The C-terminal amide is not required for the biological activity. The compounds, de-amido^{28}-VIP, VIP-Gly or [Leu^{17}]VIP-Gly-Lys [7.110], [7.111] have similar activity. Shortening the molecule at the C- and N-terminus has a negative effect on the biological activity. The fragment, VIP-(10-28) exerts an antagonistic effect on the VIP stimulated formation of cAMP and pepsinogen secretion [7.112], [7.113]. Met^{17} can be replaced by norleucine without loss of activity.

The amino acids of the hydrophobic region of the helical domain (6-28) are important for receptor binding, while the amino acids of the hydrophilic region appear to be less important. The peptide

```
            *   *   *   *   * *       * *     * * *   *
H S D A V F S D S Y S R F R R S M A V R R Y L S N V L Ta
```

with 12 substitutions (*) is about as active as VIP, while substitutions of the amino acids in positions 6, 10, 12, 17, 19, 22, 23, and 27 produce less active derivatives [7.114]. The iodization of VIP leads to at least two different iodine compounds which have full biological activity [7.115].

The displacement of [^{125}I]VIP bound to receptors on rat liver membranes by various analogs occurs in the following order: VIP > [D-Ala^4]VIP > [D-Asp^3]VIP > [D-Ser^2]VIP > [D-His^1]VIP > [D-Phe^2]VIP > [D-Arg^2]VIP.

Of these compounds, [D-Phe^2]VIP is a competitive VIP antagonist, while the other analogs still have agonistic activity (cAMP formation on rat pancreas membranes) [7.116]. Still better VIP antagonists are [D-pCl-Phe^6,Leu^{17}]VIP, micromolar amounts of which inhibit the VIP stimulated amylase [7.117], and Ac-[Tyr^1,D-Phe^2]somatoliberin-(1-29)-amide [7.118].

The C-terminal elongated h-PHV-42 relaxes the uterus and stomach of the rat and the guinea pig trachea more than h-PHM-27. VIP is four times as active as h-PHV-42 in reducing blood pressure [7.119].

[D-Phe^4]PHI-27 is a highly selective VIP agonist [7.120].

7.3. Pituitary Adenylate Cyclase Activating Polypeptide (PACAP)

Occurrence. There is considerable homology between VIP and o-PACAP-38 [7.121] and o-PACAP-27 [7.122], two peptide amides from the sheep hypothalamus. The prepro-peptide of h-, o- [7.123], and r-PACAP [7.124] has been characterized via the cDNA. h-, o-, and r-PACAP-38 are identical. Apart from the hypothalamus, the mRNA for PACAP-38 has also been observed in human testes [7.123].

	1																										27
h, o, r-PACAP-38	H	S	D	G	I	F	T	D	S	Y	S	R	Y	R	K	Q	M	A	V	K	K	Y	L	A	A	V	L-
o-PACAP-27	H	S	D	G	I	F	T	D	S	Y	S	R	Y	R	K	Q	M	A	V	K	K	Y	L	A	A	V	La.

	28										38
PACAP-38	G	K	R	Y	K	Q	R	V	K	N	Ka

Highly specific receptors have been found primarily in the hypothalamus, brain stem, cerebellum, and in the lung. PACAP-27 binds more tightly than PACAP-38.

In the rat hypophysis, o-PACAP-38 and o-PACAP-27 stimulate adenylate cyclase 1000 times more effectively than VIP [7.121], [7.122]. Like VIP, o-PACAP-27 exhibits vasodepressor activity [7.122]. Concentration of 10^{-10}M release somatotropin, prolactin and ACTH from rat hypophyseal cells and 10^{-9}M releases lutropin as well [7.125].

7.4. Helodermin and Helospectin

Occurrence. Helodermin (HD) was isolated from the poison of *Heloderma suspectum* and helospectin I and II were obtained from the poison of *Heloderma horridum* [7.126]. i.r.-HD has also been found in the brain, hypophysis, duodenum, jejunum, saliva,salivary gland [7.127], [7.128], adrenal gland [7.129], C cells of the thyroid gland [7.130], [7.131] and in medullary thyroid carcinoma [7.132] of mammals.

	1																					22
HD	H	S	D	A	I	F	T	E	E	Y	S	K	L	L	A	K	L	A	L	Q	K	Y-
HS I	H	S	D	A	T	F	T	A	E	Y	S	K	L	L	A	K	L	A	L	Q	K	Y-
HS II	H	S	D	A	T	F	T	A	E	Y	S	K	L	L	A	K	L	A	L	Q	K	Y

	23												35			
HD	L	A	S	I	L	G	S	R	T	S	P	P	Pa.			
HS I	L	E	S	I	L	G	S	S	T	S	P	R	P	P	S	S.
HS II	L	E	S	I	L	G	S	S	T	S	P	R	P	P	S.	

Receptors. HD binds to VIP and secretin receptors, leading to the activation of adenylate cyclase [7.126].

A new type of VIP receptor has been found on human SUP-T1 lymphoblasts. It binds helodermin and helospectin more tightly than VIP, and activates adenylate cyclase much better [7.133].

sa-Calcitonin binds to HD antibodies. This is probably due to the Arg-Thr sequence which is common to both sa-CT and HD [7.131].

Biological Effects. In comparison with secretin and VIP, HD causes a much greater secretion of amylase in rat pancreatic acini and a lower synthesis of cAMP [7.134]. HD stimulates the basal thyroid hormone secretion in mice [7.130] and the release of glucagon, which can in part be abolished by α and β adrenoceptor antagonists. However, it has no effect on the release of insulin [7.135].

h-Prepro-GG [7.141]

```
1                                                                                  30
R S L Q D T E E K S R S F S A S Q A D P L S D P D Q M N E D K R-
33                                     53              61
H S Q G T F T S D Y S K Y L D S R R A Q D F V Q W L M N T-
―――――――――――――――――――――――― glucagon ――――――――――――――
62    64          69    71                77
K R N R N N I A K R H D E F E R-
    ―― SpP-I ――――      ――――――――――
78                                                                        108
H A E G T F T S D V S S Y L E G Q A A K E F I A W L V K G R G-
―――――――――――――――――― glucagon like peptide I ――――――――――――――
109  111                             123  125
R R D F P E E V A I V E E L G R R-
    ―――――――― SpP-II ――――――――
126                                                                                  159
H A D G S F S D E M N T I L D N L A A R D F I N W L I Q T K I T D R.
―――――――――――――――――― glucagon like peptide II ――――――――――――――――――
```

Like VIP, HD, HS I, and HS II lower the blood pressure and relax the femoral arteries of the rat [7.136]. HD inhibits the biosynthesis of PAF stimulated leukotrienes, LTD_4, LTC_4 and LTE_4 in the rat lung [7.137].

HD suppresses the incorportion of calcium into the bones of the rat [7.138] and increases PTH induced bone resorption [7.130].

7.5. Glucagon and Related Peptides [7.139]

Glucagon (GG) is formed together with the two glucagon like peptides, GLP-1 (prepro-GG-(71-108)) and GLP-2 (prepro-GG-(126-159)) as prepro-GG (r- [7.140], h- [7.141], b- [7.142], p- [7.143], cav- [7.144], oc- [7.145]). In the pancreatic A cells, prepro-GG is processed to GG and longer C-terminal peptides, which contain GLP-1 and GLP-2. In the small intestine, however, prepro-GG is cleaved to form glycetin (prepro-GG-(1-69)), enteroglucagon (oxyntomodulin = prepro-GG-(33-69)) and GLP-1 and GLP-2 [7.146]. In particular, GLP-1 is processed in the small intestine without the N-terminal 6 amino acids as the C-terminal amide (GLP-1-(7-36)-amide) [7.147], [7.148].

I.r. GG-(1-21) is observed [7.149] in the intestinal tract of the dog and in the liver. GG is processed to GG-(19-29) by cleavage at the Arg^{17}-Arg^{18} doublet [7.150].

As far as the hyperglycemic effect is concerned, this group also includes the crustacean hyperglycemic hormone (CHH).

7.5.1. Glucagon

Occurrence. Glucagon is formed in the A cells of the pancreas. It is, however, also found in the rat hypophysis.

Release. Pulsatile GG release is observed in humans at 13-20 minute intervals [7.158]. The Ca^{2+} dependent secretion of GG is stimulated by amino acids, such as arginine, alanine, and glutamic acid, stress (through α_1- and α_2-adrenoreceptors) [7.159], β-endorphin [7.160], morphine (via the inhibition of somatostatin (SRIF)) [7.161], endotoxin [7.162], dopamine (inhibition of SRIF), carbachol, cholecystokinin [7.163], bombesin, xenopsin, Substance P, and to high doses of oxytocin and vasopressin. Parathyroid hormone potentiates the Arg stimulated release of GG.

The secretion of glucagon is inhibited by glucose via the activation of GABA-A receptor chloride channels in the A cells of the pancreas [7.164], [7.165], Mg^{2+}, thyroliberin (TRH), calcitonin, SRIF, glicentin-(1-16), glicentin-(62-69) [7.166] and by sulfonyl ureas, e.g., glibenclamide [7.167]. The plasma level of GG is higher in overweight persons [7.168] and is reduced in type 1 diabetics [7.169].

Glucagon from various species

	1 5 10 15 21 25 29
p-, b-, h-, r-, cr-GG	H S Q G T F T S D Y S K Y L D S R R A Q D F V Q W L M N T
g-, di-GG [7.151]	H S Q G T F T S D Y S K Y L D S R R A Q D F V Q W L M S T
ra-GG [7.152]	H S Q G T F T S D Y S K Y L D S R R A Q D F V Q W L M N S
an-, ps-GG [7.152]	H S Q G T F T S D Y S K Y L D T R R A Q D F V Q W L M S T
ch-GG [7.153]	H S Q G T F T S D Y S K H L D S R Y A Q E F V Q W L M N T
oc-GG [7.145]	H S Q G T F T S D Y S K F L D T R R A Q D F L D W L K N T
cav-GG [7.144]	H S Q G T F T S D Y S K Y L D S R R A Q Q F L K W L L N V
lo-GG-I [7.154]	H S E G T F S N D Y S K Y L E D R K A Q E F V R W L M N N
lo-GG-II [7.154]	H S E G T F S N D Y S K Y L E T R R A Q D F V Q W L K N S
sc-GG [7.157]	H S E G T F T S D Y S K Y M D N R R A K D F V Q W L M N T
to-GG [7.157]	H S E G T F T S D Y S K Y L D N R R A K D F V Q W L M N T
sa-GG [7.156]	H S E G T F S N D Y S K Y Q E E R M A Q D F V Q W L M N S
ic-GG [7.155]	H S E G T F S N D Y S K Y L E T R R A Q D F V Q W L M N S

Biological Effects [7.170]. Glucagon plays a role in carbohydrate, fat and protein metabolism. The liver is the most important target organ for GG. In the liver, glucagon stimulates the formation of glucose from glycogen and of ketone bodies from amino acids. In this process, the synthesis of glycogen is inactivated via a phosphorylation reaction [7.171] and the receptor binding of insulin in the liver is reduced [7.172]. On i.c.v. application, GG increases the blood glucose via cholinergic and a-adrenergic mechanisms to a much greater extent than after intravenous administration [7.173].

In hepatocytes, GG inhibits fatty acid synthesis via the phosphorylation of an acetyl CoA carboxylase by means of a cAMP dependent protein kinase [7.174]. It stimulates lipolysis in brown adipose tissue and increases the consumption of free fatty acids [7.175]. It reduces the activity of HMG CoA reductase [7.176] and lowers the plasma level of apolipoprotein E, cholesterol, triacylglycerol and phospholipids [7.177].

Other effects of GG

Stimulation of the plasma level of cAMP

Reduction of Ca^{2+} and PO_4^{-3} blood levels via increased renal excretion, but similar to PTH, increase in the reabsorption of calcium and magnesium in Henle's loop [7.178]

Similar to ANF, increase in the glomerular rate of filtration and natriuresis in the rat [7.179]

Breakdown of alcohol by activating alcohol dehydrogenase [7.180]
Spasmolytic activity in high doses
Relaxation of the esophagus
Contraction of the pylorus (delays the emptying of the stomach [7.181] and reduces the intake of food [7.182]
Stimulation of blood circulation in the intestinal mucosa [7.183]
Relaxation of Oddi's sphincter [7.184] and stimulation of bile flow [7.185], [7.186]
Bronchodilatory effect in asthmatics [7.187]
Postive inotropic and chronotropic effect on the heart
Increase in the plasma level of ammonia in patients suffering from liver cirrhosis [7.188]
In combination with insulin, protective and healing effect in hepatitis
Stimulation of SRIF release from the stomach [7.189]
In high doses, reduction of gastric acid and pepsin secretion
Stimulation of insulin secretion
Increase in somatotropin, ACTH, and cortisol plasma level [7.190], [7.191]
Increase in estradiol secretion in the placenta [7.192]
Decrease in the plasma amino acid concentration, which may result in skin rash
Inhibition of the anaphylactic release of histamine
Inhibition of morphine induced bradycardia and antinociception when applied centrally [7.193]

Structure-Activity Relationships. The shortening of the peptide chain and the replacement of the N-terminal histidine causes a drop in the glycogenolytic activity of glucagon [7.194]. However, the spasmolytic activity responds differently. Here, GG-(1-21) is fully active, whereas only about 0.1% of the cAMP releasing activity could be observed. Thus, GG-(1-21) exerts a relaxing effect on Oddi's sphincter [7.195] and an analgesic effect in biliary colic [7.196]. Further shortening of the compound, e.g., GG-(1-20) and GG-(2-21), causes a reduction of the spasmolytic activity to 2 and 1% of the GG activity. The fragment, GG-(19-29) does not activate adenylate cyclase, but is 1000 times as active as GG in inhibiting the Ca^{2+} activated and Mg^{2+} dependent ATPase [7.197] and the liver plasma membrane calcium pump [7.150].

Various substitutions in the N-terminal part of GG usually result in a reduction of the biological activity. Only [D-Ser^2]GG [7.198] and [D-Phe^4]GG had twice and six times the glycogenolytic activity in the rat [7.199]. This glycogenolytic activity could be enhanced in the combination preparation, [D-Phe^4,Phe^{25},Leu^{27}]GG-amide (7.7 times the GG activity) [7.198].

The amino acid region 10-13 is not essential for the biological activity, but is important for receptor binding [7.200].

While the S-methylation of Met^{27} reduces the glucagon activity to 1%, [S-carboxymethyl-Met^{27}]GG [7.201], [2-thiol-Trp^{25}]GG, [oxindolyl-Ala^{25}]GG, [$Met(O_2)^{27}$]GG, monoiodo-GG and [Nε-alkylamidino-Lys^{12}]GG are fully active.

GG Antagonists. Nα-Trinitrophenyl-glucagon and Nα-trinitrophenyl-[12-homoarginine]GG are weak antagonists that exert a partially agonistic effect.

A full antagonist is de-His^1-[Glu^9]GG-amide ($pA_2=7.2$) [7.202]. Still stronger competitive GG antagonists are de-His^1-[Glu^9,Orn^{12}]GG-amide (pA_2 = 7.4), de-His^1-[Glu^9,Glu^{16},Lys^{29}]GG-amide (pA_2 = 7.8) and de-His^1-[Glu^9,Lys^{29}]GG-amide (pA_2 = 8.1) [7.203]. [Asp^3,D-Phe^4,Ser^5,Lys^{17,18},Glu^{21}]GG and [D-Phe^4,Tyr^5,I_2-Tyr^{10},Arg^{12},Lys^{17,18},Glu^{21}]GG also have antagonistic activity [7.204].

Applications. Glucagon (Glucagon Novo,Glukagon Lilly) is used in the treatment of hypoglycemia. In McArdle's disease, a subcutaneous injection of 2 mg of glucagon results in an increase in the blood sugar level and, thus, improves muscle function.

The spasmolytic activity of glucagon is of importance in both endoscopic and radiological investigations of the gastrointestinal tract and in computed tomography of the liver, pancreas, and kindney. GG reliefs also esohageal cramps (caused by food) [7.205] and hiccups [7.206]. The spasmolytic activity in combination with the enhanced flow of bile has been useful in the treatment of biliary colic. In individul cases, gallstones have even been expelled.

As a result of its effect on the plasma Ca^{2+} level, glucagon has been used in combination with calcitonin and mithramycin in the treament of Paget's disease.

GG can be applied parenterally (also with a portable pump [7.207] or as a depot preparation [7.208], nasally [7.209], [7.210] or as eye drops [7.211].

7.5.2. Glicentin and Enteroglucagon

Glicentin (prepro-GG-(1-69)) and enteroglucagon (oxyntomodulin, prepro-GG-(33-69)) are formed in the small intestine from prepro-GG [7.146].

Glicetin and enteroglucagon have the glucagon sequence and the C-terminal octapeptide in common. The plasma level of glicentin rises during hypoglycemia [7.212] and physiological doses of both peptides decrease the secretion of gastric acid [7.213]. Glicentin has only 10% of the cAMP stimulating activity of GG. Thus, glicentin and enteroglucagon assume the function of a hypothetical enterogastrone in the intestine. The common C-terminal octapeptide appears to be especially important and is, in fact, capable of inhibiting the pentagastrin stimulated secretion of gastric acid (although 100 times weaker) [7.214].

p-Oxyntomodulin-(30-37): **K R N K N N I A**

Enteroglucagon is also 10 times as active as GG in inhibiting the caerulein stimulated exocrine pancreatic secretion in rats [7.215].

7.5.3. Glucagon Like Peptide-1-(7-36)-amide

Occurrence. The glucagon like peptide (GLP)-1 is processed primarily in the small intestine as the C-terminal amide, without the N-terminal six amino acids (GLP-1-(7-36)-amide) [7.147], [7.148].

glucagonlikepeptide-1-(7-36)-amide:

	7 . . . . 15 . . . . 20 . . . . 25 . . . . 30 36
h-, b-, [7.142], r-[7.140], cr-, cav-[7.144], oc-GLP-1 [7.145]	H A E G T F T S D V S S Y L E G Q A A K E F I A W L V K G Ra
lo-GLP-1-I [7.154]	H A D G T F T S D V S S Y L K D Q A I K D F V D R L K A G Q V R R E
lo-GLP-1-II [7.154]	H A D G T Y T S D V S S Y L Q D Q A A K D F V S W L K A G R G R R E
sa-GLP-1 [7.156]	H A D G T Y T S N V S T Y L Q D Q A A K D F V S W L K S G R A
ic-GLP-1 [7.155]	H A D G T Y T S D V S S Y L Q D Q A A K D F I T W L K S G Q P

Release. The release of GLP-1-(7-36)-amide in the intestine is stimulated not only by orally applied glucose, but also by D-galactose, D-glucuronic acid, 3-O-methyl-D-glucose, maltose, sucrose or by maltitol [7.216]. Fructose, D-fucose, D-mannose, D-xylose and lactose has no effect [7.216]. The postprandial increase in the plasma level of GLP-1-(7-36)-amide is enhanced in postgastrectomy dumping syndrome. This may be the cause for the hyperinsulinemia and hypoglycemia observed in this disregulation [7.217].

Receptors. GLP-1-(7-36)-amide binds specifically to a cloned insulin producing cell line from the rat, RIN 5AH. Even concentrations of 10^{-6} M of glucagon, GLP-1-(1-36)-amide, or GIP do not displace GLP-1-(7-36)-amide from the receptor [7.218]. Dexamethasone reduces the number of GLP-1-(7-36)-amide receptors on the B cells [7.219].

Biological Effects. GLP-1-(7-36)-amide is more effective in stimulating cAMP in rat fundic and antral glands [7.220] and in a rat islet cell line [7.221] than GLP-1 or glucagon.

GLP-1-(7-36)-amide is more active than GLP-1 in the glucose dependent release of insulin from isolated pancreatic islet cells of the rat [7.221]. Tests on humans have shown that GLP-1-(7-36)-amide has the properties of inkretin: its formation in the gastrointestinal tract is enhanced after an oral glucose meal,it releases insulin in a glucose dependent process, and lowers the plasma levels of glucose and glucagon [7.217]. Indeed, it is more active than GIP. GLP-1-(7-36)-amide and GIP have a synergistic effect on the release of insulin [7.222]. GLP-1-(7-36)-amide inhibits the secretion of glucagon in the pancreas through the inhibition of post-transcriptional processes [7.223] and decreases the secretion of somatostatin (SRIF), but it has no effect on the secretion of gastrin and SRIF in the stomach [7.224].

Structure-Activity Relationships. The amide structure of GLP-1-(7-36)-amide is apparently not essential for the insulin stimulating activity. GLP-1-(7-37) is as active as GLP-1-(7-36)-amide [7.225].

Applications. In type 2 diabetes (non-insulin-dependent), GLP-1-(7-36)-amide stimulates the release of insulin and inhibits the secretion of glucagon and SRIF. Both in type I diabetes (insulin dependent) as well as in type II diabetes, the isoglycemic meal related insulin requirement is lowered [7.226].

7.5.4. Crustacean Hyperglycemic and Molt Inhibiting Hormone (CHH)

The hormone, car-CHH was isolated from the sinus glands of *Carcinus maenas*. It is a peptide containing 72 amino acids and three disulfide bridges between positions 7 and 43, 23 and 39, and 26 and 52 [7.227]. The molt inhibiting hormone from the sinus gland of the lobster *Homarus americanus* (hom-MIH), which, like car-CHH, also has a hyperglycemic effect, is structurally very similar to car-CHH [7.228].

```
          1                                                                    25
car-CHH   •Q I Y D T S C K G V Y D R A L F N D L E H V C D D-
hom-MIH   •Q V F D Q A C K G V Y D R N L F K K L D R V C E D-

          26                                                                   50
car-CHH   C Y N L Y R T S Y V A S A C R S N C Y S N L V F R-
hom-MIH   C Y N L Y R K P F V A T T C R E N C Y S N W V F R-

          51                                                         72
car-CHH   Q C M D D L L M M D E F D Q Y A R K V Q M Va.
hom-MIH   Q C L D D L L L S N V I D E Y V S N V Q M
```

7.6. Gastrin Inhibiting or Glucose Dependent Insulinotropic Peptide [7.229]

Occurrence. Gastrin inhibiting or glucose dependent insulinotropic peptide (GIP), a peptide containing 42 amino acids, is found in the duodenum, jejunum as well as in the A cells of the pancreas. Thus, it is found together with glucagon.

```
        1                                                25
h-GIP   Y A E G T F I S D Y S I A M D K I H Q Q D F V N W-
p-GIP   Y A E G T F I S D Y S I A M D K I R Q Q D F V N W-
b-GIP   Y A E G T F I S D Y S I A M D K I R Q Q D F V N W-

        26                              42
h-GIP   L L A Q K G K K N D W K H N I T Q
p-GIP   L L A Q K G K K S D W K H N I T Q
b-GIP   L L A Q K G K K S D W I H N I T Q
```

The peptide sequence GIP-(2-6) occurs in chymodenin.

Release. GIP is secreted in the duodenum and jejunum particularly in response to a diet rich in fat and carbohydrate. The pancreatic enzymes also play a role in this release [7.230]. During physical exercise, however, the secretion of GIP in response to orally applied glucose is greatly reduced [7.231]. Furthermore, GIP is enhanced by bombesin or low doses of motilin.

High doses of motilin, somatostatin and dexamethasone inhibit the release of GIP [7.232]. The raised level of GIP in insulin-dependent diabetes (type I diabetes) can be normalized by treatment with insulin [7.233]. This is, however, not the case with non-insulin-dependent diabetes mellitus (NIDDM) [7.234]. The postprandial plasma level of GIP is low in obese non-insulin-dependent diabetics [7.235]. On the other hand, obese hyperglycemic mice have an increased plasma GIP level [7.236].

Biological Effects. GIP inhibits the gastrin stimulated secretion of gastric acid, secretion of pepsin, and the motorial activity of the stomach. It stimulates the intestinal secretion and increases the glucose dependent release of insulin, glucagon, pancreatic peptide (PP) and somatostatin in the pancreas. cAMP is activted in this process. The GIP mediated release of insulin is dependent on phospholipase A_2 [7.237] and is linked to a cholinergic mechanism [7.238].

In adipocytes, GIP increase the receptor affinity for insulin [7.239] i.e., it potentiates the inhibitory effect of insulin on hepatic glycogenolysis [7.240].

The physiological importance of GIP is probably based more on its insulinotropic activity and not on the reduction of the gastric acid concentration. Apart from GLP-1-(7-36)-amide, GIP is an important candidate for the role of the postulated inkretin (a substance that stimulates the pancreatic endocrine secretion depending on the concentration of glucose).

Structure-Activity Relationships. The synthetic work carried out in the seventies was based on an incorrect structural formula. It was assumed that Gln^{29} was followed by another Gln. The activity and the immunological behavior of this endo-29a-Gln-GIP was very similar to that of GIP. Shortening the molecule at the N-terminal by only two amino acids reduces the insulin releasing effect to about 2%. Endo-29a-Gln-GIP-(1-38) and b-GIP-(1-39) [7.241] still possess full insulinotropic activity. GIP-(1-28) and GIP-(3-24) are no longer capable of releasing insulin [7.242].

7.6.1. Chymodenin

Chymodenin was isolated from the small intestine of the pig. It increases the content of chymotrypsinogen in the pancreatic juice [7.243] and appears to be identical to the b-heart cytochrome oxidase subunit PP VII [7.244], [7.245]. There is some similarity between the C-terminal region and the N-terminal end of GIP.

	1 25
b-PP VII [7.230]	A E D I Q A K I K N Y Q T A P D S R F P N Q N Q T-
	26 50
b-PP VII	R N C W Q N Y L D F H R C E K A M T A K G G D V S-
	51 75
b-PP VII	V C E W Y R R V Y K S L C P I S W V S T W D D R R-
	76 84
b-PP VII	A E G T F P G K I
p-GIP-(2-6)	A E G T F

7.7. Somatoliberin and the Growth Hormone Cascade

Somatoliberin (SRF) and galanine (GAL) stimulate somatotropin (STH), which, in trun, releases the somatomedins (SMe). Somatostatin (SRIF), which inhibits SRF and STH, is of great importance in the regulation of growth hormone. GAL inhibits the release of SRIF, which favours the release of SRF and STH. On the other hand, SRF and STH stimulate the synthesis of SRIF and thus permit a negative feedback,

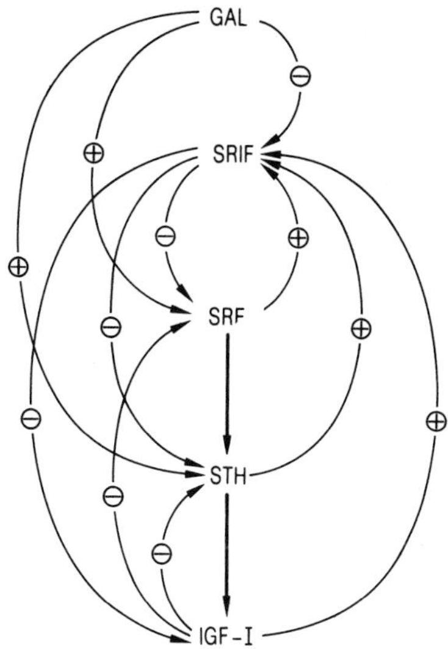

Figure 5. The growth hormone cascade
GAL = galanine; STH = growth hormone (somatotropin); SRF = growth hormone releasing hormone (somatoliberin); IGF-I = insulin-like growth factor I; SRIF = somatotropin release inhibiting hormone (somatotropin)

see Figure 5. A synthetic hexapeptide (SK&F 110679) also releases STH from the hypophysis, independently of SRF receptors.

There is great structural similarity between STH and placental lactogen, somatolactin, prolactin, proliferin and prolactin related or like proteins. The insulin like growth factors, IGF-I (somatomedin C) and IGF-II, and somatomedin B, which is structurally similar to vitronectin and trypsin inhibitors, belong to the group of somatomedins. Insuline and relaxin are structurally related to IGF-I and IGF-II. Apart from somatostatin and galanine, pancreastatin is an inhibitor of insulin release.

7.7.1. Galanine [7.246]

Occurrence. Galanine (GAL), was isolated from the intestine of the pig. The structure of this peptide has been elucidated [7.247]. It was named for its N- and C-terminal amino acids (glycine and alanine). It is a 29 amino acid peptide amide with a C-terminus that is similar to the neurokinins, e.g., neurokinin A(NKA) or Substance P (SP).

The peptides, p- and b-GAL are processed in the from of a prepro-peptide containing 123 amino acids [7.248].

	1									10										20									29
p-GAL	G	W	T	L	N	S	A	G	Y	L	L	G	P	H	A	I	D	N	H	R	S	F	H	D	K	Y	G	L	Aa
b-GAL	G	W	T	L	N	S	A	G	Y	L	L	G	P	H	A	L	D	S	H	R	S	F	Q	D	K	H	G	L	Aa
NKA																			H	K	T	/	D	S	F	V	G	L	Ma
SP																		R	P	K	P	/	Q	Q	F	F	G	L	Ma

In humans, GAL is not only found in the nerve cells of the small intestine, but also in the brain, spinal cord, pancreas, bladder, hypophysis [7.249], adrenal gland [7.250], and in the genital tract of men and women [7.251]. In the central nervous system, the highest concentrations are found in the hypothalamus [7.252] and in the median eminence. The GAL containing nerve cells in the pancreas are in the direct vicinity of the islet cells [7.253]. Unusally high concentrations of GAL are found in pheochromocytomas [7.254].

The calcitonin gene related peptide (CGRP) and SP have also been found in cell bodies with material similar to GAL. GAL is colocalized with gonadoliberin in neurons of the hypothalamus [7.255].

Release. The release of GAL in the anterior hypophysis [7.249], [7.256] and in the hypothalamus [7.257] is stimulated by estrogen [7.249], [7.257], [7.256] and the thyroid hormones, e.g., T_4 [7.256]. The formation of GAL decreases after the withdrawal of food [7.258] and treatment with capsaicin.

Biological Effects. GAL directly contracts preparations from the rat fundus, ileum, colon, bladder [7.247], jejunum [7.260], and from the lower esophageal spincter of the oppossum [7.261]. It potentiates the contraction of the rat vas deferens that is induced by electrical stimulation or by norepinephrine [7.262], [7.263]. In strips of guinea pig colon, however, GAL inhibits the contraction induced by electrical stimulation and caused via SP and acetylcholine. It also inhibits the neurally evoked circular muscle contractions in the guinea pig ileum [7.264] and the motility of the small intestine of the rabbit [7.265]. GAL inhibits the release of acteylcholine and SP from motorial nerves [7.263] and the synaptic transmission at the cholinergic nicotine synapses in the myenteric plexus [7.266]. In the rat lung, GAL, like NPY, increases the level of LTD_4 [7.267].

In humans, GAL inhibits the gastrointestinal motility, delays gastric emtying, and increases the mouth to cecum transit time. The postprandial increase in glucose, PYY, insulin, neurotensin, enteroglucagon, glucagon, pancreatic peptide (PP), and somatostatin (SRIF) is suppressed by GAL [7.268]. The GAL stimulated inhibition of the release of insulin has also been observed *in vitro* [7.253], [7.269], [7.270]. GAL directly inhibits the exocytosis of insulin via G proteins [7.271], [7.272]. Glucagon is stimulated *in vitro.* In mice, GAL inhibits the basal and stimulated insulin level, increases the plasma glucagon level temporarily and causes hyperglycemia [7.273]. In the perfused porcine pancreas, however, GAL increases the release of insulin and inhibits the secretion of SRIF [7.274].

GAL inhibits the release of gastrin and somatostatin from isolated perfused rat stomachs [7.275]. It also inhibits the secretion of gastric acid which is stimulated by pentagastrin in rats [7.276] and by bombesin and 2-deoxy-D-glucose in dogs that are awake [7.277].

The i.c.v. application of GAL to rats or the i.v. infusion in humans results in a great increase in the plasma levels of somatotropin (STH) and prolactin (PRL) (lower) [7.252] via the stimulation of epinephrine [7.278], somatoliberin (SRF) [7.279] or the inhibition of SRIF. GAL also releases STH and PRL from cultivated hypophyseal cells [7.280], [7.281] and acts synergistically with SRF [7.280]. GAL increases the secretion of gonadoliberin from the nerve ends of the median eminence.

In rats, injections of GAL into the paraventricular nucleus cause an increase in the food uptake. While NPY primarily stimulates the carbohydrate uptake via α_2 noradrenergic receptors, GAL favor the uptake of fat. After the withdrawal of fat, it favours the uptake of carbohydrate [7.282]. GAL protects hippocampal neurons from the functional effects of anoxia [7.283]. In mice, GAL exerts an antinociceptive effect via endogenous opioides when applied intrathecally [7.284].

Structure-Activity Relationships. It is possible to shorten the molecule at the C-terminus without loss of activity, while N-terminal deletions result in complete loss of activity.

GAL-(1-10) strongly contracts the longitudinal muscle of the rat jejunum, but it does not inhibit the electrically stimulated contraction of guinea pig colon strips [7.260]. The action of GAL-(1-20) is species and muscle dependent, similar to that of GAL or weaker [7.285]. While GAL-(1-16) has an agonistic effect on the rat hippocampal GAL receptor, GAL-(3-29) [7.286] and GAL-(17-29) [7.287] are inactive. The effect on the guinea pig ileum is as follows:

GAL > GAL-(2-9) > GAL-(1-15) [7.288]. In rats, the pentagastrin stimulated secretion of gastric acid is inhibited by GAL-(9-29) with 5% of the GAL activity, while GAL-(15-29) is devoid of activity [7.276].

7.7.2. Somatoliberin [7.289], [7.290]

Occurrence. The characterization of human prepro-somatoliberin (h-prepro-SRF) confirmed that h-SRF is a peptide amide containing 44 amino acids. It was isolated from a pancreatic tumor along with smaller peptides (1-40 and 1-37).

Although SRF is mainly synthesized in the hypothalamus, i.r.-SRF has also been found in the gastrointestinal tract (especially in the jejunum), placenta, human and rat milk and in spinal fluid [7.291].

	1																								25
h-SRF	Y	A	D	A	I	F	T	N	S	Y	R	K	V	L	G	Q	L	S	A	R	K	L	L	Q	D-
b-, cap-SRF	Y	A	D	A	I	F	T	N	S	Y	R	K	V	L	G	Q	L	S	A	R	K	L	L	Q	D-
o-SRF	Y	A	D	A	I	F	T	N	S	Y	R	K	I	L	G	Q	L	S	A	R	K	L	L	Q	D-
r-SRF	H	A	D	A	I	F	T	S	S	Y	R	R	I	L	G	Q	L	Y	A	R	K	L	L	H	E-

	26			29															44
h-SRF	I	M	S	R	Q	Q	G	E	S	N	Q	E	R	G	A	R	A	R	La
b-, cap-SRF	I	M	N	R	Q	Q	G	E	R	N	Q	E	Q	G	A	K	V	R	La
o-SRF	I	M	N	R	Q	Q	G	E	R	N	Q	E	Q	G	A	K	V	R	La
r-SRF	I	M	N	R	Q	Q	G	E	R	N	Q	E	Q	R	S	R	F	N.	

In the neurons of the arcuate nucleus, SRF is associated with NPY [7.292], dopamine [7.293] and the dopamine marker, tyrosine hydroxylase [7.294].

During the processing of prepro-h-SRF, the cryptic peptide, prepro-h-SRF-(78-

107)-amide with a pyroglutamyl residue at the N-terminus (CTPG or anorectin) is formed. This peptide reduces the intake of food when injected into the third ventricle [7.295].

h-CTPG: *Q V D S M W A E Q K Q M E L E S I L V A L L Q K H S R N S Q

Release. The release of SRF is stimulated by central α_2-adrenergic stimuli (clonidin), high potassium concentrations, GABA, opioids (FK 33-824, β-endorphin and morphine), galanine [7.296], dopa (ral) [7.297], thyroid hormones and by a glucose meal [7.298]. More SRF is found in the hypothalamus of male rats than in that of female rats [7.299]. The plasma level of i.r. SRF increases greatly during puberty (more so in girls than in boys) [7.300].

Somatostatin (SRIF) and pentobarbital anesthesia have an inhibitory effect on the secretion of SRF. Somatotropin (STH) [7.301], insulin like growth factor IGF I [7.302], [7.303] and IGF II [7.303], SRF itself [7.304] and dopamine also inhibit the release of SRF via the stimulation of SRIF. It is assumed that this SRF stimulated secretion of SRIF is regulated by β-endorphin containing neurons [7.305].

The plasma level of SRF is lower in children suffering from dwarfism and in acromegaly [7.306]. Less SRF is found in the hypothalamus of old rats than of young rats [7.307].

Receptors. Chronic infusion (*in vivo*) of SRF [7.308] and adrenalectomy [7.309] lead to the down regulation of SRF receptors in the rat.

Biological Effects. Especially on pulsatile application, SRF stimulates both the transcription of the STH gene as well as the release of STH [7.310] in the hypophysis via an increase in cAMP, arachidonic acid and PGE_2. Chronic application does not produce a total inhibition of the release of STH [7.311]. It is assumed that melatonin [7.312], cholinergic agonists (e.g., pilocarpines) [7.313] and β_1 adrenergic receptor blockers (e.g., atenolol [7.314]), hypoglycemia [7.315], testosterone [7.290] and VIP increase the SRF stimulated STH level via the inhibition of SRIF. The SRF stimulated release of STH is inhibited by SRIF, cholinergic antagonists (pirenzepine or atropine) [7.313], α-receptor blockers [7.316], dopamine [7.317], hyperglycemia [7.318] and by IGF I and II [7.303]. The SRF induced release of STH is reduced in children with thalassemia major [7.319], in patients suffering from noninsulin-dependent diabetes (type II) [7.320], hypothyroidism [7.321], Cushing's syndrome [7.322] or depression [7.323], in test persons with an increased plasma level of free fatty acids (induced by a lipid-heparin infusion) [7.324] and in obesity [7.290]. On the other hand, the SRF stimulated secretion of STH is raised in streptozotocin diabetic rats [7.325], in diabetics with retinopathy [7.326], in acromegaly and in young Alzheimer patients [7.290]. In patients with hypothyroidism, T_4 potentiates the SRF induced release of STH [7.321].

Glucocorticoids exert a biphasic effect on the release of STH. They potentiate SRF activity by stabilizing the SRF receptors and enhancing the transcription rate of the STH gene. In cases of chronic misuse of glucocorticoids, however, an inhibition of the release of STH is observed in humans as well as in experimental animals [7.327], [7.328].

Other Effects of SRF

Release of protein 7B2 from GH_1 cells [7.329]
Release of calcitonin and neurotensin from rat C-cells [7.330]
Inhibition of TRH release from the rat hypothalamus [7.331]
Acceleration of glucose transport in isolated rat adipocytes [7.332]
Proliferation of endothelial cells in the mucosa of the fundus and antrum [7.333]
Increase in the production of milk in cows and sheep
Positive inotropic effect [7.334]
Increase in the REM sleep (i.c.v. applied) [7.335]
Stimulation of histamine release [7.336]

Structure-Activity Relationships. SRF can be shortened at the C-terminus. In rats, h-SRF-(1-29)-amide still has 50% of the activity of h-SRF-(1-44)-amide. However, h-SRF-(1-29)-amide has the same activity as h-SRF-(1-44)-amide in children [7.337]. Further shortening at the C-terminus results in a significant reduction of activity [7.338]. The N-terminal amino group is not important for the biological activity because its acetylation gives more potent compounds.

Structure-activity relationships show that position 1 should be occupied by an aromatic amino acid [7.339]. Nle^{27} and D-Ala^2 were found to increase activity [7.340], [7.341]. Although [D-Ala^2]SRF-(1-29)-amide had 50 times the activity in some tests and Ac-[D-Tyr^1,D-Ala^2]-SRF-(1-29)-amide had 38 times the SRF-(1-29)-amide activity in the rat, these compounds did not exhibit a higher activity in humans [7.342]. The replacement of Arg^{29} by agmatin is of advantage for the activity. Potent analogs that act for a long time are [D-Ala^2, Nle^{27}]h-SRF-(1-28)-Agm [7.343] and [de-amino-Tyr^1,Ala^{15},Nle^{27}]SRF-(1-28)-Agm [7.344].

[D-Arg^2]h-SRF-(1-29)-amide) exhibits SRF antagonistic activity in concentration of 10^{-7} M [7.340]. More potent antagonists are [D-$Arg^{2,29}$,Arg^{30}]h-SRF-(1-30)-amide (3.64 times as active as [D- Arg^2]h-SRF-(1-29)-amide) and Ac-[His^1,D-Arg^2,Ala^{15}]r-SRF-(1-29)-amide (27.7 times as active as [D-Arg^2]h-SRF-(1-29)-amide [7.345]).

An antagonist that acts in the 10 μM range is [1-Ser^9-ψ-(CH_2NH)-Tyr^{10}]SRF-(1-29)-amide [7.346]. Acetyl-[D-Phe^2]h-SRF-(1-29)-amide is a VIP antagonist [7.347]. It reduces the SRF and VIP stimulated adenylate cyclase, but not the secretin stimulated adenylate cyclase and inhibits the receptor binding of VIP and SRF to rat pancreatic membranes.

The activity of h-SRF-(1-27)-amide can also be increased drastically. [D-Asp^3,D-Asn^8,Leu^{27}]h-SRF-(1-27)-amide is especially active, having twice the activity of h-SRF-(1-29)-amide [7.341].

SRF peptides with more amphiphilic and α-helical regions have up to twice the biological activity [7.348], [7.349]. The activity of SRF is species specific: human sequences are more active in humans than in the rat and vice versa [7.350].

Applications. It is possible to differentiate between hypothalamic and hypophyseal dependent STH deficiency diseases by using SRF as a diagnostic aid. Patients who have a defective hypothalamus usually react to SRF. The older the patient, i.e., the longer the endogenous SRF deficiency disease exists, the poorer the reaction of the hypophysis to SRF.

The application of SRF in the treatment of children with STH deficiency is of great interest. The linear growth of children could be enhanced in a study involving the s.c. application of 250 μg of SRF-(1-29)-amide twice a day [7.351] or after pulsatile treatment (s.c. application of 1-3 μg/kg every 3 hours) [7.352]. However, children suffering from dwarfism responded better to h-STH than to h-SRF-(1-29)-amide [7.353]. In women after menopause, SRF has a bone forming, plasma osteocalcin increasing effect [7.354].

7.7.2.1. SK&F 110679

Of the large number of smaller synthetic peptides [7.355] one of the most active is the hexapeptide amide SK&F 110679, which has about 10% of the somatoliberin (SRF) activity [7.356], [7.357], [7.358]. There is no structural similarity between these peptides and SRF.

SK&F 110679 : H w A W f Ka

It releases somatotropin (STH) from hypophyseal cell cultures *in vitro* and acts synergistically with SRF. Thus, it evidently does not bind to the same receptor as SRF [7.359]. SK&F 110679 bound to membrane fragments of the rat hypothalamus and the rat hypophysis can be displaced by r-SRF, VIP, VIP antagonists, SRIF, dynorphin A and ACTH-(1-24) [7.360]. In humans, doses as low as 0.25-1 μg/kg/30 min stimulate the release of STH. It appears to be more effective here than SRF [7.361], [7.362]. In primates oral doses (ED_{50} = 0.75 mg/kg) also cause STH release [7.363].

7.7.3. Somatotropin, Placenta Lactogen and Prolactin [7.364]

Prolactin (PRL) and somatotropin (STH) from the hypophysis, and placenta lactogen (PL) and two STH variants (STH-V and STH-V2) [7.365] from the placenta are structurally related, see Table 17. PL, which is more similar to STH than to PRL, possesses not only its lactogenic properties, but also about 13% of the STH activity in the Tibia test. On the other hand, PRL exhibits no somatotropic activity. However, STH has weak PRL activity. STH-V2 contains an intron of 84 amino acids after Trp^{126} of STH-V and a reading frame which is shifted by one base [7.365].

Another peptide of similar structure is somatolactin (pl-SL), which is isolated from the hypopysis of the flounder. Its structure is midway betwen that of STH and PRL [7.366]. Other substances related to prolactin are proliferin (m-PLF), prolactin related cDNA (b-PRC I, II, and III), proliferin-related protein (m-PRP) and prolactin-like proteins (r-PLP A and B, b-PLP I, II, and III) [7.367].

There is great similarity in the structures of secretin (SEC), h-STH- and h-PL-(1-16).

h-STH-(1-16)	F	P	T	I	P	L	S	R	L	F	D	N	A	M	L	R
H-PL-(1-16)	V	Q	T	V	P	L	S	R	L	F	D	H	A	M	L	Q
p-SEC-(5-20)	T	F	T	S	E	L	S	R	L	R	D	S	A	R	L	Q

On the other hand, there is amino acid homology between h-STH-(30-38), gastrin (GT) and the C-terminal end of the insulin B chain.

h-GT-11	E	E	E	E	A	Y	G	W	M	D	Fa
h-STH-(30-38)	E	F	E	E	A	Y	I	P	K	E	Q
h-INS-B-(21-30)	E	R	G	F	F	Y	T	P	K	T	

7.7.3.1. Somatotropin [7.364]

Occurrence. Somatotropin is formed in the hypophysis as pro-STH. Circulating STH occurs in several forms: monomeric (little STH), dimeric (big STH) and oligomeric (big-big STH, MW > 60000) [7.368]. In addition, 20 K-h-STH, a de-(32-46)-STH, and the two STH variants from the placenta, STH-V and STH-V2, [7.365] have also been found.

Table 17. STH sequences of various species [7.368a], [7.368b], [7.366]

Protein	Amino acid sequence
	1 … 25
h-STH	F P T I P L S R L F D N A M L R A H R L H Q L A F-
h-STH-V [7.365]	F P T I P L S R L F D N A M L R A R R L Y Q L A Y-
cap-STH	F P A M S L S G L F A N A V L R A Q H L H Q L A A-
b-STH	F P A M S L S G L F A N A V L R A Q H L H Q L A A-
el-STH [7.368c]	F P A M P L S S L F A N A V L R A Q H L H Q L A A-
r-STH	F P A M P L S S L F A N A V L R A Q H L H Q L A A-
mu-STH [7.368d]	F P A M P L S S L F A N A V L R A Q H L H Q L A A-
pl-STH	Q P I T E N Q R L F S I A V G R V Q Y L H L V A K-
se-STH	Q P I T D S Q H L F S I A V S R I Q N L H L L A Q-
sco-STH	Q P I T D S Q R L F S I A V S R V Q H L H L L A Q-
sa-STH	I E N Q R L F N I A V S R V Q H L H L L A Q-
g-STH	T F P A M P L S N L F A N A V L R A Q H L H L L A A-
h-PL	V Q T V P L S R L F D H A M L Q A H R A H Q L A I-
	26 … 50
h-STH	D T Y Q E F E E A Y I P K E Q K Y S F L Q N P Q T-
h-STH-V	D T Y Q E F E E A Y I L K E Q K Y S F L Q N P Q T-
cap-STH	D T F K E F E R T A I P E G Q R Y S T / Q N T Q V-
b-STH	D T F K E F E R T A I P E G Q R Y S T / Q N T Q V-
el-STH	D T Y K E F E R A Y I P E G Q R Y S I / Q N A Q A-
r-STH	D T Y K E F E R A A I P E G Q R Y S T / Q N A Q A-
mu-STH	D T Y K D F E R A Y I P E G Q R Y S I / Q N A Q A-
pl-STH	K L F S D F E N S L Q L E D Q R L L N / K I A S K-
se-STH	R L F S N F E S T L Q T E D Q R Q L N / K I F L Q-

Protein	Amino acid sequence
	26 … 50
sco-STH	R L F S D F E S S L Q T E E Q R Q L N L / I F L Q-
sa-STH	K M F N D F D G T L L P D E R R Q L N / K I F L L-
g-STH	E T Y K E F E R T Y I P E D Q R Y T N / K N S Q A-
h-PL	D T Y Q E F E E T A I P K D Q K Y S F L H D S Q T-
	51 … 75
h-STH	S L C F S E S I P T P S N R E E T Q Q K S N L E L-
h-STH-V	S L C F S E S I P T P S N R V K T Q Q K S N L E L-
cap-STH	A F C F S E T I P A P T G K N E A Q Q K S D L E L-
b-STH	A F C F S E T I P A P T G K N E A Q Q K S D L E L-
el-STH	A F C F S E T I P A P T G K D E A Q Q R S D V E L-
r-STH	A F C F S E T I P A P T G K E E A Q Q R T D M E L-
mu-STH	A F C F S E T I P A P T G K D E A Q Q R S D M E L-
pl-STH	E F C H S D N F L S P I D K H E T Q G S S V Q K L-
se-STH	D F C N S D Y I I S P I D K H E T Q R S S V L K L-
sco-STH	D F C N S D Y I I S P I D L H E T Q R S S V L K L-
sa-STH	D F C N S D S I V S P V D K H E T Q K S S V L K L-
g-STH	A F C Y S E T I P A P T G K D D A Q Q K S D M E L-
h-PL	S F C F S D S I P T P S N M E E T Q Q K S N L E L-
	76 … 100
h-STH	L R I S L L L I Q S W L E P V Q F L R S V F A N S-
h-STH-V	L R I S L L L I Q S W L E P V Q L L R S V F A N S-
cap-STH	L R I S L L L I Q S W L G P L Q F L S R V F T N S-
b-STH	L R I S L L L I Q S W L G P L Q F L S R V F T N S-
el-STH	L R F S L L L I Q S W L G P V Q F L S R V F T N S-
r-STH	L R F S L L L I Q S W L G P V Q F L S R I F T N S-
mu-STH	L R F S L L L I Q S W L G P V Q F L S R V F T N S-
pl-STH	L S V S Y R L I E S W E F F S R F L V A S F / / A-
se-STH	L S I S Y R L V E S W E F S S R F L S G G S / / A-
sco-STH	L S I S Y R L V E S W E F P S R S L S G G S / / A-
sa-STH	L H I S F R L I E S W E Y P S Q T L / / I I S N S-
g-STH	L R F S L V L I Q S W L T P V Q Y L S K V F T N N-
h-PL	L R I S L L L I E S W L E P V R F L R S M F A N N-
	101 … 125
h-STH	L V Y G A S D S N V Y D L L K D L E E G I Q T L M-
h-STH-V	L V Y G A S D S N V Y R H L K D L E E G I Q T L M-
cap-STH	L V F G T S D / R V Y E K L K D L E E G I L A L M-
b-STH	L V F G T S D / R V Y E K L K D L E E G I L A L M-
el-STH	L V F G T S D / R V Y E K L K D L E E G I Q A L M-
r-STH	L M F G T S D / R V Y E K L K D L E E G I Q A L M-
mu-STH	L V F G T S D / R V Y E K L K D L E E G I Q A L M-
pl-STH	V / / R / / T Q V T / S K L S E L K M G L L K L I-
se-STH	L / / R / / N Q I S / P R L S E L K T G I Q L L I-
sco-STH	P / / R / / N Q I S / P K L S E L K T G I H L L I-
sa-STH	L M V R N A N Q I S E / K L S D L K V G I N L L I-
g-STH	L V F G T S D / R V F E K L K D L E E G I Q A L M-
h-PL	L V Y D T S D S D D Y H L L K D L Q E G I E T L M-

Protein	Amino acid sequence
	126 150
h-STH	G R L E D G / / S P R T G Q I F K Q T Y S K F D T N S-
h-STH-V	W R L E D G / / S P R T G Q I F N Q S Y S K F D T K S-
cap-STH	R E L E D V / / T P R A G Q I L K Q T Y D K F D T N M-
b-STH	R E L E D G / / T P R A G Q I L K Q T Y D K F D T N M-
el-STH	R E L E D G / / S P R P G Q V L K Q T Y D K F D T N M-
r-STH	Q E L E D G / / S P R I G Q I L K Q T Y D K F D A N M-
mu-STH	R E L E D G / / S P R A G P I L K Q T Y D K F D T N L-
pl-STH	E A N Q D G A G G F S E S S V L Q L T P Y G / / / / /-
se-STH	T A N Q D G A E M F S D V S A L Q L A P Y G N F Y Q S-
sco-STH	R A N Q D G A E M F A D S S A L Q L A P Y G N Y Y Q S-
sa-STH	T G S Q D G V L S L D D N D S Q Q L P P Y G N Y Y Q N-
g-STH	R E L E D R / / S P R G P Q L L R P T Y D K F D I H I-
h-PL	G R L E D G / / S R R T G Q I L K Q T Y S K F D T N S-
	151 175
h-STH	H N D D A L L K / N Y G L L Y C F R K D M D K V E T-
h-STH-V	H N D D A L L K / N Y G L L Y C F R K D M D K V E T-
cap-STH	R S D D A L L K / N Y G L L S C F R K D L H K T E T-
b-STH	R S D D A L L K / N Y G L L S C F R K D L H K T E T-
el-STH	R S D D A L L K / N Y G L L S C F K K D L H K A E T-
r-STH	R S D D A L L K / N Y G L L S C F K K D L H K A E T-
mu-STH	R S D D A L L K / N Y G L L S C F K K D L H K A E T-
pl-STH	/ / / / / / / / / N S E L F A C F K K D M H K V E T-
se-STH	L G G E E L L R R N Y E L L A C F K K D M H K V E T-
sco-STH	L G A D E S L R R S Y E L L A C F K K D M H K V E T-
sa-STH	L G G D G N V R R N Y E L L A C F K K D M H K V E T-
g-STH	R N E D A L L K / N Y G L L S C F K K D L H K V E T-
h-PL	H N H D A L L K / N Y G L L Y C F R K D M D K V E T-
	176 191
h-STH	F L R I V Q C R / S V E G S C G F.
h-STH-V	F L R I V Q C R / S V E G S C G F.
cap-STH	Y L R V M K C R R F G E A S C A F.
b-STH	Y L R V M K C R R F G E A S C A F.
el-STH	Y L R V M K C R R F V E S S C A F.
r-STH	Y L R V M K C R R F A E S S C A F.
mu-STH	Y L R V M K C R R F V E S S C A F.
pl-STH	Y L T V A K C R L F P E A N C T L.
se-STH	Y L T V A K C R L S P E A N C T L.
sco-STH	Y L T V A K C R L S P E A N C T L.
sa-STH	Y L T V A K C R K S L E A N C T L.
g-STH	Y L K V M K C R R F G E S N C T I.
h-PL	F L R M V Q C R / S V E G S C G F.

Release [7.369]. STH is regulated primarily by the hypothalamic hormones galanine (GAL), somatoliberin (SRF), and somatostatin (SRIF). GAL and SRF stimulate and SRIF inhibits the release of STH from the hypopysis. If the SRIF infusion in rats is subjected to short interruptions, STH bursts are observed which lead to an increase in weight [7.370]. Prior treatment wih SRIF also results in a greater SRF stimulated release of STH [7.371]. In addition, STH is released in response to stimulation of α_1 adrenergic receptors [7.372] and cholinergic muscarin receptors [7.373], inhibition of α_2-adrenergic receptors [7.374], SKF-110679, insulin, hypoglycemia, pharmacological doses of glucagon, arginine infusion, arginine pyroglutamate [7.375], physical stress [7.376], α-melanotropin (α-MSH), opioids, e.g., dermorphin [7.377] or guanidyl-DAMME [7.378], angiotensin II [7.379], pentagastrin [7.380], GIP, neurotensin, combined application of thyroliberin (TRH) and ACTH [7.381] or of TRH/gonadoliberin (GnRH) and histamine [7.382], slow wave sleep [7.383], i.c.v. application of delta sleep inducing peptide (DSIP) via cholinergic [7.383], and dopaminergic [7.384] mechanisms, interleukin-1β (IL-1β) [7.385], serotonin [7.386] dopamine and dopamine agonists [7.387] and to indomethacin [7.388].

In women, the release of STH is dependent on estrogen. Thus, the plasma level of STH falls during menopause [7.389] and in suppression therapy with high doses of GnRH agonists [7.390]. In hypogonadal young males, testosterone stimulates the formation of STH and growth [7.391]. Glucocorticoids also regulate the synthesis of STH. Both hypoadrenalism [7.392] as well as high doses of glucocorticoids [7.393] reduce the release of STH. However, glucocorticoids and triiodothyronine (T_3) enhance the release of STH *in vitro* through SRF and increase the mRNA for STH.

The release of STH is inhibited by recombinant Met-STH [7.394], centrally applied gastrin releasing peptide (GRP) [7.395], corticoliberin [7.396], calcitonin [7.397], calcitonin gene related peptide (CGRP) [7.398] and by TRH [7.399], mainly by the secretion of SRIF.

While plasma STH levels in persons with acromegaly and type I diabetes [7.400] are higher than in healthy persons, they are reduced in depressive patients with an increased level of somatomedin C [7.401]. While TRH has no influence on the release of STH in normal test persons, it causes a paradoxical increase in the plasma level of STH in patients suffering from acromegaly [7.402], depressions [7.403] and metastatic testicular cancer [7.404]. In depressive patients, this increase can be inhibited by the dopamine antagonist domperidon [7.403] or by the cholinergic muscarin receptor antagonist pirenzepine [7.405]. In type I and type II diabetics, Gn-RH also raises the plasma STH level [7.406].

Receptors. The amino acid sequences of the STH receptors (STHR) from humans, rabbits [7.407], and rats [7.408] have been elucidated via the cDNA.

STH binding proteins are mainly the extracellular fragments of STHR [7.409]. In patients with Laron type dwarfism (high plasma STH level and low IGF I level), the STH receptor is defective [7.410] and the high affinity STH binding protein is missing [7.411]. African Pygmies have low levels of STH binding protein [7.412].

Biological Effects. STH is species specific and promotes the longitudinal growth and regeneration of bone. It exerts a strong mitogenic effect on osteoblasts [7.413] and on pancreatic β cells [7.414], raises the osteocalcin level [7.415], and enhances DNA/RNA and protein synthesis. It possesses lipolytic, diabetogenic and insulinotropic properties, and has an antidiuretic and antinatriuretic effect via the activation of renin and the increase in the plasma adosterone level [7.416], [7.417]. Together with relaxin and estrogen, STH also plays a role in the relaxation of the interpubic disc.

In a negative feedback mechanism, STH inhibits the release of SRF in the hypothalamus [7.418] via the release of SRIF [7.419].

In particular, pulsatile intravenous STH stimulates the formation of somatomedin C [7.420] not only in the liver but also in skeletal tissue. This substance is identical to the insulin like growth factor I (IGF I) and is largely responsible for the growth effect of STH [7.421].

In vitro, STH stimulates the release of insulin from pancreatic islet cells [7.422] and the uptake and oxidation of glucose in adipose and muscle tissues in an insulin dependent process [7.423]. The insulin like effect of STH is found in vivo only in STH deficiency models, e.g., in hypophysectomized animals or in children suffering from STH deficiency. However, chronic therapy with STH or very high doses of STH produce a diabetogenic effect in the tissues of hypophysectomized animals. Thus, after longer STH treatment the fasting plasma glucose level in humans is raised [7.424] and in patients with acromegaly, glucose metabolism is reduced in spite of the increased release of insulin [7.425]. The somatomedins are apparently not involved in the insulin like or diabetogenic activity of STH [7.425].

STH potentiates the menopausal gonadotropin stimulated release of inhibin and estradiol [7.426] and the h-CG stimulated synthesis of cAMP and testosterone in Leydig cells of the rat [7.427].

Similar to the marcophage activating interferon-γ, physiological doses of STH also stimulate the production of peroxide aninons in macrophages. STH increases the activity of cytolytic T cells and of natural killer cells, stimulates the formation of antibodies and the synthesis of thymulin [7.428], and regenerates the thymus in dogs [7.429].

Structure-Activity Relationships. Similar to h-STH, recombinant Met-h-STH, which acts like natural h-STH in children with STH deficiency [7.430], has both diabetogenic and insulin like activity [7.431]. Only the tendency to antibody formation appears to be somewhat raised in Met-h-STH [7.432]. More acidic STH derivatives (deamidated h-STH) are as acive as h-STH [7.433].

Various fragments of STH exhibit biological activity in the tibia test in hypophysectomized animals, e.g., STH-(88-124), STH-(125-156), b-STH-(96-133), b-STH-(151-191), b-STH-(123-131), h-STH-(1-135) carboxymethylated at Cys^{53} and its cyanobromide fragment h-STH-(15-125) (but weak), de-(32-46)-h-STH (20K-h-STH) which can be isolated from the hypophysis, and de-(135-145)-h-STH (formed by trypsin treatment of h-STH at pH 6 in 90% glycerol). h-STH, but not 20K-h-STH, reduces the concentration of free fatty acids in children deficient in STH [7.434]. The binding of 20K-h-STH to lactogenic receptors is about 500 times less than that of STH [7.435]. The lactogenic activity of h-STH and o-PRL could be competitively inhibited by [Met^{14}]h-STH-(14-191) [7.436].

The peptides, h-STH-(4-15) and h-STH-(6-13) stimulate glucose oxidation, glucose uptake into the adipose tissue, the transport of 2-deoxyglucose in adipocytes, and the synthesis of glycogen [7.437]. The minimal STH sequence which has this insulin potentiating effect is h-STH-(8-13). An artefact, the Asp-imide variant of these peptides, appears to be responsible for this effect [7.438].

h-STH-(4-15)	I	P	L	S	R	L	F	D	N	A	M	L
h-STH-(6-13)			L	S	R	L	F	D	N	A		
h-STH-(8-13)					R	L	F	D	N	A		

Other insulinotropic STH fragments are h-STH-(31-44), h-STH-(32-38) [7.439] and h-STH-(32-46) [7.440], [7.441]. h-STH-(31-34) has no influence on blood glucose and plasma insulin levels, but chronic application of this compound to rats decreases the subcutaneous, perirenal, mesenteral and epididymal fat cotent. Lys-Asp (STH-(115-116)) also has lipolytic activity [7.442].

h-STH-(31-46)	F	E	E	A	Y	I	P	K	E	Q	K	Y	S	F	L	Q
h-STH-(31-34)	F	E	E	A												

Both h-STH-(177-191) and h-STH-(178-191) have diabetogenic activity.

h-STH-(177-191)	L	R	I	V	Q	C	R	S	V	E	G	S	C	G	F

Applications. Met-h-STH and h-STH (Genotropin [7.443]) are produced by genetic engineering and are now used instead of natural h-STH in the therapy of hypophyseal dwarfism.

h-STH has also been used to activate growth in persons with Turner's syndrome where growth is reduced in spite of normal STH formation [7.444]. h-STH also finds application for the healing of wounds [7.445] and for increasing the mass [7.446] and density of bones [7.447]. In older patients with a reduced plasma STH level, STH improves the psyche [7.448], increases the bone density, lean body mass and skin thickness and reduces the amount of adipose tissue [7.449].

7.7.3.2. Prolactin [7.450]

Occurrence. Prolactin (PRL) is formed in the hypophysis, partially in the phosphorylated form [7.451] glycosylated at Asn^{31} [7.452] or dimerized via disulfide bridges [7.452]. Other sites of synthesis are the placenta, myometrium, ovaries [7.453], and the hypothalamus. Apart from prolactin, proliferin (m-PLF), prolactin related cDNA (b-PRC I, II and III), proliferin related protein (m-PRP), and prolactin like proteins (r-PLP A and B, b-PLP I, II, and III) have been found [7.454]

Somatolactin (SL), first isolated from the flounder (pl-SL), is a zwitter between STH and PRL [7.455]

Release [7.458]. The pulsatile secretion of PRL from the hypophysis is regulated by hypothalamic factors. The release of PRL is stimulated by TRH and VIP and inhibited by dopamine. The C-terminal glycopetide of vasopressin-neurophysin, which contains 39 amino acids, acts as the specific PRL releasing factor [7.459]. PRL is also released in response to K^+ [7.460], hypocalcemia [7.461], 1,25$(OH)_2D_3$, [7.462], [7.463] helodermin and PHI-27 [7.464], cholecystokinin, insulin, arginine (via insulin?) [7.465], epidermal growth factor (EGF) [7.466], thymosin fraction 5 [7.467], beer [7.468], the sucking stimulus through endogenous opioids [7.469] (activation of μ-

Prolactin Sequences of various species and pl-somatolactin [7.456]

	1																												25
h-PRL	L	P	I	/	C	P	G	G	A	A	/	/	/	R	C	Q	V	T	L	R	D	/	L	F	D	R	A	V	V	L-
b-PRL	T	P	V	/	C	P	N	G	P	G	/	/	/	N	C	Q	V	S	L	R	D	/	L	F	D	R	A	V	M	V-
o-PRL	T	P	V	/	C	P	D	G	P	G	/	/	/	N	C	Q	V	S	L	R	D	/	L	F	D	R	A	V	M	V-
r-PRL	L	P	V	/	C	S	G	G	/	/	/	/	/	D	C	Q	T	P	L	P	E	/	L	F	D	R	V	V	M	L-
m-PRL	L	P	I	/	C	S	A	G	/	/	/	/	/	D	C	Q	T	S	L	R	E	/	L	F	D	R	V	V	I	L-
g-PRL	L	P	I	/	C	P	I	G	S	V	/	/	/	N	C	Q	V	S	L	G	E	/	L	F	D	R	A	V	K	L-
sa-PRL [7.457]																	I	G	L	S	D	/	L	M	E	R	A	S	Q	R-
cy-PRL																	V	G	L	N	D	/	L	L	E	R	A	S	E	L-
pl-SL	I	P	L	D	C	K	E	E	Q	G	S	L	S	R	C	P	S	I	S	Q	E	K	L	L	D	R	V	I	Q	H-

	26																								50
h-PRL	S	H	Y	I	H	N	L	S	S	E	M	F	S	E	F	D	K	R	Y	T	H	G	R	G	F-
b-PRL	S	H	Y	I	H	N	L	S	S	E	M	F	N	E	F	D	K	R	Y	A	Q	G	K	G	F-
o-PRL	S	H	Y	I	H	N	L	S	S	E	M	F	N	E	F	D	K	R	Y	A	Q	G	K	G	F-
r-PRL	S	H	Y	I	H	T	L	Y	T	D	M	F	I	E	F	D	K	Q	Y	V	Q	D	R	E	F-
m-PRL	S	H	Y	I	H	T	L	Y	T	D	M	F	I	E	F	D	K	Q	Y	V	Q	D	R	E	E-
g-PRL	S	H	Y	I	H	Y	L	S	S	E	I	F	N	E	F	D	E	R	Y	A	Q	G	R	G	F-
sa-PRL	S	D	K	L	H	S	L	S	T	S	L	T	K	D	L	D	S	H	F	P	P	M	G	R	V-
cy-PRL	S	D	K	L	H	S	L	S	T	S	L	T	N	D	L	D	S	H	F	P	P	V	G	R	V-
pl-SL	A	E	L	I	Y	R	V	S	E	E	S	C	S	M	F	E	E	M	F	V	P	F	P	L	R-

	51																										75
h-PRL	I	T	K	A	I	N	S	/	/	C	H	T	S	S	L	A	T	P	E	D	K	E	Q	A	Q	Q	M-
b-PRL	I	T	M	A	L	N	S	/	/	C	H	T	S	S	L	P	T	P	E	D	K	E	Q	A	Q	Q	T-
o-PRL	I	T	M	A	L	N	S	/	/	C	H	T	S	S	L	P	T	P	E	D	K	E	Q	A	Q	Q	T-
r-PRL	I	A	K	A	I	N	D	/	/	C	P	T	S	S	L	A	T	P	E	D	K	E	Q	A	Q	K	V-
m-PRL	M	V	K	V	I	N	D	/	/	C	P	T	S	S	L	A	T	P	E	D	K	E	Q	A	L	K	V-
g-PRL	I	T	K	A	V	N	G	/	/	C	H	T	S	S	L	T	T	P	E	D	K	E	Q	A	Q	Q	I-
sa-PRL	M	M	P	R	P	S	M	/	/	C	H	T	S	S	L	E	T	P	K	D	K	E	Q	A	L	K	V-
cy-PRL	M	M	P	R	P	S	M	/	/	C	H	T	S	S	L	Q	V	P	N	D	K	D	Q	A	L	K	V-
pl-SL	L	Q	R	N	Q	A	G	Y	A	C	I	T	K	A	L	P	I	P	S	S	K	S	E	I	Q	Q	I-

76 … 99

h-PRL N Q K D F L V S I L I L / R S W N E P L Y H L V T-
b-PRL H H E V L M S L I L G L L R S W N D P L Y H L V T-
o-PRL H H E V L M S L I L G L / R S W N D P L Y H L V T-
r-PRL P P E V L L N L I L S L V H S W N D P L F Q L I T-
m-PRL P P E V L L N L I L S L V Q S S S D P L F Q L I T-
g-PRL H H E D L L N L V V G V L R S W N D P L I H L A S-
sa-PRL S E N E L I S L A R Y L L L A W N D P L L L L S S-
cy-PRL P E D P L L S L A R S L L L A W S D P L A L L S S-
pl-SL S D T W L L H S V L M L V Q S W I E P L V Y L Q T-

100 … 149

h-PRL E V R G B Q E A P E A I L S K A V E I E E Q T K R-
b-PRL E V R G M K G A P D A I L S R A I E I E E E N K R-
o-PRL E V R G M K G V P D A I L S R A I E I E E E N K R-
r-PRL G L G G I H E A P D A I I S R A K E I E E Q N K R-
m-PRL G V G G I Q E A P E Y I L S R A K E I E E Q N K Q-
g-PRL E V Q R I K E A P D T I L W K A V E I E E Q N K R-
sa-PRL E A P T L P H T P S N G D I S S K I R E L Q D Y S-
cy-PRL E A S S L A H / P E R N T I D S K T K E L Q E N I-
pl-SL T L D R Y D N A P D M L L N K T K W V S D K L I S-

125 … 149

h-PRL L L E G M E L / I V S Q V H P E T K E D E I Y P V W-
b-PRL L L E G M E M / I F G Q V I P G A K E T E P Y P V W-
o-PRL L L E G M E M / I F G Q V I P G A K E T E P Y P V W-
r-PRL L L E G I E K / I I S Q A Y P E A K G N E I Y L V W-
m-PRL L L E G V B K / I I S Q A Y P E A K G N G I Y F V W-
g-PRL L L E G M E K / I V G R V H S G H A G N E I Y S H S-
sa-PRL K S L G D G L D I M V N K M G P S S Q Y I S S I P F-
cy-PRL N S L G A G L E H V F N K M D S T S D N L S S L P F-
pl-SRL L E Q G V V V / L I R K M L D E G M L T A T Y N E Q-

	150																								174
h-PRL	S	G	L	P	S	L	Q	M	A	D	E	S	E	R	L	S	A	Y	Y	N	L	L	H	C	L-
b-PRL	S	G	L	P	S	L	Q	T	K	D	E	D	A	R	Y	S	A	F	Y	N	L	L	H	C	L-
o-PRL	S	G	L	P	S	L	Q	T	K	D	E	D	A	R	H	S	A	F	Y	N	L	L	H	C	L-
r-PRL	S	Q	L	P	S	L	Q	G	V	D	E	E	S	K	D	L	A	F	Y	N	L	I	R	C	L-
m-PRL	S	Q	L	P	S	L	Q	G	V	D	E	E	S	K	I	L	S	L	R	N	T	I	R	C	L-
g-PRL	D	G	L	P	S	L	Q	L	A	D	E	D	S	R	L	F	A	F	Y	N	L	L	H	C	H-
sa-PRL	K	G	G	D	/	L	G	N	D	K	T	/	S	R	L	I	N	F	H	F	L	M	S	C	F-
cy-PRL	Y	T	N	S	/	L	G	E	D	K	T	/	S	R	L	V	N	F	H	F	L	L	S	C	F-
pl-SL	G	L	F	Q	Y	D	A	Q	P	D	M	L	E	S	V	M	R	D	Y	T	L	L	S	C	F-

	175																								198	
h-PRL	R	R	D	S	H	K	I	D	N	Y	L	K	L	L	K	C	R	I	I	/	/	H	N	N	N	C.	
b-PRL	R	R	D	S	S	K	I	D	T	Y	L	K	L	L	N	C	R	I	I	/	/	Y	N	N	N	C.	
o-PRL	R	R	D	S	S	K	I	D	T	Y	L	K	L	L	N	C	R	I	I	/	/	Y	N	N	N	C.	
r-PRL	R	R	D	S	H	K	V	D	N	Y	L	K	F	L	R	C	Q	I	V	/	/	H	K	N	N	C.	
m-PRL	R	R	D	S	H	K	V	D	N	F	L	K	V	L	R	C	Q	I	A	/	/	H	Q	N	N	C.	
g-PRL	R	R	D	S	H	K	I	D	N	Y	L	K	V	L	K	C	R	L	I	/	/	H	D	S	N	C.	
sa-PRL	R	R	D	S	H	K	I	D	S	F	L	K	V	L	R	C	R	A	T	N	M	R	P	E	T	C.	
cy-PRL	R	R	D	S	H	K	I	D	S	F	L	K	V	L	R	C	R	A	K	K	/	R	P	E	M	C.	
pl-SL	K	K	D	A	H	K	M	E	I	F	L	K	L	L	K	C	R	Q	T	D	/	K	Y	/	N	C	A.

[7.470] and κ-receptors [7.471] and VIP [7.472], stress and physical exercise [7.473], [7.474] through serotoninergic mechanisms [7.475], β-endorphin (central effect [7.476], [7.477]), AT II [7.478], [7.479], after a longer enalapril therapy [7.480], bradykinin (similar to TRH) [7.481], gonadoliberin (Gn-RH) [7.482] through the formation of AT II [7.483], neurotensin [7.484], tryptoline (gonadal dependent) [7.485], and to glutamate [7.486]. It is also released through the stimulation of α_2-adrenergic receptors [7.487] or histamine H_1 receptors [7.488]. In rats, hyperprolactinemia is induced by the transplantation of the hypophysis below the kidney.

Dopamine from the hypothalamus and hypophysis [7.489] which is released by PRL should be one of the most important inhibitors of PRL. Ascorbic acid potentiates the inhibitory effect of dopamine on the release of PRL in hypophyseal cell cultures [7.490] and δ-opiates stimulate the synthesis of dopamine in the brain of rats [7.491], i.e., they inhibit the release of PRL.

The secretion of PRL is also inhibited by hypercalcemia [7.492], the PRL inhibition factor (PIF) PRL, itself [7.493], stimulation of α_2-[7.487], [7.494], GABA-A [7.495] or histamine H_2 receptors [7.488], chronic nicotine intake [7.496], somatostatin (SRIF), SRIF analogs (e.g., RC-160) [7.497], calcitonin [7.498], [4.499], IL-1 [7.500], T_3 [7.501], Ni^{2+} ions [7.502], melatonin [7.503], glucorticoids [7.462], centrally [7.504] or subcutaneously [7.505] applied vasopressin (presumably via a dopaminergic pathway [7.506]), centrally applied α-MSH, CLIP [7.507] or by neurotensin. A competitive antagoist of PRL is a recombinant h-somatotropin, which lacks the 13 amino acids at the N-terminus [7.508].

The sexual steroid hormones modulate the release of PRL. Estrogen potentiates the secretion of PRL in experimental animals [7.509], [7.510], [7.511], but it can also potentiate the dopamine stimulated inhibition of PRL in humans. Even in cells of the hypophysis, 17-β-estradiol potentiates the SRIF stimulated inhibition of PRL [7.512]. However, prolactinomas can be induced in men by massive estrogen therapy [7.513]. The release of PRL from human tumors is also stimulated by estrogen [7.514]. In postmenopausal women, estradiol potentiates the release of PRL stimulated by i.v. Gn-RH [7.515] and TRH [7.516], but not by VIP. Testosterone appears to have an inhibitory effect on the secretion of PRL because the TRH stimulated release of PRL is greatly enhanced in delayed puberty (males) [7.517].

A period of breast feeding is followed by a reduced PRL level for a time span of at least 12-13 years [7.518], which could be connected with the reduced risk of getting breast cancer after pregnancies. Women who are not lactating have a pronounced PRL deficiency [7.519]. In humans, the PRL response to TRH decreases with age [7.520]. Patients who have chronic kidney damage and rats with experimentally induced kidney damage develop hyperprolactinemia. The plasma level of PRL is raisd in spontaneously hypertensive rats and in patients with essential hypertension. The PRL response to TRH in pre-menopausal women who suffer from fibrocystic breast disease is greaer than in healthy women [7.521].

Receptors. Receptors for PRL (PRLR) are widely distributed. They are found, e.g., in the mammary gland, liver, kidneys, brain (particularly in the hypothalamus [7.522]), prostata, testes, ovaries and in the lymphocytes [7.523]. Until now, the structures of r-PRLR [7.524] and cu-PRLR [7.525] have been elucidated. In particular, the structure of the longer cu-PRLR is very similar to that of STHR.

In the rat, the number of PRLR in the mammary gland increases on the day of estrus. Together with estradiol, LH stimualtes the formation of PRLR [7.526].The number of testicular PRL receptors in the golden hamster is increased by PRL, FSH and LH [7.527]. Both ovariectomy as well as hypophysectomy reduce the synthesis of PRLR [7.522], [7.527].

PRL is displaced from the receptors on lymphocytes by cyclosporine [7.523].

Biological Effects. In female mammals, PRL increases the production of milk and initiates maternal behaviour. The PRL stimualted biosynthesis of casein and lipid is reduced by inhibitors of phospholipase A_2 [7.528].

PRL stimulates the synthesis of dopamine and increases the density of dopamine receptors [7.529]. Thus, both the dopaminergic tone in the central nervous system and in the hypothalamus as well as the dopaminergic inhibition of aldosterone are increased in patients suffering from hyperprolactinism [7.530].

Similar to placenta lactogen, PRL also has luteotropic properties and stimulates the synthesis of progesterone [7.531], [7.532], [7.533] but inhibits the formation of estradiol and testosterone [7.533]. Thus, in rats suffering from hyperprolactinemia, the activities of 3-β-hydroxysteroid dehydrogenase, 17-hydroxylase, 17,20-desmolase, 17-ketosteroid reductase and aromatase are reduced [7.534]. The synthesis of gonadotropin in hyperprolactinemic rats is also decreased [7.535] because of the inhibition of estrogen [7.536] and the down regulation of the GnRH receptors in the hypopysis [7.537]. Even in men with chronically raised PRL levels, the plasma level of testosterone is lowered and the excretion of LH in the urine is reduced [7.538]. In women, hyperprolactinemia leads to irregular menstruation or even amenorrhea [7.539].

PRL causes an increase in the blood pressure of experimental animals which is dependent on the adrenal glands [7.540] and presumably occurs via the potentiation of noradrenaline activity. As a real stress hormone, PRL potentiates the antidiuretic activity of vasopressin [7.541] and the AT II induced uptake of water and retention of fluids [7.542]. It also inhibits the bromocryptin or dopamine stimulated natriuresis in rats [7.543], [7.544] and reduces the salt content of sweat [7.545].

In cultivated rat pancreatic cells, o-PRL increases the basal and glucose stimulated secretion of insulin and reduces the threshold dose of glucose [7.546]. PRL increases dose dependently the food intake and weight of rats [7.547].

Other effects of hyperprolactinemia are an increase in the production of ascorbic acid [7.548] and androgens [7.549] in the adrenal gland, enhanced DNA synthesis in the liver [7.550], an increase in the susceptibility to chemically induced tumors [7.551] and spontaneous mammacarcinoma, increase in the estrogen receptors in the mammary gland [7.552], protection from stress induced stomach ulcers [7.553], decreased i.r.-β-endorphin in the hypothalmus and in the hypophysis [7.554], hypercalcemia in several species [7.555] (presumably via the release of parathyroid hormone [7.556] and a reduction of the mineral content of the bone [7.557], [7.558] (via reduced estrogen synthesis [7.559]).

PRL has also immunostimulating activity. Thus, the reactivity of the lymphocytes is reduced both *in vivo* and *in vitro* by blocking the release of PRL in the hypophysis with bromocryptin. The immunosuppressive effect of cyclosporine, which displaces PRL from the T lymphocytes, can be abolished by PRL [7.523].

Structure-Activity Relationships. De-Tyr^{28}-Met-b-PRL is devoid of activity [7.560].

7.7.3.3. Placenta Lactogen

Placenta lactogen (PL) is formed in the placenta as early as in the first trimester of pregnancy and can be detected in the blood (35 times more than prolactin (PRL)) and urine of pregant females. The secretion of PL decreases after birth.

PL combines the function of somatotropin (STH) and prolactin (PRL). It is assumed that h-PL frist stimulates the development of the breast tissue without causing the secretion of milk. When the level of h-PL decreases after birth, hypophyseal h-PRL stimulates the secretion of milk.

The hormone h-PL has luteotropic properties. It stimulates the release of progesterone and estrogen from the corpus luteum. This effect is potentiated by h-choriogonadotropin (h-CG). The growth promoting effect of h-PL is somewhat weaker than that of h-STH, but appears to play a more important role in the development of the fetus. For intance, it stimulates the formation of IGF II in the rat fetus.

7.7.4. Somatomedins, Insulin and Relaxin

Somatomedins of the IGF type (insulin like growth factors I and II) are structurally related to insulin and relaxin. They are synthesized as single-chain prepropeptides, crosslinked by three disulfide bridges. In the case of insulin and relaxin, the C-peptide (connecting peptide) between the C-terminus of the N-terminal B-chain and and the N-ternimus of the C-terminal A-chain is enzymatically removed to give a peptide consisting of two chains joined together by two disulfide bridges. In the case of IGF I and IGF II, the C-peptide is not cleaved. In spite of the great structural similarity between relaxin and insulin, relaxin has no influence on the blood glucose level. On the other hand, IGF I, IGF II and insulin are functionally related.

7.7.4.1. Somatomedins [7.561], [7.562], [7.563]

Occurrence [7.563], [7.564]. The somatomedins (SMs, SM C and SM A) were previously also called the sulfation factor, thymidine factor, non suppressible insulin-like activity in serum (NSILA-S) or the multiplication stimulating activity (MSA). Later, after elucidation of the structure, these peptides were called insulin like growth factors (IGF I and IGF II). IGF I is identical to SM C and SM A appears to be a deamidated IGF I [7.565]. On the other hand, SM B has a completely different structure, which is similar to the N-terminal end of vitronectin [7.566], [7.567].

The SMs are found in the liver, lung, prostata, testes, seminal and follicular fluid

[7.568], thymus, thyroid gland [7.569], skeleton, brain, heart, kidneys [7.570], gastrointestinal tract, saliva [7.571], milk [7.572], fibroblasts, chondrocytes [7.573], skeletal muscles and in fat pads. A large number of tumors produce the mRNA for IGF I and/or IGF II: Wilms tumor, rhabdomyo sarcoma, colon carcinoma, pheochromocytoma, liposarcoma, neuroblastoma, hepatoblastoma, leiomyosarcoma, leiomyoma and breast cancer [7.562].

IGF I and IGF II consist of a peptide chain which, like in pro-insulin, is linked by disulfide bridges. Analogous to proinsulin, the peptide chain is divided into the N-terminal B chain, the connecting C peptide, the A chain and the C-terminal D chain. The propeptides also contain the C-terminal E peptide and the prepro-compounds contain, in addition, the N-terminal prepeptide. The following structures have been elucidated: h-, b-, o-, p-, m-, and r-IGF I [7.574], h-, b-, o-, p-, m-, and two r-IGF II [7.575], h-, and r-prepro-IGF I and II, one h-pro-IGF II variant (10 kD-IGF II var [7.576]), one h- and r-prepro-IGF I variant (h-prepro-IGF IB [7.577], r-prepro-IGF IB [7.574]). An h-IGF I variant (h-IGF I var) which is shortened at the amino end by three amino acids has been isolated from the brain of the human fetus [7.578]. The peptides h-, o-, p-, and b-IGF I are identical [7.575].

```
                     1          10           20                 30   35
h, b, o, p-IGF I: G P E T L C G A E  L V D A L Q F V C G D R G F Y F N K P T G Y G S S

                 36    40                50                   60         70
                 R R A P Q T G I  V D  E C C F R S C D L R R L E M Y C A P L K P A K S A.
```

B chain

				1																		20
h, b, p, o-IGF I				G	P	E	T	L	C	G	A	E	L	V	D	A	L	Q	F	V	C	G	D-
r, m-IGF I				G	P	E	T	L	C	G	A	E	L	V	D	A	L	Q	F	V	C	G	P-
h, b, p-IGF II	A	Y	R	P	S	E	T	L	C	G	G	E	L	V	D	T	L	Q	F	V	C	G	D-
r-IGF II	A	Y	R	P	S	E	T	L	C	G	G	E	L	V	D	T	L	Q	F	V	C	S	D-
m-IGF II	A	Y	G	P	G	E	T	L	C	G	G	E	L	V	D	T	L	Q	F	V	C	S	D-

B-chain

	21								29
h, b, p, o-IGF I	R	G	F	Y	F	N	K	P	T-
r, m-IGF I	R	G	F	Y	F	N	K	P	T-
h-IGF II	R	G	F	Y	F	S	R	P	A-
b, o-IGF II	R	G	F	Y	F	S	R	P	S-
r, m-IGF II	R	G	F	Y	F	S	R	P	S-

C-peptide

	30											41
h, b, p, o-IGF	G	Y	G	S	S	S	R	R	A	P	Q	T-
r, m-IGF I	G	Y	G	S	S	I	R	R	A	P	Q	T-
h-IGF II	S	R	V	S	/	/	R	R	S	R	/	/-
r-IGF II	S	R	A	N	/	/	R	R	S	R	/	/-
b, o-IGF II	S	R	I	N	/	/	R	R	S	R	/	/-

A chain

	42																				62
h, b, p, o, r, m-IGF I	G	I	V	D	E	C	C	F	R	S	C	D	L	R	R	L	E	M	Y	C	A-
h, b, p, o, r, m-IGF II	G	I	V	E	E	C	C	F	R	S	C	D	L	A	L	L	E	T	Y	C	A-

D chain

	63							70
h, b, p, o-IGF I	P	L	K	P	A	K	S	A-
r-IGF I	P	L	K	P	T	K	S	A-
m-IGF I	P	L	K	P	T	K	A	A-
h, b, p, r, m-IGF II	T	/	/	P	A	K	S	E-
o-IGF II	A	/	/	P	A	K	S	E.

Release. The release of SM is stimulated by somatotropin (STH) (particularly in the liver), prolactin (in the liver [7.579]), insulin, injuries [7.580], physical exercise [7.581] increased vascular strain [7.582], gonadoliberin (GnRH), follitropin (FSH), lutropin (LH) and h-choriogonadotropin (h-CG) [7.583], estradiol (especially in the uterus [7.584] and in the osteoblasts [7.585]) [7.586], [7.587] ACTH (in fetal adrenal cells [7.562]), androgens, parathyroid hormone (in bone cells [7.562]) and by cAMP (in osteoblast cultures [7.588]). Prior to birth, neither IGF I nor IGF II appears to be dependent on STH [7.589]. In this case, placenta lactogen (PL), but not STH, insulin or other growth factors stimulate the sythesis of IGF II.

The secretion of SM is reduced by insufficient food intake [7.590], low protein diet [7.589], [7.591], chronic alcohol consumption [7.592], suppression treatment with Gn-RH agonists [7.593] and by dexamethasone (in rat neural and glial cells [7.594]).

Serum SM levels decrease significantly after the age of sixty and in patients with hypophyseal insufficiency and diabetes mellitus. In comparison, women taking contraceptives and patients with hyperthyroidism exhibit an increased level of SMs [7.595]. In Cushing's syndrom, the level of SMs in the spinal fluid is raised [7.595].

Receptors [7.562]. IGF I and IGF II bind with high affinity to specific receptors and with low affinity to receptors of other type [7.596].

Like the insulin receptor, the IGF I receptor (IGF-I-R) also consist of two α and two β subunits. The α subunit binds the hormones in the cysteine rich domain between His^{223} and Met^{274} [7.597], while the β-subunit contains the tyrosine kinase which phosphorylates intracellular proteins after ligand binding. The two tyrosines, B25 and C2 of IGF I are important for the binding to IGF-I-R [7.598].

The IGF-II-R is very similar to the mannose-6-phosphate receptor and also binds mannose-6-phosphate [7.599], [7.600]. It consists of a single polypeptide chain, has no kinase activity , does not bind insulin, and binds only slightly to IGF I.

In bovine breast tissue, the IGF I and IGF II receptors are increased during lactation [7.601]. The formation of IGF-I-R is stimulated by IGF II [7.602], by h-CG in Leydig cells [7.603], by T_3 in hypophyseal cells [7.604], and by reduced food uptake in the median eminence [7.605]. STH increases the density of the IGF II receptors in rat adipocytes [7.606].

IGF Binding Proteins (IGFBPs) [7.562]. An IGF binding component in the serum is the extracellular part of the IGF II recptor [7.607]. In addition, four structurally related IGF binding proteins have been characterized until now: h-IGFBP-1, h-, b-, and r-IGFBP-2, h- and r-IGFBP-3 [7.608] and h-, and r-IGFBP-4 [7.609].

IGFBP-1 binds IGF I and IGF II with similar affinity. In comparison, IGFBP-2 binds preferably to IGF II. The majority of serum IGF is bound by a complex which consists of IGF I or IGF II (γ subunit), IGFBP-3 (β subunit) and an α subunit that does not bind IGF.

Both IGFBP-1 and IGFBP-2 contain the RGD sequence at the C-terminal end. This sequence could be responsible for the binding phenomena.

h-IGFBP-1-(218-220)	P	E	I	R	G	D	P	N	C	Q	M
h-IGFBP-2-(262-272)	P	T	I	R	G	D	P	E	C	H	L

The synthesis of r-IGFBP-1 is stimulated by STH deficiency [7.610], relaxin [7.611], hypoglycemia or by the withdrawal of food [7.612]. r-IGFBP-1 is decreased by STH [7.610], IGF I [7.611], insulin [7.611], and by oral glucose [7.612].

On the other hand, IGFBP-3, like IGF I, is formed in a STH dependent process and circulates in amounts equimolar to the IGFs.

The serum IGF bioactivity is inhibited by h-IGFBP-1 and h-IGFBP- 2-[7.613], [7.614]. IGFBP-2-(148-270), a C-terminal fragment, binds IGF I and IGF II like IGFBP-2, but with a lower affinity [7.615]. On the other hand IGFBP-3 has a more stimulating effect on the activity of IGF I [7.562].

Biological Effects. IGF I stimulaes cellular 1,2-diacylglycerol and induces the degradation of phosphatidyl choline, a process which involves protein kinase [7.616].

In hypophysectomized animals, i.e., even without STH, IGF I, like STH, results in increased body weight and tibial epiphyseal width and longitudinal bone growth. In streptozotocin diabetic rats which do not respond to STH, IGF I also increases the body weight and skeletal growth [7.617]. STH stimulates the proliferation of osteoblasts at least in part via the local synthesis of IGF I [7.618]. The STH dependent increase in calcitriol after a low phosphate diet is mediated by IGF I [7.619].

Both IGF I as well as IGF II, which bind 100 to 200 times less tighly to insulin receptors than insulin, imitate the effect of insulin on adipose and muscle tissue (glucose uptake and glcose oxidation to CO_2). In humans, both have about 6% of the hypoglycemic activity of insulin [7.620] and inhibit the endogenous release of insulin [7.621]

The basal and somatoliberin (SRF) stimulated release of STH is inhibited by a negative feedback mechanism by IGF I. It acts in part directly on the hypophysis, inhibiting the transcription of the STH gene [7.622], and partly by releasing somatostatin in the hypothalamus (e.g., on i.c.v. application it reduces the food uptake of rats) [7.589]. On the other hand, IGF II exerts no negative feedback effect on the basal or SRF stimulated STH [7.563]. IGF I also inhibits the sercetion of PRL in the hypophysis [7.623], but it stimulates the release of PRL in cells of the human decidua [7.624].

In Leydig cells, IGF I stimulates the growth of h-CG receptors [7.603], [7.625] and the basal [7.626] as well as the h-CG stimulated secretion of testosterone [7.625],

[7.626], [7.627] in a LH dependent process [7.627]. IGF I stimulates, above all, progesterone and pregnenolone in ovarian cells [7.628], [7.629].

IGF I and IGF II also play a role in muscle development as autocrine growth factors [7.630], in the proliferation of h-small cell lung cancer cells *in vitro* [7.631], in neuroectodermal tumors with a t(11,22) translocation [7.632] and in MCF-7 breast cancer cell lines [7.633]. They increase the motility of melanoma cells and, therefore, could also increase the metastatic potential [7.634].

IGF I also has a modulating effect on the immune system since it stimulates the growth of the thymus [7.635], but it also inhibits the IL-2 induced proliferation and antibody formation of splenocytes *in vitro* [7.636].

Structure-Activity Relationships. While the A chain domains of IGF I are of great importance for receptor binding and the growth effect [7.637], the B chain sequences play a role in the interaction with IGFBPs [7.638], [7.639].

In IGF I, Tyr^{24}, Tyr^{31} and Tyr^{60} are important for binding to the IGF I receptor and Tyr^{60} for binding to the IGF II receptor [7.640].

While IGF I-(6-70) has only one hundredth of the activity of IGF I (protein synthesis in rat myoblasts), IGF I-(4-70) and IGF I-(5-70) are ten times and twice as active as IGF I in stimulating protein synthesis, respectively [7.641]. Like IGF I, de-(1-3)-IGF I binds to receptors, but not to IGFBP-2 [7.614].

Applications. IGF I reduces the blood sugar and insulin level in Mendenhall's syndrome with severe insulin resistance [7.642] and in Laron type dwarfism with a defective STH receptor [7.643].

7.7.4.2. Somatomedin B

Occurrence. Human somatomedin B (h-SM-B), a peptide consisting of 44 amino acids, was isolated from human plasma [7.644]. The N-terminal end of h-vitronectin (h-VN) is almost identical to SM-B [7.645]. SM-B is also similar to trypsin inhibitors [7.644].

```
        1                                                 25
h-SM-B  D Q E S C K G R C T E G F N V D K K C Q C D E L C-
h-VN    D Q E S C K G R C T E G F N V D K K C Q C D E L C-
        26                                  44
h-SM-B  S Y Y Q S N C T C Y T A E C K P Q V T.
h-VN    S Y Y Q S C C T D Y T A E C K P Q V T
```

The formation of SM-B depends on somatotropin. The serum SM-B level is higher in acromegaly and lower in hypophyseal dwarfs than in normal persons. High SM-B levels were also observed during pregnancy and during estrogen treatment [7.644].

Biological Effects. SM-B inhibits trypsin, but not plasmin, thrombin or kallikrein [7.644]. SM-B alone has no effect on the growth of chicken fibroblasts or adipocytes.

7.7.4.3. Insulin [7.646]

Occurrence. Insulin (INS) which was used for the first time in 1921 as a therapeutic agent in diabetes mellitus, is synthesized in the *β*-cells of the pancreatic islets of Langerhans. However, immunoreactive INS has also been found in other tissues, e.g., in the brain, testes and liver.

INS consists of two peptide chains which are joined together by two disulfide bridges. It is formed from single-chain proinsulin after enzymatic removal of the C-peptide (connecting peptide) [7.647] between the C-terminal A-chain and the N-terminal B-chain. The A-chain (acid chain) almost always contains 21 amino acids, while the B- chain (basic chain) usually consists of 30 amino acids. The C-chain with its N- and C-terminal basic amino acids contains 35 amino acids. *In vivo*, the C-chain is found without the N- and C-terminl basic amino acids.

Insulin like molecules are also observed in the silkworm *Bombyx mori* (Bombyxin [7.649]), in the mollusc *Lymnaea stagnalis* (prothoracicotropic hormone) [7.649] and in the brain of the insect *Locusta migratoria* [7.650].

In physiological concentrations, insulin has been found in human plasma as a monomeric compound which does not bind to any serum protein [7.651].

Release [7.661]. Signals for the release of INS are an increased plasma level of glucose or amino acids, cAMP, and Ca^{2+} (Mg^{2+}). The plasma INS level increases after the intake of food. The inkretins stimulate the secretion of INS from the pancreas in a glucose dependent process. The release of the inkretins from the intestine is increased by oral glucose and inhibited by INS. The criteria for the postulated inkretin correspond best with the glucagon like peptide 1-(7-36)-amide (GLP 1-(7-36)-amide) [7.662], the glucose dependent insulinotropic peptide (GIP), and with oxyntomodulin [7.663], [7.664], which are released from the small intestine in response to oral glucose and stimulate the INS secretion from the B cells of the pancreas. The glucose dependent secretion of INS is also stimulated by gastrin releasing peptid (GRP), neurotensin, Substance P, gastrin and pentagastrin (via inhibition of somatostatin). The inkretin activity of cholecystokinin discussed previously could not be confirmed. Only high non-physiological doses of cholecystokinin cause the release of INS. The glucose dependent release of INS is also stimulated by CCK-8, sulfated or unsulfated, the C-terminal tetrapeptide of CCK, cholinergic substances (e.g., carbachol and acetylcholine), α_2-adreno-receptor antagonists [7.665], TGF-β_1, TGF-β_2 [7.666], oxytocin [7.667], ACTH, CLIP, *β*-cell-tropin, glucagon, secretin [7.668], calcitriol [7.669], i.c.v. applied calcitonin [7.670], estradiol [7.671], thyroid hormones (via a *β* adrenergic mechanism [7.672]), prostaglandin D_2, lipoxygenase products [7.673], [7.674], and by phospholipase A_2 products [7.675], [7.676].

Insulin A-chains of various species [7.646]

	1				5					10					15						21
h, p, cu, ca, ph, ba_1, mac [7.652]	G	I	V	E	Q	C	C	T	S	I	C	S	L	Y	Q	L	E	N	Y	C	N
sai [7.653]	G	V	V	D	Q	C	C	T	S	I	C	S	L	Y	Q	L	Q	N	Y	C	N
e	G	I	V	E	Q	C	C	T	G	I	C	S	L	Y	Q	L	E	N	Y	C	N
ba_2	G	I	V	E	Q	C	C	A	S	T	C	S	L	Y	Q	L	E	N	Y	C	N
el	G	I	V	E	Q	C	C	T	G	V	C	S	L	Y	Q	L	E	N	Y	C	N
b	G	I	V	E	Q	C	C	A	S	V	C	S	L	Y	Q	L	E	N	Y	C	N
cap, o	G	I	V	E	Q	C	C	A	G	V	C	S	L	Y	Q	L	E	N	Y	C	N
g, ps, al [7.654]	G	I	V	E	Q	C	C	H	N	T	C	S	L	Y	Q	L	E	N	Y	C	N
an	G	I	V	E	Q	C	C	E	N	P	C	S	L	Y	Q	L	E	N	Y	C	N
cro [7.654]	G	I	V	E	Q	C	C	E	N	T	C	S	L	Y	Q	L	E	N	Y	C	N
'r, m-1 m-2	G	I	V	D	Q	C	C	T	S	I	C	S	L	Y	Q	L	E	N	Y	C	N
di [7.655]	G	I	V	E	Q	C	C	N	S	I	C	S	L	Y	Q	L	E	T	Y	C	N
ch	G	I	V	D	Q	C	C	T	N	I	C	T	L	Y	Q	L	E	N	Y	C	N
hy	G	I	V	D	Q	C	C	T	G	V	C	S	L	Y	Q	L	E	N	Y	C	N
pro	G	I	V	D	Q	C	C	T	N	I	C	S	R	N	Q	L	L	T	Y	C	N
myo	G	I	V	D	Q	C	C	T	N	I	C	S	R	N	Q	L	M	S	Y	C	N
cav	G	I	V	D	Q	C	C	T	G	T	C	T	R	H	Q	L	Q	S	Y	C	N
Cuis [7.656]	G	I	V	D	Q	C	C	T	R	I	C	T	S	Y	Q	L	R	N	Y	C	N
oc [7,657]	G	I	V	D	Q	C	C	N	N	I	C	T	A	N	Q	L	Q	N	Y	C	N
sq	G	I	V	E	H	C	C	H	N	T	C	S	L	Y	D	L	E	G	Y	C	N
my	G	I	V	E	Q	C	C	H	K	R	C	S	I	Y	N	L	Q	N	Y	C	N
cy	G	I	V	E	Q	C	C	H	K	P	C	S	I	F	E	L	Q	N	Y	C	N
th	G	I	H	E	Q	C	C	H	K	P	C	D	I	F	Q	L	E	N	Y	C	N
lo	G	I	V	E	Q	C	C	H	R	P	C	N	I	F	D	L	Q	N	Y	C	N
on_2 [7.658]	G	I	V	E	Q	C	C	H	R	P	C	N	I	F	D	L	Q	N	Y	C	N
sco	G	I	V	E	Q	C	C	H	K	P	C	N	I	F	D	L	Q	N	Y	C	N
on_1 [7.658]	G	I	V	E	Q	C	C	H	K	P	C	N	I	F	D	L	Q	N	Y	C	N
co [7.659]	G	I	V	E	Q	C	C	H	R	P	C	N	I	R	V	L	E	N	Y	C	N
ant I	G	I	V	E	Q	C	C	H	R	P	C	D	I	F	D	L	Q	S	Y	C	N
ant II	G	I	V	E	Q	C	C	H	R	P	C	D	K	F	D	L	Q	S	Y	C	N
ga	G	I	V	D	Q	C	C	H	R	P	C	D	I	F	D	L	Q	N	Y	C	N
br [7.660]	G	L	V	E	E	C	C	Y	N	V	C	D	Y	S	Q	L	E	S	Y	C	N

Insulin B-chains of various species [7.646]

```
                                1       5         10        15        20        25        30
h, el, mac                      F V N Q H L C G S H L V E A L Y L V C G E R G F F Y T P K T
p, ca, e                        F V N Q H L C G S H L V E A L Y L V C G E R G F F Y T P K A
b, o, cap, ph, ba2, sai             F V N Q H L C G P H L V E A L Y L V C G E R G F F Y A P K T
cu, cr                          F V N Q H L C G S H L V E A L Y L V C G E R G F F Y T P K S
g, ps                           A A N Q H L C G S H L V E A L Y L V C G E R G F F Y S P K A
an                              A A N Q H L C G S H L V E A L Y L V C G E R G F F Y S P K A
al                              A A N Q R L C G S H L V D A L Y L V C G E R G F F Y S P K G
cro                             A A N Q R L C G S H L V E A L F L I C G E R G F F Y S P R S
r, m-1                          F V K Q H L C G P H L V E A L Y L V C G E R G F F Y T P K S
r, m-2                          F V K Q H L C G S H L V E A L Y L V C G E R G F F Y T P M S
di                              L V N Q H L C G S H L V E A L Y L V C G E R G F F Y T P K A
ch                              F V N K H L C G S H L V D A L Y L V C G D R G F F Y T P M A
hy                              F V N Q H L C G S H L V E A L Y L V C G N D G F F Y R P K A
pro                             Y V G Q R L C G S Q L V D T L Y S V C K H R G / F Y R P S E
myo                             Y V S Q R L C G S Q L V D T L Y S V C R H R G / F Y R P N D
cav                             F V S R H L C G S N L V E T L Y S V C Q D D G F F Y I P K D
Cuis                            F F N R H L C G S N L V D A L Y V V C K D K G F F S R P D /
oc                              Y S S Q H L C G S N L V E A L Y M T C G R S G / F Y R P H D
sq                              L P S Q H L C G S H L V E A L Y F V C G P K G F Y Y L P K B Z V
my                              R T T G H L C G K D L V N A L Y I A C G V R G F F Y D P T K M
cy                          N A G A P Q H L C G S H L V E A L Y L V C G P T G F F Y N P K /
th                              A A N P H L C G S H L V E A L Y L V C G E R G F F Y Q P K A
on1                             A A A Q H L C G S H L V D A L Y L V C G E K G F F Y N P K /
on2                             A A A Q H L C G S H L V D A L Y L V C G E K G F F Y T P K /
co                              A D P Q H L C G S H L V D A L Y L V C G D R G F F Y N P K /
lo                            V A P A Q H L C G S H L V D A L Y L V C G D R G F F Y N P K /
sco                           V A P P Q H L C G S H L V D A L Y L V C G D R G F F Y N P K /
ant I                         M A P P Q H L C G S H L V D A L Y L V C G D R G F F Y N S K /
ant II                        M A P P Q H L C G S H L V D A L Y L V C G D R G F F Y N S / /
ga                            M A P P Q H L C G S H L V D A L Y L V C G D R G F F Y N P K /
br                                Q A E Y L C G S T L A D V L S F V C G N R G Y N S Q P R /
```

Endogenous opioids and exorphins, especially in low doses, stimulate the secretion of INS, while high doses have a more inhibitory effect. The hyperinsulinism observed after an endotoxin shock can be reduced by naloxone, i.e., it is partially caused by endogenous opioids [7.677].

The glucose stimulated release of insulin is inhibited by adrenaline (α_2 and β adrenergic mechanisms [7.678]), prostaglandin E_2 [7.679], [7.680], somatostatin (SRIF) (via inhibition of adenylate cyclase), pancreastatin, pancreatic peptide (PP), galanine [7.681], calcitonin gene related peptide (CGRP) [7.682] and amylin, calcitonin, chronically raised parathyroid hormone [7.683], exogenous INS [7.684], somatotropin (STH), interferon [7.685], interleukin 1β *in vitro* [7.686], [7.687] and in adrenalectomized rats [7.688] and by interleukin 6 [7.689].

The cell specific destruction of the pancreatic β cells, which precedes INS dependent diabetes mellitus (IDDM), is mediated by an autoimmune mechanism. A 64 kD β-cell autoantigen is the target of the autoantibodies, which have been found in more than 80% of recently diagnosed IDDM patients. This 64 kD β-cell autoantigen has been identified as glutamic acid decarboxylase [7.690] or as heat shock protein 65 [7.691].

Insulin Receptor (INSR) [7.692]. The INSR consists of two α- and two β-subunits joined together by disulfide bridges to give the configuration:

(β-chain-S-S-α-chain)-S-S-(α-chain-S-S-β-chain). The cysteine rich domain between Asn^{230} and Ile^{285} of the α-chain is important for the binding of INS [7.693]. The β-subunit is a transmembrane protein and contains a tyrosine specific protein kinase. INS stimulates the phosphorylation of the β-subunit. At least 6 of the 13 posssible tyrosines are phosphorylated. These are, among others, the tyrosines of the following intracellular sequences [7.694]:

Y E T D Y Y R
T Y D E H I P Y T

The autophosphorylation of INSR is restricted by triiodo-thyronine [7.695].

INSR is synthesized in the form of a prepro-compound. There is sequence homology between INSR and the epidermal growth factor receptor and members of the src family of oncogene products [7.696].

Diabetes resistant to INS can result from receptor defects, e.g., a deletion of the tyrosine kinase domain [7.697] or the substitution of valine for Gly^{996} [7.698]. Trp^{1200} is also important for the normal function of insulin receptor kinase [7.699]. In Hodgkin's disease, antibodies against the insulin receptor can cause hypoglycemia because of their INS like activity [7.700], [7.701].

For the receptor mediated internalization of INS, a sequence of 12 amino acids in the juxtamembrane region of the INSR β-subunit is required:

^{954}A S S N P E Y L S A S D^{965}.

The NPEY sequence also occurs in a slightly modified form in the LDL receptor

(NPVY). However, receptor endocytosis is not necessary for INS activity [7.702].

The formation of INSR is stimulated by glucose [7.703] and down regulated by INS and pro-INS [7.704]. Human retina express neural INSR which are upregulated in diabetes [7.705]. In cancerous breast tissue, the concentration of INS receptors is more than six times that in normal tissue and correlates with the size of the tumor and with the estrogen receptor concentration [7.706].

The following positions in INS appear to be important for receptor binding: $GlyA^{1}$, $GlnA^{5}$, $TyrA^{19}$, $AsnA^{21}$, $ValB^{12}$, $TyrB^{16}$, $ArgB^{22}$, $GlyB^{23}$, $PheB^{24}$, $PheB^{25}$, and $TyrB^{26}$.

While estrogen, progesterone [7.707], h-STH, h-STH-(7-13) [7.708] and relaxin [7.709] increase the binding of INSR, prolactin [7.709], heparin [7.710] or contraceptive therapy in women [7.711] reduce the INSR binding.

Biological Effects. INS lowers the blood glucose level by promoting the uptake of glucose into the skeletal muscle, heart and adipose tissue, by the activation of glycogen synthesis [7.712], and by the inhibition of glycogenolysis in the liver [7.713]. Glucose transporters, a family of structurally related proteins, play a role in the uptake of glucose [7.714], 7.715], [7.716].

INS rapidly increases the diacylglycerol level in monocytes by activating *de novo* phosphatidic acid synthesis [7.717].

Taurin [7.718], the basic peptides Arg-Lys and Arg-Arg [7.719], and SRIF (via glucagon inhibition) potentiate the insulin effect on the blood glucose level.

Apart from stimulating glucose and amino acid transport, INS also increases the secretion of somatomedins, the affinity of the IGF II receptors [7.720], the proliferation and differentiation of cells (healing of wounds), ion transport, renal Na^{+}/K^{+} ATPase [7.721], and lipogenase, normalizes platelet aggregation [7.722], potentiates the FSH stimulated synthesis of estrogen and progestin [7.723] and the h-CG stimulated synthesis of testosterone [7.724], increases food intake [7.725], stimulates the secretion of gastric acid [7.726] and the release of pancreatic peptide, enhances the bile flow, and is important for the postprandial CCK or secretin stimulated exocrine pancreatic secretion [7.727].

Hypoglycemia induced by INS is a stress component which leads to the release of ACTH and cortisol (via β adrenergic mechanisms [7.728]). INS also increases the K^{+} stimulated production of aldosterone [7.729] and potentiates the activity of aldosterone [7.730]. In type I diabetics, insulin lowers the level of cortisol raised by fasting [7.731].

Insulin inhibits the synthesis and release of somatostatin and enteroglucagon [7.732], reduces the basal and PTH stimulated excretion of phosphate in the urine [7.733] as well as the renal Ca^{2+} reabsorption, inhibits alkaline phosphatase in osteoblast like cells [7.734], stimulates the synthesis of collagen [7.735], potentiates the PTH stimulated synthesis of calcitriol in the kidneys [7.736] and reduces skeletal damage in streptozotocin diabetic rats [7.737].

Insulin inhibits the norepinephrine or AT II stimulated contraction of arteries and veins, i.e., it has a hypotensive effect [7.738], increases the heart rate [7.739], and is positively inotropic [7.740].

The i.v.c. application of insulin to rats causes the release of TRH, TSH and triiodothyronine [7.741]. Insulin has a central analgetic effect on rats [7.742] and increases the rectal temperature [7.743].

The inability of a given concentration of insulin to produce an expected biological effect is called insulin resistance. This resistance can be caused by circulating antagonists of insulin activity (antibodies, prostaglandins of the E series and glucocorticoids [7.744]), down regulation of the receptors (hyperinsulinism in type IIb diabetics) or by receptor defects.

Structure-Activity Relationships. The biological acticity of insulin is expressed in international units (I.U.). The activity is measured in the mouse cramp test or by determining the fall in blood sugar in the rabbit. Crystal insulin has 26-28 I.U./mg.

Proinsulin is only about 1/3 as active as insulin in lowering the blood sugar level, but its action is signicantly prolonged [7.745]. However, the antilipolytic activity of h-proinsulin is higher than that of h-insulin [7.746].

The hydrophobic surfaces of insulin are responsible for the formation of insulin dimers. It is assumed that especially B12, B16, B21, B22, and B24-26 are involved in dimerisation. The hexamer of insulin is formed from three dimers with two or four zinc atoms [7.747]. The formation of hexamers is influenced by B6, B10, B14, B17, B18, A13 and A14. It is assumed that monomeric insulin is reponsible for the biological activity, i.e., for binding to the receptor. In particular, the regions around A1, A2, A19, A21 and B22-B25 are important for the biological activity.

A fully active insulin is obtained when the B-chain is shortened at the N-terminus by two amino acids and at the C-terminus by three amino acids. De-(B26-30)-[PheB^{25}-amide]insulin is also completely active [7.748].

[AspB^{10}]h-INS is about 4-5 times as active as h-INS [7.749]. An analog which is about 11-13 times more active is de-(B^{26}-B^{30})- [AspB^{10},Tyr-amideB^{25}]INS [7.750]. GlyB^{23} can be replaced by D-Ala without loss of activity, but [AlaB^{23}]insulin has a greatly reduced biological activity [7.751].

Both [D-PheB^{24}]b-INS and [D-AlaB^{24}]b-INS still possess insulin activity [7.752]. Substitution of Glu for TyrB^{26} or ThrB^{27} or of Asp for ProB^{28} leads to monomeric insulin analogs having a slightly higher activity [7.753]. ThrB^{29} can be replaced by many different amino acids without loss of activity [7.754].

The N-terminal Gly of the A-chain can only be replaced by D-amino acids without loss of activity. [D-LeuA^{1}]INS possesses full insulin activity and [D-TrpA^{1}]INS is even slightly more potent than native insulin. TyrA^{19} is of significance for INS activity. The hydroxyl group of TyrAr^{19} is linked to the carboxyl group of GlyA^{1} by means of a hydrogen bridge, which stabilizes the molecule. Indeed, [PheA^{19}]p-INS has only about 23% and [LeuA^{19}]o-INS less than 0.1% of the p-INS activity [7.755]. Modifications of AsnA^{21} also lead to less active compounds. In all insulins found until now, the cysteines are invariant and extremely important for the tertiary structure.

Less soluble analogs with prolonged activity are obtained by introducing Arg into the molecule or by reducing the acidity by amidating the Glu-γ-carboxyl groups in positions A4, A17, A13, and B21 and the C-terminal carboxyl group of the B-chain [7.756], and by the palmitoylation of the A-chain [7.757]. Long-acting analogs of insulin are [GlyA^{21}, ArgB^{27}, ThrB^{30}-amide]h-INS (NovoSol-Basal) [7.758], h-INS-B^{31}-Arg-amide [7.759] and h-INS-B$^{31\text{-}32}$-Arg-Arg-OH [7.760]. Fast acting derivatives of INS are [AspB^{9}, GluB^{27}]h-INS [7.761] and [AspB^{10}]h-INS [7.761], [7.762], which occur in the monomeric form.

An insulin fragment of the structure:

B-chain	15										26
	L	Y	L	V	C	G	E	R	G	F	F	Y
					|						
				Y	C	N						
A-chain				19		21						

with the essential B-chain sequence and the important AsnA,^{21} had 0.3% of the INS activity in the mouse cramp test. Thus, it is considerably more active than the peptides of the INS-B-chain-(22-26) region, which were also said to have insulin like activity.

Applications [7.763],[7.764]. A large number of parenterally applicable insulin preparations are available for the treatment of diabetes mellitus. In general, bovine (b), porcine (p) and human (h) insulin are used for therapy. In the case of the h-insulin preparations that have been put on the market recently, it can be distinguished between the enzymatically modified p-insulin (emp-h-INS, H-insulin Hoechst, Depot-H-insulin Hoechst, Insulin Monotard HM, Insulin Actrapid HM) and the chain recombinant bacterial insulin (crb-h-INS, Huminsulin).

De-PheB^{1}-insulin (insulin defalan) derivatives have full insulin activity, but are more soluble than the starting insulin products and are, in addition, less antigenic. In all cases of insulin allergy, de-PheB^{1}-p-insulin could be used successfully [7.765]. However, de-Phe-insulins are no longer on the market because the problem of a stable preparation could not be solved.

Insulin pumps have been developed for better insulin adjustment. In this connection, pulsatile infusion appears to be more efficient than continuous infusion [7.766].

Apart from the treatment of diabetes mellitus, treatment with insulin is also recommended in total parenteral feeding [7.767], in cancer induced anorexia and loss of weight [7.768]. Furthermore, topically applied insulin promotes the healing of wounds [7.769].

7.7.4. 3.1. Pancreastatin

Occurrence. Porcine pancreastatin (p-PST), a peptide containing 49 amino acids, was isolated from the pancreas of the pig [7.770].

The precursor of pancreastatin is chromogranin A, an acidic glycoprotein which is widely distributed in the neuroendocrine system and released in large amounts from the chromaffin cells of the adrenal medulla [7.771]

A comparison with the sequence of h-chromogranin A led to the assumption that h-PST consists of 52 amino acids. Chromogranin A-(210-301) and h-PST-29 (h-PST-(24-52)) were isolated from human carcinoid liver metastasis [7.772] and chromogranin A-(116-301) from the liver metastasis of a patient with insulinoma. The latter is cleaved with trypsin to h-PST-48 (h-PST-(5-52)) [7.773].

PST contains the penta-Glu sequence of gastrin (GT) and has a C-terminal end that is similar to that of gonadoliberin (Gn-RH).

In patients with non-insulin-dependent diabetes mellitus (NIDDM), the plasma PST level increases after the digestion of glucose, which is not the case in healthy persons [7.774].

PST sequences of various species [7.771]

	1 14 25
p-PST	G W P Q A P A M D G A G K T G A E E A Q P P E G K-
	1 5 24 27
h-PST	G E S R S E A L A V D G A G K P G A E E A Q D P E G K-
b-PST	A A P G W P E / / / / D G A G K M G A E E A K P P G K-
r-PST	D D D G Q S E S Q A V N / / G K T G A S E A V P S E G K-
	26 33 49
p-PST	G A R E H S R Q / E E E E E / / T A G A P Q G L F R Ga.
	28 52
h-PST	G E Q E H S Q Q K E E E E E / / M A V V P Q G L F R Ga.
b-PST	G E W A H S R Q / E E E E / / / M A R A P Q V L F R Ga.
r-PST	G E L E H S Q Q / E E D G E E A M A G P P Q G L F P Ga.
p-GnRH	*QH W S Y G L R P Ga.
p-GT	... K K Q G P W M / E E E E E A Y G W M D Fa

Biological Effects. PST inhibits the first phase of the glucose or arginine induced release of insulin [7.770], [7.775], [7.776] and the arginine induced release of somatostatin (SRIF) [7.775], but potentiates the arginine induced secretion of glucagon [7.775] from the isolated perfused rat pancreas. These PST induced changes in the pancreas are reminiscent of non-insulin-dependent diabetes.

In the rat, PST lowers the plasma insulin level and increases the plasma level of glucose in response to intragastric glucose [7.777] and it raises the plasma glucagon level in response to intravenous arginine [7.778]. PST, like glucagon, stimulates hepatic glycogenoloysis [7.779].

PST also inhibits the CCK stimulated secretion of protein and fluid from the pancreas [7.778] and the CCK stimulated growth of human pancreatic adenocarcinoma [7.780]. In canine parietal cells, PST inhibits the secretion of acid and potentiates the inhibitory effect of SRIF [7.781]. On the other hand PST increases the meal stimulated gastric acid secretion in the dog [7.782].

Structure-Activity Relationships. The inhibitory effect on the glucose stimulated release of insulin is restricted to the C-terminal end of the molecule because even the fragments p-PST-(14-49) (p-PST-36) and p-PST-(33-49) (p-PST-17) (stronger) exert this effect [7.770]. The shortest, still active C-terminal fragment is p-PST-(35-49) [7.783]. The C-terminal amide function is required for the biological activity [7.783], [7.784].

The compounds, h-PST-29, h-PST-52, de-Glu^{37}-h-PST-29, de-Lys^{36}-h- PST-(29-52) [7.785], and r-PST-26 [7.786] also inhibit the glucose stimulated secretion of insulin and increase the plasma level of glucose.

7.7.4.4. Relaxin [7.787], [7.788]

Occurrence. Relaxin (RLX) is found primarily in the serum and tissues of pregnant mammals. It occurs in especially high concentrations in the corpus luteum of pregnant animals, and is isolated from this source. In addition, RLX is found in the decidua, breast [7.789], testes, prostata and in seminal fluid.

There is considerable similarity in the structures of RLX, insulin and the insulin like growth factors. RLX, like insulin, consists of two chains (A- and B-chain) that are connected by disulfide bridges.

Relaxin, like insulin, is synthesized as single-chain pro-relaxin. Two different genes, H1 and H2 have been found for h-RLX. H2 is formed in the ovaries during pregnancy and is reponsible for RLX in the plasma [7.790]. p-RLX occurs in three forms. The B-chain can be shortened by three amino acids at the C-terminus and the A-chain can be extended by Phe at the N-terminus [7.791].

h-Relaxin-2

```
          ┌───────────┐
*Q L Y S A L A N K C C H V G C T K R S L A R F C
                    \                         |
     S W M E E V I K L C G R E L V R A Q I A I C G M S T W S K R S L
```

Relaxins from various species [7.790]

A-chains	
h-RLX-1	R P Y V A L F E K C C L I G C T K R S L A K Y C
h-RLX-2	•Q L Y S A L A N K C C H V G C T K R S L A R F C
r-RLX	•Q S G A L L S E Q C C H I G C T R R S I A K L C
p, ba-RLX [7.792]	R M T L S E K C C Q V G C I R K D I A R L C
pr-RLX	A T S P A M S I K C C I Y G C T K K D I S V L C
sc-RLX [7.793]	E G S P G M S S K C C T Y G C T R K D I S I L C

B-chains	1 … 14
h-RLX-1	K W K D D V I K L C G R E L-
h-RLX-2	S W M E E V I K L C G R E L-
r-RLX	R V S E E W M D Q V I Q V C G R G Y-
p-RLX	•Q S T N D F I K A C G R E L-
ba$_3$-RLX	•Q S T N D L I K A C G R E L-
ba$_4$-RLX	•Q S T N D L I K A C G R E L-
pr-RLX	•Q S L S N A G S G I K L C G R G F-
sc-RLX	•Q S F K N A E P G I K L C G R E F-

	15 … 32
h-RLX-1	V R A Q I A I C G M S T W S K R S L.
h-RLX-2	V R A Q I A I C G M S T W S K R S L.
r-RLX	A R A W I E V C G A S / V G R L A L.
p-RLX	V R L W V E I C G S V S W G R T A.
ba$_3$-RLX	V R L W V E I C G S V R W G Q S A L.
ba$_4$-RLX	V R L W V E I C G S V S W G R T A L.
pr-RLX	I R A I I F A C G G S R.
sc-RLX	I R A V I Y T C G G S R W.

Release. The plasma level of RLX increases greatly at the end of pregnancy and falls rapidly after birth. In hypophysectomized, but not in hysterectomized animals, endogenous RLX can be activated by estrogens and progesterone. The release of RLX is also increased by h-CG [7.794], oxytocin and prostaglandin F_2 (PGF_2). On the other hand, indomethacin prevents the increase in the RLX level and delays birth.

In trophoblastic disease, which is accompanied by an increase in the plasma h-CG level, RLX is detectable in the plasma [7.795].

Biological Effects. Purified RLX (NIH-RXN-P1 standard) has about 3000 U/mg (measured on the guinea pig interpubic disc). One unit is about the threshold dose which causes a softening of the interpubic disc in 2/3 of the guinea pigs tested.

In vitro, RLX stimulates the formation of cAMP and increases the Ca^{2+} efflux from mature rat myometrical cells [7.796].

In advanced pregnancy, the contraction of the uterus is reduced by the action of RLX, which is supported by progesterone. This effect is of significance in the maintenance of pregnancy. RLX has a growth promoting and relaxing effect on the uterus and cervix and a growth promoting and lactation inhibiting effect on the mammary gland. Balanced proportions of progesterone, estradiol and RLX are required for the development of an endometrium similar to that in pregnancy. RLX alone lowers the formation of progesterone and increases the sythesis of estrogen [7.797]. RLX also appears to prepare the myometrium for parturition because without RLX oxytocin cannot induce birth. RLX inhibits the formation of PGE during pregnancy, but stimulates it during birth [7.798].

The relaxing effect of RLX on the interpubic disc (pelvic symphysis) is caused by the conversion of the compact cartilaginous tissue between the two pubic bones to a more flexible substance. The water content of the tissue is increased and the secretion of proteolytic enzymes (plasminogen activator, collagenase, and proteoglycanase [7.799] is stimulated, loosening the network of collagen and glycosaminoglycan [7.800]. The increased flexibility and the enlargement of the birth canal shorten the time of delivery and increase the number of live births. Estrogen and hypophyseal hormones (e.g., somatotropin) are also involved in this deaggregation of collagen. Intravaginal RLX (2mg/patient) causes cervical maturation in humans and induces birth.

RLX increases the motility of sperm and exerts a hypotensive effect [7.801]. It inhibits the pressor activity of AT II, vasopressin and norepinephrine [7.802].

Structure-Activity Relationships. Both r-RLX and p-RLX are active in the mouse. However, RLX of the shark (pr-RLX) is not active in the mouse, but it exhibits full activity in the guinea pig.

N-terminal shortening of the A-chain of p-RLX by two amino acids does not affect the biological activity. However, if 5 or 6 N-terminal amino acids of the A-chain of p-RLX are deleted, the biological activity is halved [7.803].

The oxidation of $TrpB^{27}$ in p-RLX with N-bromosuccinimide to oxindol has no influence on the biological activity. However, p-RLX becomes inactive when $TrpB^{18}$ is oxidized.

7.7.5. Somatostatin [7.804], [7.805]

Occurrence. Somatostatin (SRIF) is formed as prepro-SRIF. The main product of gene expression in the stomach and pancreas of mammals is pro-SRIF-(1-64), which is processed at the C-terminus to form SRIF-28 and SRIF-14 [7.806].

SRIF-14 and SRIF like substances have been found, for instance, in the hypothalamus, central and peripheral nervous system, pancreatic islet D cells, mucosa of the gastrointestinal tract, thymus, tongue, extrahepatic bile ducts, plasma [7.807], along with calcitonin in the C cells of the thyroid gland, ovaries and in plants (e.g., tobacco [7.808] and spinach [7.809]. In cutivated neurons from the rat cerebrum, SRIF occurs with acetylcholine esterase.

SRIF-like urotensin II (UT II), which has until now been found in fish, does not exert an SRIF like effect in mammals.

Release. SRIF is released in the antrum in response to fasting [7.814], acid in the antrum, intraduodenal infusion of fat and HCl [7.807], stress [7.815], cholinergic and β-adrenergic mechanisms, dopamine D_2 agonists [7.816], [7.817], estrogen and androgens [7.818], [7.817], and to various peptides. These peptides include the gastrin releasing peptide (GRP), glucose dependent insulinotropic peptide (GIP), insulin [7.807], Substance P (SP), cholecystokinin (CCK), gastrin (GT) (via activation of protein kinase C [7.819], secretin (SEC), neurotensin (NT), glucagon (GG), somatoliberin (SRF) (via the release of arachidonic acid and cyclooxygenase as well as cytochrome P450 metabolites [7.820], thyroliberin (TRH), somatotropin (STH) [7.821], somatomedins (SMs), interleukin 1, tumor necrosis factor α, and interleukin 6 [7.822].

The secretion of SRIF is reduced shortly after meals [7.814], by electrical stimulation of the vagus of sham feeding, by α-adrenergic stimuli, dexamethasone [7.823], serotonin [7.824], cysteamine, and by various peptides. These peptides include VIP, endogenous opioids (β-endorphin, enkephalins, and β-casomorphines), galanine [7.825], pancreatic peptide (PP), and insulin.

The plasma level of SRIF is lower at night [7.826]. The plasma SRIF level in children below the age of two is raised and increases continuously with age [7.827].

The plasma SRIF level is increased in patients with chronic renal failure, cirrhosis [7.828], and primary hypothyroidism [7.829]. The level of SRIF is raised in the cerebrospinal fluid of patients with brain tumors. metabolic encephalopathy, spinal cord compression, and intracranial hypertension [7.830]. In Huntington's disease, the level of SRIF in the basal ganglia is increased [7.831].

The plasma level of SRIF is reduced in patients with pernicious anemia, achlorhydria [7.827], cluster headache [7.832], psoriasis (correlates with hyperinsulinism) [7.833], overweight [7.834], and in women who are taking estrogen containing contraceptives [7.835]. The level of SRIF in the cerebrospinal fluid is reduced in patients with Alzheimer's disease [7.835], [7.836], unipolar depression and during active multiple sclerosis [7.815]. The level of SRIF is reduced in the cortex and hippocampus of patients suffering Parkinson's disease [7.837] (destruction of SRIF containing neurons is an early event in the course of the disease [7.838].

Somatostatin and urotensin II of various species [7.804], [7.810]

```
h, r, o, b-SRIF-28                      S A N S N P A M A P R E R K A G C K N F F W K T F T S C
my-SRIF-34 [7.811]
                            A V E R P R Q D G Q V H E P P G R E A K A G C K N F F W K T F T S C
pet-SRIF-37 [7.812]
                      A L R A A A V A G S P Q Q L L P L G Q R E R K A G C K N F F W K T F S S C
lo-SRIF-I-28                            A A S G G P L L A P R E R K A G C K N F F W K T F T S C
lo-SRIF-II-28                           S V D S T N N L P P R E R K A G C K N F Y W K G F T S C
co-SRIF-28                              S V D P P N N I P L R E R K A G C K N F Y W K G F T S C
sa-SRIF-25                                    S V D N L P P R E R K A G C K N F Y W K G F T S C
sil-SRIF-I                                          D N T V R S K P L N C M N Y F W K S S T A C
ic-SRIF-II-22                                       D N T V T S K P L N C M N Y F W K S R T A C
mammalian SRIF-14                                                   A G C K N F F W K T F T S C
                                                                    1                     12
go-UT-II                                                            A G T A D C F W K Y C V
cy-UT-II-α                                                          G G G A D C F W K Y C V
cy-UT-II-β1                                                         G G N T E C F W K Y C V
cy-UT-II-β2                                                         G S N T E C F W K Y C V
cy-UT-II-γ                                                          G G G A D C F W K Y C I
cat-UT-II-A                                                         G S G A D C F W K Y C V
cat-UT-II-B                                                         G S N T E C F W K Y C V
pl-UT-II [7.813]                                                    A G T T E C F W K Y C V
```

Receptors. Several SRIF receptors are found in the brain. ^{125}I-[Tyr^{3}]octreotide is a selective radioligand for one of these receptors [7.839].

Receptors for SRIF are detected in most STH and TSH producing pituitary adenomas, carcinoids and islet cell carcinomas, EGF receptor negative glia tumors, and small cell lung carcinomas [7.840]. SRIF receptors are reduced in the cerebral cortex of patients with Alzheimer's disease [7.841].

Biological Effects:

Actions of SRIF-14 [7.842], [7.843]

Stimulation of phosphodiesterase, inhibition of cAMP synthesis and blocking of calcium increase and the turnover of phospho-inositides in the cytosol [7.844].

Inhibition of hormone release and action: somatotropin, prolactin, thyreotropin, 7B2 [7.845], stress induced corticotropin (by the inhibition of CRH release [7.846], adenylate cyclase activity and cAMP-dependent protein kinases [7.847]), hypoglycemia-induced oxytocin via opioid-receptors [7.848], calcitonin, insulin, glucagon, GIP, VIP, secretin, pancreatic polypeptide, gastrin releasing peptide [7.807], gastrin, cholecystokinin and motilin. The effect is stronger in fat persons than in thin persons [7.849]. The somatotrophs are the most sensitive to SRIF-inhibition, followed by the thyrotrophs and B-cells followed by the A-cells [7.850].

Increase in extrahepatic glucose clearance [7.851], [7.852]. But decreased INS-stimulated glucose utilisation in adipocytes [7.853] and skeletal muscle [7.854] as well as insulin-stimulated amino acid uptake into cultured rat myoblasts [7.855] and enhanced glucagon stimulated heptatic glucose output [7.856].

Inhibition of exocrine secretion: gastric acid, pepsin (anti-ulcer effect), pancreatic enzymes (improvement of experimentally induced pancreatitis in rats [7.857] and dogs [7.858]), pancreatic bicarbonate, colonic fluid (reduction of secretory diarrhea [7.859]), bile flow (patients with somatostatinoma usually have gallstone disease [7.860]).

Inhibition of splanchnic and mesenteric blood flow and stimulation of vascular resistance.

Potentiates *in vitro* collagen- and arachidonic acid-stimulated human platelelt aggregation [7.861]. (Control of upper gastrointestinal hemorrhage.

Inhibition of motor functions (via inhibition of acetylcholine? [7.862]), inhibition of postprandial rise in lower esophageal sphincter pressure (LESP) [7.863], gall bladder contraction [7.864] and antral motility [7.865].

But increases contraction of sphincter of oddi.

Regulation of the migratory motorial complex by local SRIF: The start of an migrating activity front in the upper duodenum was accompanied by plasma-SRIF-peaks [7.866]. Enhances gastric emptying [7.867] and intestinal transit [7.868]. SRIF may be useful in the management of dumping syndrome [7.869].

Inhibition of absorption of: calcium, glucose and other sugars, amino acids, triglycerids, ions, water.

Reduces the effective renal plasma flow, the glomerular filtration rates, the urinary flow [7.870], sodium and creatinine excretion [7.871] and vasopressin-stimulated adenylate cyclase in kidney cells [7.872]. Inhibits corticosteroidogenesis *in vitro* [7.873].

Prevents postprandial hypotension [7.874].

Stimulates acetylcholine from canine heart [7.875].

Inhibits CNS activities, acetylcholine release and neuronal activity.

Inhibits cell proliferation: exocrine pancreas [7.876], rabbit jejunum [7.877], liver after hepatectomy [7.878] and tumor growth. The antitumor effect of SRIF is mediated by the stimulation of tyronsine phosphatase [7.879].

Reduces CCl_4-induced liver damage [7.880].

Stimulates histamine release from rat peritoneal mast cells comparable to substance P [7.881].

Immunomodulatory actions: stimulation of leukocyte migration inhibitory factor [7.882], enhancement of human lymphocyte natural killer activity [7.883] and mouse spleen lymphocyte proliferation [7.884].

Inhibits stimulatory effects of testosterone on bone formation in the mouse [7.885]. Octreotide stimulates osteoblastic activity and inhibits PTH-induced blockade of osteoblastic activity [7.886].

Central application: hypotension and bradycardia [7.887], blocks the electroconvulsive shock induced retrograde amnesia in rats [7.888], administered into the cerebrospinal fluid or epidural and intrathecal space, SRIF acts as potent analgesic [7.889], [7.890], activates opioid-receptors, enhances acetylcholine-induced excitations in rat hippocampus and cortex [7.891], reduces jejunal absorption of water and electrolytes (via opioide receptors?) [7.892], i.c.v.-SRIF partially reverses suppression of food intake induced by i.c.v. CRH [7.893] and increases food intake in fastig rats [7.894] (The octreotide stimulated intake of food is linked with a loss of body weight [7.894]), hypothalmic SRIF levels mediate normal temperature responses in rats (rise in colon temperature) [7.895] and stimulate prolactin secretion [7.896].

Activity of UT II. UT II contracts smooth muscles and exerts hypertensive and osmoregulatory effects in fish and birds. UT II causes endothelium dependent relaxation and endothelium independent contraction of rat aortic strips [7.897].

Structure-Activity Relationships. The substitution of D-Trp for Trp^8 produced a SRIF-14 analog that had 8 times the activity.

The long acting and potent SRIF analogs of the urotensin II type are of tremendous interest. For instance, octreotide (SMS 201- 995) is 70 times as effective as SRIF-14 in inhibiting STH *in vivo* 15 minutes after administration and it exhibits a long duration of action when given intramuscularly.

UT-II:	A G T A D C F W K Y C V
Octreotide [7.898]	f C F w K T C T-ol
Vapreotide (RC-160)[CAS-103222-11-3] [7.899], [7.900]	f C F w K T C Wa
RC-98-I [7.901]	w C F w K T C Ta
RC-121 [7.899]	f C Y w K Y C Ta
BM-23014 [7.902]	D-Nal C Y w K T C Ta

Cyclic peptides without a disulfide structure are also potent and long acting compounds:

I	cyclo-(Pro-Phe-D-Trp-Lys-Thr-Phe)
Seglitide, MK-678	cyclo-(N-Me-Ala-Tyr-D-Trp-Lys-Val-Phe)
III	cyclo-(Aoc-Phe-D-Trp-Lys-Val-Phe)

The cyclic peptide I is about 20 to 25 times as active *in vivo* as SRIF-14 and acts for about 5 hours when applied subcutaneously. Seglitide has about 10 times the activity of I. Cyclic peptide III inhibits, above all, glucagon secretion without affecting the release of insulin.

With regard to insulin, glucagon and STH, cyclo-(7-aminoheptanoyl-Phe-D-Trp-Lys-Thr(Bzl)) has the properties of a competitive SRIF inhibitor. However, it could not abolish the inhibitory effect of SRIF-14 on the secretion of gastric acid, but it stimulates the growth of young rats [7.903].

Cytoprotective SRIF Analogs. Cyclic retro-peptides of type I, e.g., cyclo-(Phe-Thr-Lys-Trp-Phe-D-Pro) (008) inhibit the uptake of cholate and phallotoxin in liver cells [7.904] and protect against taurocholate and ceruletide induced pancreatitis [7.905], and against ethanol induced stomach lesions [7.906] (more effective than SRIF), without possessing the real SRIF activity.

Opioid Antagonists with SRIF Structure. Analogs of octreotide which usually contain pennicillamine instead of Cys^7 are poor ligands for SRIF receptors, but exhibit stronger binding to opioid receptors [7.907].

D-Phe-Cys-Tyr-D-Trp-Lys-Thr-Pen-Thr-NH_2 (Cys–Pen bridged)	(CTP) [7.908]
D-Phe-Cys-Tyr-D-Trp-Orn-Thr-Pen-Thr-NH_2	(CTOP) [7.910]
D-Tic-Cys-Tyr-D-Trp-Lys-Thr-Pen-Thr-NH_2	(TCTP) [7.911]
D-Tic-Cys-Tyr-D-Trp-Orn-Thr-Pen-Thr-NH_2	(TCTOP) [7.911]
D-Tic-Cys-Tyr-D-Trp-Arg-Thr-Pen-Thr-NH_2	(TCTAP) [7.911]
D-Nal-Cys-Tyr-D-Pal-Lys-Val-Cys-Thr-NH_2	(DC-13-161) [7.912]

The derivatives listed above act *in vivo* and *in vitro* as selective antagonists for μ opioid receptors, i.e., they inhibit morphine induced analgesia [7.912].

Applications. SRIF-14 infusions (Stilamin and Somatofalk) have proved useful for the treatment of bleeding peptic ulcers and gastrointestinal lesions, for preventive treatment against stress ulcers, and in the healing of fistulae of the small intestine and gallbladder.

Octreotide [7.913], [7.914] is used in the treatment of somatotropin and thyrotropin secreting pituitary tumors, carcinoid tumors, VIPomas, glucagonomas, insulinomas, and gastrinomas. The use of octreotide in acute esophageal variceal bleeding, pancreatic pseudocysts, gastrointestinal and pancreatic external fistulae, short bowel syndrome, dumping syndrome and in acquired immunodeficiency syndrome related refractory hyper-secretory diarrhea has provided encouraging results. Preliminary reports indicate the efficiency of octerotide in the treatment of psoriasis, autonomic neuropathy (postprandial and orthostatic hypotension), panic like attacks [7.915] and of carcinoid flush [7.916], and its ability to reduce height velocity in tall adolescents. Other potential application include the control of pain [7.917], headaches, rheumatoid arthritis, and diabetic microangiopathy as well as the prevention of post operative pancreatic complications (fistulae, pancreatitis and abscess).

Vapreotide and BM 23014 should find similar applications, e.g., in cancer [7.918], [7.919] and as an analgetic [7.920].

8. Neurotensin [8.1], [8.2]

8.1. Occurrence

Neurotensin (NT), a peptide containing 13 amino acids, was first isolated from bovine hypothalamic extracts and characterized. Later, immunoreactive NT was found primarily in the gastrointestinal tract, the highest concentrations being in the mucosa of the jejunum and ileum. The structurally and functionally related p-neuromedin N was isolated from the porcine spinal cord [8.3]. This is a peptide consisting of 6 amino acids which, except for one amino acid, is identical to g-LANT-6 [8.4] from the chicken intestine. NT and neuromedin N are coded on the same gene. In the meantime, the prepro-compounds from the rat, cow and dog have been characterized [8.5]. N-terminal shortened Neurotensin [8.1] and high molecular neuromedin N (17 and 8 kDa) [8.6] have been isolated from the intestine of the dog.

In human beings, immunoreactive NT has been isolated primarily from the nervous system of the hypothalamus, median eminence, substantia nigra and from brain stem structures. It occurs with dopamine in dopaminergic neurons [8.7]. NT-(1-13) and the main metabolite NT-(1-8) are also found in the adrenal medulla [8.8]. g-NT has been found in the chicken rectum where it stimulates the contraction [8.9].

The level of i.r.-NT is raised by about 250% in the hot flushes experienced during menopause. This i.r.-NT is not identical to NT or one of its known metabolites. It may be formed in the blood, which contains large amounts of proteins that can release an i.r.-NT by the action of pepsin [8.10]. This i.r.-NT may be h-kintensin, a nonapeptide that is formed by enzymatic degradation with acidic proteases from human plasma [8.11]. It is very similar to h-serum albumin and α-fetoprotein sequences. Identical and similar peptides were also found in other species and were known as neurotensin or xenopsin related peptides (NRP) [8.12], [8.13] or neurotensin related substances (NRS) [8.14].

Xenopsin, a peptide that is structurally and functionally related to NT, was isolated from the skin of the frog, *Xenopus laevis*. This peptide is also synthesized as a precursor molecule [8.15]. There is structural similarity between the tryptophyllins from the skin of *Phyllomedusa rhodei* and xenopsin.

The structure Arg-Arg-Pro also occurs in the histamine stimulating granuliberin R, a peptide from the skin of the frog, *Rana rugosa*. The histamine releasing peptides, mastoparan M, X, and C from wasp poison also possess structural characteristics of NT. As far as the histamine releasing activity is concerned, the mast cell degranulating peptide (MCD peptide) from bee poison also belongs to this group.

Comparison of the sequences of neurotensin, neuromedin N, kintensin, and xenopsin with those of tryptophyllins, granuliberin R, mastoparans and MCD peptide

r-, ca-, b-, h-NT	•Q L Y E N K P R R P Y I L
ca-NT-(4-13) [8.1]	E N K P R R P Y I L
cav-NT [8.16]	•Q L Y E N K S R R P Y I L
g-NT	•Q L H E V K A R R P Y I L
g-LANT 6	K N P Y I L
p-Neuromedin N	K I P Y I L
ca, b [8.12], h-Kintensin [8.11]	I A R R H P Y F L
r-Kintensin [8.12]	V A R R H P Y F L
h-Serum albumin-	
(138-150)	Y L Y E I A R R H P Y F Y
(330-342)	F L Y E Y A R R H P D Y S
α-Fetoprotein-	
(143-155)	F I Y E I A R R H P F L Y
(335-347)	F V H E Y S R R H P Q L A
g-Xenopsin-2 [8.13]	F H P K R P W I L
g-Xenopsin-1 [8.14]	H P K R P W I L
g-NRS [8.14]	F E R F Q G M R T R G P Y F L
g-NRS-8 [8.14]	R T R G P Y F L
xe-Xenopsin	•Q G K R P W I L
Tryptophyllins	•Q P W La
	•Q P W Ma
	•Q P W Va
	•Q A W M
	•Q P W M
	F P P W La
	F P P W Ma
	F P P W Va
	F P P W
	F P P W L
	V P P L G W M
Tryptophyllin-13 [8.17]	•Q E K P Y W P P P I Y P M
Granuliberin R	F G F L P I Y R R P A Sa
Mastoparan M	I N L K A L A A L A K K / / I La
Mastoparan X	I N W K G I A A M A K K / / I La
Mastoparan C	L N L K A L L A V A K K / / I La
MCD-peptide	I K C N C K R H V I K P H I C R K I C G K Na

8.2. Release

NT is released from hypothalamic preparations by depolarizing K^+ and dopaminergic mechanisms. Both the dopamine releaser methamphetamine and dopamine itself [8.19] as well as the dopamine antagonist haloperidol (via D_2 receptors) [8.18] and neuroleptic agents [8.20] release NT in various areas of the brain and increase the number of NT containing neurons.

NT is released from the N cells of the small intestine (particularly the ileum) into the blood in response to meals rich in fat and oleic acid, bombesin infusions, adrenaline [8.21], hydrocortisone [8.22], nerve growth factor (NGF) [8.22], insulin [8.22], and to i.v. $CaCl_2$ [8.23]. While hexamethonium and morphine prevent this oleic acid induced release of NT, it is potentiated by haloperidol [8.24] and cigarette smoking [8.25]. The secretion of NT from the intestine is inhibited by somatostatin (SRIF) [8.23] and dexamethasone [8.26].

NT is released by NGF and dexamethasone [8.8] from the adrenal medulla and by somatoliberin from cloned rat C cells [8.27].

The plasma level of NT can be raised in various tumor patients, e.g., with liver tumors [8.28], small cell cancer [8.29] and tumor of the pancreas [8.30]. On the other hand, the spinal fluid of children suffering from febrile aseptic meningitis has a lower content of i.r.-NT [8.31].

Neuroleptic agents increase the NT receptor binding sites in the substantia nigra [8.20].

8.3. Biological Effects

NT and C-terminal fragments such as NT-(8-13) raise the level of cytosolic calcium [8.32] and cGMP [8.33] and stimulate the formation of inositol phosphates [8.34].

NT has neuroleptic properties when applied centrally. It reduces the synthesis of dopamine within the nerve ends, increases the turnover of dopamine [8.35], and inhibits the hyperactivity of dopamine receptor agonists [8.36]. Depending on the site of the central injection, sometimes NT and sometimes neuromedin N exerts a stronger inhibitory effect on dopamine induced hyperactivity [8.37]. The affinity of D_2 dopamine receptors in the neostriatum and in the nucleus accumbens is also reduced by NT [8.38]. However, NT (10^{-4}-10^{-9} M) increases *in vitro* the electrically caused or K^+ stimulated release of dopamine from the cat and rat striatum [8.38]-[8.40] and amygdala [8.35].

NT reduces the food uptake when injected into the paraventricular hypothalamic nucleus [8.21] or into the ipsilateral paraventricular nucleus [8.41]. Apart from this reduced food uptake, aphagia (inability to swallow) is also observed on ventral tegmental injection [8.42].

When centrally applied, NT stimulates the release of somatostatin (SRIF), the secretion of lutropin [8.43] via estrogen dependent [8.44] dopaminergic and α adrenergic mechanisms [8.45], and the release of corticosterone [8.46]. It also enhances the effet of barbiturates [8.30], exerts an antinociceptive effect [8.47], reduces the secretion of thyrotropin, somatotropin and of prolactin, inhibits the carbachol, 2-Deoxy-D-glucose and pentagastrin stimulated gastric acid secretion [8.48], exerts a cytoprotective effect on stress induced gastric ulcers through the prostaglandins, and has a hypothermic effect [8.49].

The most sensitive target organs for intravenously applied NT are the upper gastrointestinal tract and the adipose tissue [8.50]. NT reduces sphincter tone, inhibits gastric secretion, gastric emptying and intestinal motility. It has a vasoconstrictive effect in adipose tissue.

In the stomach, NT inhibits the pentagastrin stimulated release of pepsin and the secretion of gastric acid (via SRIF release [8.51]). It increases the exocrine $NaHCO_3$ secretion in the pancreas and the endocrine secretion of PP, insulin and glucagon.

Relatively high, unphysiological doses of NT (>100 ng/kg.min) raise the plasma glucose concentration and reduce blood pressure. The hypotensive and blood-glucose increasing effects are probably achieved through the release of histamine and other vasoactive substances from mast cells, substances which exert an inhibitory effect on the sympathetic nervous system. The NT induced formation of edema is also caused by mediators of the mast cells [8.52]. In concentrations of 10^{-11} to 10^{-7}M, NT enhances phagocytosis of macrophages and neutrophils and, consequently, could play a role in inflammatory processes.

On intradermal application, the kintensins increase vascular permeability and are more effective in releasing histamine from rat mast cells than NT and bradykinin [8.12]. They may be mediators of the inflammatory reactions in the second phase of inflammation in which acid proteases are released [8.12].

The tryptophyllins have no effect on smooth muscles, but stimulate protein synthesis in the liver [8.53]. The caerulein like tryptophyllin heptapeptide, Val-Pro-Pro-Leu-Gly-Trp-Met has a sedative and sleep promoting effect when applied i.c.v. to doves [8.53]. Like the smaller peptides, however, it also stimulates protein synthesis in the liver and increases body weight when applied subcutaneously [8.54].

8.4. Structure-Activity Relationships

The peptides, [Gln^{4}]NT, g-NT and [monoiodo-Tyr^{3}]NT [8.55] appear to be identical to b-NT with respect to biological activity. *In vitro* studies have shown that the C-terminal hexapeptide NT-(8-13) has almost the same activity as the entire molecule. The activity of the hexapeptide is enhanced by N-acetylation. The C-terminl pentapeptide and the free C-terminal carboxyl group are essential for activty. The hypotensive effect of the NT fragments shows a similar pattern. NT-(6-13) is about as active as NT [8.56]. NT-(8-13) and NT-(9-13) still have 51% and 3.8% of the NT activity, respectively. Tyr^{11} can be replaced by Trp without loss of activity.

The C-terminal carboxyl group is not required for the hypotensive, vasodilating activity since cyclopeptides, e.g.,

cyclo-(Gly-Arg-Arg-Pro-Tyr-Ile-Leu) [8.57] and
Phe-Lys-Pro-Arg-Arg-Pro-Tyr-Ile-Leu [8.56] (Lys side chain linked to C-terminal Leu)

are potent hypotensive agents.

The C-terminal region is responsible for cGMP formation in the murine neuroblastoma clone N1E-115. In fact, NT-(8-13) is 50 times as active as NT [8.58]. NT-(9-13) is less active than NT and NT-(1-12) is devoid of activity [8.59].

The entire molecule appears to be required for the inhibition of the pentagastrin stimulated secretion of gastric acid and pepsin and for the motility inhibiting effect.

NT-(8-13) has the same central antinociceptive effect as NT. However, even shorter sequences such as NT-(9-11), NT-(8-10) or NT-(8-11)-NH-C_2H_5 exhibit this activity [8.60].

The cytoprotective activity of i.c.v. applied NTs is not based on the C-terminal but on the N-terminal sequence. Thus, NT-(8-13), xenopsin, and NT-(1-6) have no cytoprotective activty. In comparison, NT-(1-8) and NT-(1-10) exhibit significant cytoprotection [8.61].

In miosis, the NT analog [D-Trp^{11}]NT and [D-Tyr^{11}]NT and the fragment NT-(1-12) have a similar (only slightly weaker) effect as NT. The C-terminal fragments NT-(8-13) amd NT-(9-13) as well as the N-terminal fragment NT-(1-6) had no effect on miosis [8.62].

The central hypothermic activity is especially pronounced in [D-Tyr^{11}]NT and [$dopa^{11}$]NT [8.63]. [D-Tyr^{11}]NT also inhibits the hyperactivity induced by dopamine agonists when applied centrally [8.64].

Depending on the dose, both the histamine releasing and the hypotensive effect of NT [8.65] as well as the NT stimulated formation of inositol phosphate [8.66] and secretion of glucagon can be inhibited by [D-Trp^{11}]NT. In contrast to NT, which reduces motorial activity on i.c.v. application, [D- Tyr^{11}]NT and [D-Phe^{11}]NT increase this effect.

The N- and C-terminal fragments of NT stimulate the phagocytosis of macrophages (10^{-7}M) [8.67].

9. Motilin [9.1], [9.2]

9.1. Occurrence

Porcine motilin (MT), a peptide containing 22 amino acids, was first isolated from side fractions of p-cholecystokinin. The precursor of h-MT has been characterized via the cDNA [9.3].

h,p-MT	F	V	P	I	F	T	Y	G	E	L	Q	R	M	Q	E	K	E	R	N	K	G	Q
ca-MT	F	V	P	I	F	T	H	S	E	L	Q	K	I	R	E	K	E	R	N	K	G	Q

Motilin is found primarily in the EC_2 cells of the jejunum. But i.r. motilin has also been found in the mucosa of the esophagus, stomach, duodenum, ileum and colon, in the pancreas, kidneys, adrenals and in the gallbladder. It is also widely distributed in the brain with the highest concentrations in the cerebellum.

Erythromycin A [9.4] and the erythromycin derivative EM-523 [9.5] are macrolide antibiotics which greatly increase the gastrointestinal motor activity and are motilin receptor agonists.

9.2. Release

The release of MT is stimulated by oral fat, exogenous and endogenous bile acids, bombesin, alkalization of the duodenum (in the dog), acidification of the duodenum (in human and pigs), stimulation of the vagus (via muscarinic and nicotinic receptors), exogenous motilin and by atropine [9.6].

The release of MT is inhibited by oral glucose, insulin, somatostatin, pancreatic peptide, VIP, and by sham feeding [9.7].

In human beings a raised level of MT was observed in diarrhea, kindney failure, in the first week after abdominal surgery and in alcoholics who do not have liver diseases [9.8].

9. 3. Biological Effects

Motilin increases intracellular Ca^{2+} by stimulating cGMP. On a molar basis, motilin has 50 times the contracting effect of acetylcholine.

The motilins appear to be species specific. *In vitro* p-MT and ca-MT contract strips of duodenum from the rabbit, but not from the guinea pig. The duodenum of the dog is contracted by ca-MT, but not by p-MT [9.9]. Motilin accelerates gastric emptying and intestinal transit by initiating phase III of the migratory motorial complex (MMC), which consists of four phases. Phase 1 is the resting phase, phase 2 consists of irregular contractions, phase 3 is the period of rhythmic contractions of maximal frequency and in phase 4 the motorial activity subsides. In humans with an empty stomach, these phases are repeated in cycles of approximately 92 minutes. The i.r.-MT peaks correlate with phase III in the duodenum of the dog, opossum and rat and with phase III in the stomach of humans. In comparison, MT has no effect on the MMC of the pig. The MT induced contraction in the small intestine of the dog may be mediated by endogenous opioides [9.10]. However, the contraction in the colon are not induced by opioides [9.11]. Intravenous corticoliberin inhibits the MT induced MMC in dogs [9.12]. MT also contracts the lower esophageal sphincter and the gallbladder (only between periods of digestion) and, like CCK, activates Oddi's sphincter [9.13]. The cyclic pancreatic secretion of water, bicarbonate, and protein is increased by MT by the shortening of the cycle length [9.14], [9.15].

In fasting, but not in well fed rats, intraperitoneally applied MT (5-10 μg/kg) increases the food uptake. This effect can be abolished by naloxone or pentagastrin. Thus, MT appears to be a hunger hormone during fasting [9.16]. An infusion of low doses of motilin accelerates the digestion of glucose, increasing the plasma levels of glucose, GIP and insulin as expected. An infusion of high doses of motilin has the opposite effect. Motilin stimulates the release of somatostatin in dogs evidently via the mediation of endogenous opiates [9.17].

MT exerts a hypotensive effet in pithed rats. It reduces the pressor activity of vasopressin and leukotriene D_4 and suppresses the release of epinephrine which is caused by stimulating the spinal cord [9.18].

9.4. Structure-Activity Relationships

p-MT-(1-16) still has 3% of the p-MT activity. The active region should be between position 6 and 16. Gly^8 can be replaced by D-Ala without loss of activity. Met^{13}, Gln^{14} and Glu^{15} appear to be of no great importance for the biological activity [9.19]. Phe^1 is important for full biological activity. The substitution of Ser or Lys for Phe^1 produces compounds that are 250 and 1000 times less active [9.20].

10. Pancreatic Spasmolytic Peptide [10.1]

10.1. Occurrence

p-Pancreatic spasmolytic peptide (PSP), a pyroglutamyl peptide containing 106 amino acids and seven disulfide bridges [10.2], was discovered in a side fraction of p-insulin. Immunoreactive PSP is found in low concentrations in a large number of organs. However, substantial amounts are located only in the pancreas. There is about 50% sequence homology between p-PSP-(14-49) and p-PSP-(63-98). There is great similarity in the sequences of p-PSP and pS_2. The latter was isolated from human gastric juice [10.3] and is induced by estrogen in the human breast cancer cell line MCF-7 [10.4], and is identical to the human epidermal growth factor like immunoreactive factor [10.5]. The structure of a protein from the skin gland of *Xenopus laevis* has been established via the cDNA. This protein has 396 amino acids, contains 33 copies of the sequence $(Ala\text{-}Pro)_{2\text{-}4}$-Ala-Glu and is known as the APEG protein [10.6]. There is a sequence at the C-terminal end of this APEG protein which is very similar to p-PSP and h-pS_2. Another protein from *Xenopus laevis*, which was characterized via the cDNA structure, contains, after possible processing two peptides which are very similar to PSP: spasmolysin I and spasmolysin II [10.7].This domain contains three disulfide bridges which have a characteristic clover leaf like structure [10.3].

10.2. Biological Effects

p-PSP inhibits the activity of adenylate cyclase in the intestinal mucosal cell membranes of the rat [10.8]. p-PSP inhibits the gastrointestinal motility and the secretion of gastric acid in laboratory animals after both parenteral (like glucagon) and oral application. It also reduces the pressure of Oddi's sphincter and the frequency of the pressure waves [10.9].

PSP acts like a growth factor on MCF-7 cells [10.3].

```
                1                                                 25
p-PSP           *Q K P A A C R C S R Q D P K N R V N C G F P G I T-

                26                                                50
p-PSP           S D Q C F T S G C C F D S Q V P G V P W C F S P L-
                    -24                                           -2
h-pS2               M A T M E N K V I C A L V L V S M L A L G T L-

                51                                                75
p-PSP           P A Q E S E E C V M Q V K A R K N S G Y P G I C P-
                -1 1                                              24
h-pS2           A E A Q T E T C T V A P R E R Q N C G F P G V T P-
                      326                                         346
APEG                  T E D C K G D P F K R T D C G Y P G I T E-
Spasmolysin I           Q D C S V A P N M R V N C G Y P T V T E-
Spasmolysin II          A E C T V D P S V R T D C G Y P G I T D-

                76                                                100
p-PSP           E D C A A R N C C F S D T I P E V P W C F F P M S-
                25                                                49
h-pS2           S Q C A N K G C C F D D T V R G V P W C F Y P N T-
                347                                               371
APEG            G Q C K A K G C C F D S S I V G V K W C F F P A Q-
Spasmolysin I:  A D C R A V G C C F D S S I L N T K W C F Y N A T-
Spasmolysin II: K E C R E K G C C Y D E C I P D V I W C F E K A V-

                101       106
p-PSP           V E D C H Y.
                50                  60
h-pS2           I D V P P E E E C E F.
                372                                               396
APEG            L V P S V Y S A P E T G R T V V I Q A L R P W N A.
Spasmolysin I:  A G P I.
Spasmolysin II: P V N N S.
```

11. References

References for Chapter 1

[1.1] A.R. Green, *Lancert I* (1989) 705.
[1.2] W. König in A. Kleemann et al. (eds.): *Arzneimittel, Fortschritte 1972 bis 1985,* VCH Verlagsgesellschft Weinheim 1987, 572-772.
[1.3] R. Geiger, *Coll. Soc. Fr. Etudes Fertil.* **26** (1988) 63.
[1.4] E. Bürger, *Arzneim.-Forsch./Drug Res.* **38** (1988) 754.
[1.5] V. Dourtoglou, B. Gross, *Synthesis* **1984,** 572.
[1.6] W. König, G. Breipohl in E. Giralt, D. Andreu (eds.): *Peptides 1990*, ESCOM Science Publishers B.V., Leiden 1991, 143
[1.7] K.M. Zsebo et al., *J. Biol. Chem.* **261** (1986) 5858.
[1.8] S.D. Emr et al., *Proc. Natl. Acad. Sci. USA* **80** (1983) 7080.
[1.9] J. Engels, E. Uhlmann in A. Fiechter (ed.): *Advances in Biochemical Engineering/Biotechnology 37,* Springer Verlag, Berlin-Heidelberg 1988, 75.
[1.10] IUPAC-IUB Commission on Biochemical Nomenclature, *J. Biol. Chem.* **247** (1972) 977.
[1.11] IUPAC-IUB Commission on Biochemical Nomenclature, *Eur. J. Biochem.* **5** (1968) 151.
[1.12] IUPAC-IUB Commission on Biochemical Nomenclature, *J. Biol. Chem.* **250** (1975) 3215.
[1.13] IUPAC-IUB Commission on Biochemical Nomenclature, *Hoppe-Seyler's Z. Physiol. Chem. 348 (1967) 262.*

References for Chapter 2

[2.1] W. König in A. Kleemann et al. (eds.): *Arzneimittel, Fortschritte 1972 bis' 1985,* VCH Verlagsgesellschaft Weinheim 1987, 672-674, 677-679.
[2.2] J.P. Adelman et al., *Proc. Natl. Acad. Sci. USA* **83** (1986) 179.
[2.3] K. Nikolics et al., *Nature (London)* **316** (1985) 511.
[2.4] S.S. Smith, S.R. Ojeda, *Endocrinology (Baltimore)* **115** (1984) 1973.
[2.5] J.A. King, R.P. Millar, *Peptides (N.Y.)* **6** (1985) 689.
[2.6] J.A. King, R.P. Millar, *Peptides (N.Y.)* **7** (1986) 827-834.
[2.7] R.C. Powell et al., *Peptides (N.Y.)* **6** (1985) 223.
[2.8] N.M. Sherwood et al., *J. Bio. Chem.* **261** (1986) 4812.
[2.9] E.G. Stopa et al., *Peptides (N.Y.)* **9** (1988) 419.
[2.10] D.D. Rasmussen, *J. Endocrinol. Invest.* **9** (1986) 427.
[2.11] B.A. Eipper et al., *Endocrinology (Baltimore)* **116** (1985) 2497.
[2.12] P.A. Melrose, *Endocrinology (Baltimore)* **121** (1987) 200.
[2.13] S.V. Drouva et al., *Neuroendocrinology* **43** (1986) 32.
[2.14] D.E. Dluzen, V.D. Ramirez, *Endocrinology (Baltimore)* **118** (1986) 1110.
[2.15] M.J. Kelly et al., *Neuroendocrinology* **49** (1989) 88.
[2.16] C.R. Parker, Jr., J.C. Porter, *J. Clin. Endorcrinol. Metab.* **58** (1984) 488.
[2.17] W.K. Samson et al., *Regul. Pep.* **2** (1981) 253.
[2.18] S. Ohtsuka et al., *Acta Endocrinol. (Copehnagen)* **117** (1988) 399.
[2.19] L.G. Allen et al., *Endocrinology (Baltimore)* **121** (1987) 1953.
[2.20] K. Gerozissis et al., *Neuroendocrinology* **40** (1985) 272.
[2.21] T. Aono et al., *Endocrinol. Jpn.* **31** (1984) 571.
[2.22] P. Bouchard et al., *J. Clin. Endorcinol. Metab.* **60** (1985) 258.
[2.23] A.K. Dubey, T.M. Plant, *Biol. Reprod.* **33** (1985) 423.
[2.24] T.J. Cicero et al., *Life Sci.* **37** (1985) 467-474.

[2.25] F. Petraglia et al., *Endocrinology (Baltimore)* **120** (1987) 1083.
[2.26] H. Rosen et al., *Endocrinology (Baltimore)* **122** (1988) 2873.
[2.27] K.E. Nikolarakis et al., *Neuroendocrinology* **44** (1986) 314.
[2.28] Z. Naor, *Endocr. Rev.* **11** (1990) 326.
[2.29] A. Jagannadha Rao, N.R. Moudgal, *IRCS Med. Sci.* **12** (1984) 1105.
[2.30] T.A. Bramley et al., *J. Endocrinol.* **108** (1986) 323.
[2.31] T.A. Bramley et al., *J. Endocrinol.* **113** (1987) 317-327.
[2.32] N. Dekel et al., *Endocrinology (Baltimore)* **123** (1988) 1205.
[2.33] M. Fekete et al., *J. Clin. Lab. Analysis* (1988) in press.
[2.34] D.R. Pieper et al., *Endocrinology (Baltimore)* **115** (1984) 1190.
[2.35] P.M. Conn et al., *Fed. Proc. Fed. Am. Soc. Exp. Biol.* **43** (1984) 2351.
[2.36] P.C. Wynn et al., *Endocrinology (Baltimore)* **119** (1986) 1852.
[2.37] A. Starzec et al., *Endocrinology (Baltimore)* **119** (1986) 561.
[2.38] G.A. Schuiling et al., *Acta Endocrinol. (Copenhagen)* **114** (1987) 488-496.
[2.39] G.A. Schuiling et al., *J. Endocrinol* **115** (1987) 469.
[2.40] W.S.A. Wun, I.H. Thorneycroft, *Mol. Cell. Endocrinol.* **54** (1987) 165.
[2.41] S.R. Swartz, G.P. Moberg, *Endocrinology (Baltimore)* **118** (1986) 2425.
[2.42] W.R. Crowley et al., *Endocrinology (Baltimore)* **120** (1987) 941.
[2.43] H. Dobson, *Acta Endocrinol. (Copenhagen)* **115** (1987) 63.
[2.44] D.E. Suter, N.B. Schwartz, *Endocrinology (Baltimore)* **117** (1985) 849.
[2.45] Pi-hsueh S. Li, *Life Sci.* **41** (1987) 2493.
[2.46] R.D. Koos, W.J. LeMaire, *Endocrinology (Baltimore)* **116** (1985) 628.
[2.47] N. Eckstein et al., *Mol. Cell. Endocrinol.* **47** (1986) 91.
[2.48] J. Wang, P.C.K Leung, *Biochem. Biophys. Res. Commun.* **146** (1987) 939.
[2.49] R.M Sharpe, I. Cooper, *J. Endocrinol.* **113** (1987) 89-96.
[2.50] A.P.N. Themmen et al., *J. Endocrinol.* **108** (1986) 431.
[2.51] B.A. Cooke, M.H.F. Sullivan, *Mol. Cell. Endocrinol.* **41** (1985) 115-122.
[2.52] S. Malozowski et al., *Horm. Metab. Res.* **17** (1985) 321.
[2.53] S.Y. Bin et al., *Contraception* **35** (1987) 79.
[2.54] T.W. Redding und A.V. Schally, *Proc. Natl. Acad. Sci. USA* **81** (1984) 248.
[2.55] T. diPaolo, P. Falardeau, *Eur. J. Pharmacol.* **102** (1984) 383.
[2.56] T. Torres-Aleman et al., *Proc. Natl. Acad. Sci. USA* **82** (1985) 1252.
[2.57] J.A. Foekens et al., *Biochem. Biophys. Rers. Commun.* **140** (1986) 550.
[2.58] B.D. Greenstein et al., *J. Endocrinol.* **112** (1987) 345.
[2.59] M.J Karten und J.E. Rivier, *Endocr. Rev.* **7** (1986) 44.
[2.60] J.J. Nestor, Jr., B.H. Vickery, *Ann. Rep. Med. Chem.* **23** (1988) 211.
[2.61] K. Folkers et al., *Proc. Natl. Acad. Sci. USA* **82** (1985) 1070.
[2.62] J.J. Nestor, Jr., et al., *J. Med. Chem.* **27** (1984) 320.
[2.63] T.L. Ho et al., *Int. J. Pept. Protein Res.* **24** (1984) 79.
[2.64] W. König et al. in T. Shiba, S. Sakakibara (eds.): *Peptide Chemistry 1987,* Protein Research Found. Osaka 1988, 591.
[2.65] J. Seprodi et al., *Biochem. Biophys. Res. Commun.* **144** (1987) 1214.
[2.66] J.J. Nestor, Jr., et al. in F. Labrie et al. (eds.): *LH-RH and its analogues,* Elsevier, Excerpta Medica, Amsterdam-New York-Oxford 1984, 24.
[2.67] J.J. Nestor, Jr., et al., *J. Med. Chem.* **31** (1988) 65.
[2.68] W. König et al., in G. Jung, E. Bayer (eds.): *Peptides 1988,* Walter de Gruyter, Berlin-New York 1989, 334.
[2.69] J.E. Rivier et al., *J. Med. Chem.* **29** (1986) 1846.
[2.70] J.E. Rivier et al., The Salk Institute for Biological Studies, EP 0 102 492, 1985.
[2.71] A. Ljungqvist et al., *Biochem. Biophys. Res. Commun.* **148** (1987) 849.
[2.72] S. Bajusz et al., *Int. J. Pept. Protein Res.* **32** (1988) 425.
[2.73] M.V. Nekola et al., *Int. Arch. Allergy Appl. Immunol.* **84** (1987) 3161.
[2.74] K. Folkers et al., *Z. Naturforsch. B. Anorg. Chem. Org. Chem.* **42B**. (1987) 101.
[2.75] S. Bajusz et al., *Proc. Natl. Acad. Sci. USA* **85** (1988) 1637-1641.
[2.76] A. Phillips et al., *Life Sci.* **41** (1987) 2017.
[2.77] A. Phillips et al., *Life Sci.* **43** (1988) 883.
[2.78] A.V. Schally et al., *Proc. Natl. Acad. Sci. USA* **84** (1987) 851-855.

[2.79] S.M. de Muinck Keizer Schrama et al., *Eur. J. Pediatr.* **146** (1987) Suppl. 2, S35.
[2.80] A. Bertelsen et al., *Eur. J. Pediatr.* **146** (1987) Suppl. 2, S40.
[2.81] S. Zgliczynski et al., *Acta Endocrinol (Copenhagen)* **105** (1984) 161.
[2.82] K.E. Anderson et al., *N. Engl. J. Med.* **311** (1984) 643.
[2.83] *Drugs of the Future* **9** (1984) 295.
[2.84] *Drugs of the Future* **3** (1978) 645; **11** (1986) 812.
[2.85] *Drugs of the Future* **8** (1983) 1038; **9** (1984) 930.
[2.86] *Drugs of the Future* **10** (1985) 355; **12** (1987) 404.
[2.87] *Drugs of the Future* **11** (1986) 623-624; **12** (1987) 720-722.
[2.88] G.C. Doelle et al., *Horm. Metab. Res.* **18** (1986) 201.
[2.89] *Drugs of the Future* **11** (1986) 745.
[2.90] A. Lemay, N. Faure, *J. Clin. Endocrinol. Metab.* **63** (1986) 1379.
[2.91] A. Lemay et al., *Fertil. Steril.* **47** (1987) 60.
[2.92] M.W. van Maarschalkerweerd, M. Gons, *Eur. J. Pediatr.* **146** (1987) 272.
[2.93] B. Bentick et al., *Br. J. Obstet. Gynaecol.* **93** (1986) 1194.
[2.94] R. Maheux et al., *Fertil. Steri.* **42** (1984) 644.
[2.95] W. Matta et al., *Br. J. Obstet. Gynaecol.* **93** (1986) 1194.
[2.96] R. Maheux et al., *Fertil. Steril.* **47** (1987) 229.
[2.97] V. Perl et al., *Fertil. Steril.* **48** (1987) 383.
[2.98] J.L. Andreyko et al., *J. Clin. Endocrinol. Metab.* **63** (1986) 854.
[2.99] K.E. Anderson et al., *Clin. Pharmacol. Ther. (St. Louis)* **41** (1987) 180.
[2.100] J.L. Gabrilove et al., *N. Engl. J. Med.* **318** (1988) 580.
[2.101] J.P. Devogelaer et al., *Lancet-II* (1987) 1498.
[2.102] W.H. Matta et al., *Br. Med. J.* **294** (1987) 1523.
[2.103] C.H. Li et al., *Proc. Natl. Acad. Sci. USA* **84** (1987) 959-962.
[2.104] W. König in A. Kleemann et al., (eds.): *Arzneimittel, Fortschritte 1972 bis 1985*, VCH Verlagsgesellschaft, Weinheim 1987, 674.
[2.105] M. Bulant et al., *J. Biol. Chem.* **263** (1988) 17189.
[2.106] A. Dutour et al., *Peptides (N.Y.)* **10** (1989) 523.
[2.107] J.F. Wilber et al., *Clin. Res.* **34** (1986) 272A.
[2.108] C.W. Hilton et al., *Neuropeptides (Edinburgh)* **15** (1990) 55.
[2.109] H. Steiner et al., *Neuropeptides (Edinburgh)* **14** (1989) 185.
[2.110] S.M. Cockle et al., *FEBS Lett.* **252** (1989) 113-117.
[2.111] M.J. Kubek et al., *Life Sci.* **36** (1985) 315.
[2.112] T.M. Engber et al., *Regul. Pept.* **12** (1985) 51.
[2.113] B.M. Lewis et al., *J. Endocrinol.* **115** (1987) 419.
[2.114] M.T. Lin et al., *Neuroendocrinology* **50** (1989) 177.
[2.115] M. Mori and M. Yamada, *J. Endocrinol.* **114** (1987) 443.
[2.116] T. Mitsuma et al., *Horm. Metab. Res.* **21** (1989) 301.
[2.117] T. Mitsuma et al., *Peptides (N.Y.)* **8** (1987) 473.
[2.118] S. Bhasin et al., *Endocrinology (Baltimore)* **114** (1984) 946.
[2.119] P. Lamberton et al., *Endocrinology (Baltimore)* **117** (1985) 1834.
[2.120] S. Aratan-Spire et al., *Acta Endocrinol. (Copenhagen)* **106** (1984) 209.
[2.121] T. Mitsuma et al., *Horm. Res.* **25** (1987) 223.
[2.122] S. Wilk, *Life Sci.* **39** (1986) 1487.
[2.123] J.L Charli et al., *Neuropeptides (Edinburgh)* **14** (1989) 191.
[2.124] Y. Oron et al., *Science (Washington, D.C.)* **238** 1406.
[2.125] M.C. Sheppard, K.I.J. Shenan, *J. Endocrinol.* **101** (1984) 101.
[2.126] L.D. Keith et al., *Neuroendocrinology* **43** (1986) 445.
[2.127] M. Mori et al., *Neuroendocrinology* **48** (1988) 153.
[2.128] H.N. Bhargava et al., *Arch. Int. Pharmacodyn. Ther.* **297** (1989) 247.
[2.129] H.N. Bhargava, *Ann. N.Y. Acad. Sci.* **553** (1989) 526.
[2.130] H.N. Bhargava, *Ann. N.Y. Acad. Sci.* **553** (1989) 528.
[2.131] A.H. Drummond, *Biochem. Biophys. Res. Commun.* **127** (1985) 63.
[2.132] R.K. Rinehart et al., *J. Pharmacol. Exp. Ther.* **238** (1986) 178.
[2.133] A.H. Drummond, P.J. Hughes, F. Ruiz Larrea, L.A. Joels, *Ann. N.Y. Acad. Sci.* **553** (1989) 197.

[2.134] E.C. Griffiths, *Clin. Science* **73** (1987) 449.
[2.135] A. Sanchez, E. Montoya, *Horm. Metab. Res.* **19** (1987) 604.
[2.136] I. Zofkova, J. Bednar, *Exp. Clin. Endocrinol.* **83** (1984) 263.
[2.137] H. Watanobe et al., *J. Endocrinol. Invest.* **8** (1985) 459.
[2.138] K. Chihara et al., *J. Clin. Endocrinol. Metab.* **58** (1984) 81.
[2.139] P. Chiodera et al., *J. Clin. Endocrinol. Metab* **59**(1984) 794.
[2.140] T. Mitsuma et al., *Exp. Clin. Endocrinol.* **89** (1987) 55.
[2.141] B. Suryaprakash et al., *IRCS Med. Sci.* **12** (1984) 895.
[2.142] W. Schäfgen et al., *Horm. Metab. Res.* **16** (1984) 615.
[2.143] J.R. Attali et al., *Endocrinology (Baltimore)* **116** (1985) 561.
[2.144] S. Aggestrup, *Scand. J. Gastroenterol.* **19** (1984) 631.
[2.145] N. Basso et al., *Gastroenterology* **84** (1983) Part 2, 1100.
[2.146] M. Bickel, personal communication.
[2.147] G. Soldani et al., *Life Sci.* **39** (1986) 595.
[2.148] L.O.D. Koskinen, *Ann. N.Y. Acad. Sci.* **553** (1989) 353.
[2.149] L.O.D. Koskinen, *Ann. N.Y. Acad. Sci.* **553** (1989) 608.
[2.150] M. Burnier et al., *Clin. Res.* **37** (1989) 249A.
[2.151] S. Reichlin, *Acta Endocrinol. (Copenhagen) Supp.* **276** (1986) 21.
[2.152] T. Suzuki et al., *Gen. Pharmacol.* **20** (1989) 239-242.
[2.153] J. Mattila and R.D. Bunag, *J. Pharmacol. Exp. Ther.* **238** (1986) 232.
[2.154] T. Garrik et al., *Life Sci.* **40** (1987) 649.
[2.155] J.P. Roussel et al., *Acta Endocrinol. (Copenhagen)* **114** (1987) 314.
[2.156] *Drugs of the Future 9* **(1984) 807-808.**
[2.157] *Drugs of the Future* **8** (1983) 1007; **11** (1986) 1059; **12** (1987) 1153; **14** (1989) 1224.
[2.158] H. Maeda et al., *Chem. Pharm. Bull.* **36** (1988) 190.
[2.159] *Drugs of the Future* **14** (1989) 203; **15** (1990) 285.
[2.160] *Drugs of the Future* **7** (1982) 21; **8** (1983) 56; **9** (1984) 62; **10** (1985) 75; **11** (1986) 6; **12** (1987) 70; **13** (1988) 73; **14** (1989) 78; **15** (1990) 92.
[2.161] H. Maeda et al., *Int. J. Pept. Protein Res.* **33** (1989) 403-411.
[2.162] *Drugs of the Future* **7** (1982) 167; **8** (1983) 282; **14** (1989) 295; **15** (1990) 319.
[2.163] *Drugs of the Future* **13** (1988) 420-423; **14** (1989) 48.
[2.164] M. Oka et al., *Arzneim. Forsch.* **39** (1989) 297.
[2.165] R. Lanzara et al., *Ann. N.Y. Acad. Sci.* **553** (1989) 559.
[2.166] R.J. Guiloff, *Ann. N.Y. Acad. Sci.* **553** (1989) 399.
[2.167] M.H. Brooke, *Ann. N.Y. Acad. Sci.* **553** (1989) 422.
[2.168] B.R. Brooks, *Ann. N.Y. Acad. Sci.* **553** (1989) 431.
[2.169] J.W. Holaday et al., *Ann. N.Y. Acad. Sci.* **553** (1989) 370.
[2.170] A.I. Faden et al., *Ann. N.Y. Acad. Sci.* **553** (1989) 380.
[2.171] A.I. Faden, *Brain Res.* **486** (1989) 228.
[2.172] M. Miyamoto et al., *Ann. N.Y. Acad. Sci.* **553** (1989) 508.
[2.173] I. Paakkari et al., *Ann. N.Y. Acad. Sci.* **553** (1989) 612.
[2.174] W. König in A. Kleemann et al. (eds.): *Arzneimittel, Fortschritte 1972 bis 1985,* VCH Verlagsgesellschaft Weinheim 1987, 741.
[2.175] J.G. Pierce, T.F. Parsons, *Ann. Rev. Biochem.* **50** (1981) 465.
[2.176] D.N. Ward et al., *J. Protein Chem.* **1** (1982) 263-280.
[2.177] F.S. Esch et al., *Proc. Natl. Acad. Sci. USA* **82** (1986) 6618.
[2.178] C.S. Liu et al., *J. Biochem.* **186** (1989) 105.
[2.179] A.M. Naylor et al., *Neuroendocrinology* **49** (1989) 531.
[2.180] S.S. Stojilkovic et al., *Science (Washington, D.C.)* **248** (1990) 1663.
[2.181] G. Saade et al., *J. Endocrinol.* **114** (1987) 95.
[2.182] G. Emons et al., *Acta Endocrinol. (Copenhagen)* **113** (1986) 219.
[2.183] G.A. Schuiling et al., *Acta Endocrinol. (Copenhagen)* **110** (1985) 329.
[2.184] L.N. Sidneva et al., *Probl. Endokrinol.* **33** (1987) 67-70.
[2.185] I. Porsova et al., *Horm. Res.* **28** (1987) 288.
[2.186] L. Kiesel et al., *Mol. Cell. Endocrinol.* **51** (1987) 1.
[2.187] L. Wogensen J. Warberg, *Acta Endocrinol. (Copenhagen)* **106** (1984) 30.
[2.188] P.M. Conn et al., *Biochem. Biophys. Res. Commun.* **126** (1985) 532.

[2.189] C.G. Brown, N.L. Poyser, *J. Endocrinol.* **103** (1984) 155.
[2.190] M. Filicori et al., *J. Clin. Endocrinol. Metab.* **62** (1986) 1136.
[2.191] P.A. Steele, S.J. Judd, *Clin. Endocrinol.* **29** (1988) 1.
[2.192] T. Nishi et al., *Acta Obstet. Gynecol. Scand.* **66** (1987) 309.
[2.193] A. Tsatsoulis et al., *Clin. Endocrinol.* **32** (1990) 73.
[2.194] F.R.A. Cagampang et al., *Horm. Metab. Res.* **22** (1990) 269.
[2.195] T. Koh et al., *Life Sci.* **34** (1984) 1635.
[2.196] E.M. McColl et al., *Clin. Endocrinol.* **31** (1989) 617.
[2.197] F. Petraglia et al., *Life Sci.* **38** (1986) 2103.
[2.198] S.E. Lenahan et al., *Neuroendocrinology* **44** (1986) 89.
[2.199] M.S. Smith et al., *Neuroendocrinology* **50** (1989) 308.
[2.200] J.H. Mendelson et al., *J. Pharmacol. Exp. Ther.* **237** (1986) 862.
[2.201] G.B. Melis et al., *J. Clin. Endocrinol. Metab.* **65** (1987) 237.
[2.202] I.R. Reid, *Ann. Intern. Med.* **106** (1987) 639.
[2.203] M. Gambacciani et al., *J. Clin. Endocrinol. Metab.* **65** (1987) 465.
[2.204] K. Seki and I. Nagata, *Acta Endocrinol. (Copenhagen)* **122** (1990) 211.
[2.205] C.R. Pohl et al., *Endocrinology (Baltimore)* **120** (1987) 849.
[2.206] S.J. Kim et al., *Biochem. Biophys. Res. Commun.* **146** (1987) 989.
[2.207] S.H Sheikha et al., *Life Sci.* **44** (1989) 1363-1369.
[2.208] S. Taleisnik and C.H. Sawyer, *Neuroendocrinology* **44** (1986) 265.
[2.209] H. Loosfelt et al., *Science (Washington, D.C.)* **245** (1989) 525.
[2.210] K.C. McFarland et al., *Science* **245** (1989) 494.
[2.211] B. Kessel et al., *Biol. Reprod* **32** (1985) 1038.
[2.212] R. Farookhi, J. Desjardins, *Mol. Cell. Endocrinol.* **47** (1986) 13.
[2.213] X.-C. Jia et al., *Endocrinology (Baltimore)* **117** (1985) 13.
[2.214] P.E. Gottschall et al., *Biochem. Biophys. Res. Commun.* **149** (1987) 502.
[2.215] J.S. Davis et al., *Biochem. J.* **238** (1986) 597.
[2.216] V.V. Patwardhan, A. Lanthier, *J. Endocrinol.* **101** (1984) 305.
[2.217] J.J. Boniface, L.E. Reichert, Jr., *Science (Washington, D.C.)* **247** (1990) 61.
[2.218] H.T. Keutmann et al., *Proc. Natl. Acad. Sci. USA* **84** (1987) 2038.
[2.219] T.A. Santa Coloma et al., *Biochemistry* **29** (1990) 1194.
[2.220] T.A. Santa Coloma, L.E. Reichert, Jr., *J. Biol. Chem.* **265** (1990) 5037.
[2.221] M.C. Charlesworth et al., *J. Biol. Chem.* **262** (1987) 13409.
[2.222] B.B. Saxena, P. Rathnam, *Biochemistry* **24** (1985) 813.
[2.223] J.-M. Bidart et al., *Science (Washington, D.C.)* **248** (1990) 736.
[2.224] L. Liu et al., *Endocrinology (Baltimore)* **124** (1989) 175.
[2.225] A. Traub, Jr., *J. Med. Sci.* **155** (1986) Suppl. 30.
[2.226] A. Lanzone et al., *Fertil. Steril.* **48** (1987) 1058.
[2.227] W.G. Rossmanith et al., *Int. J. Fertil.* **32** (1987) 460.
[2.228] Y. Nakamura et al., *Fertil. Steril.* **51** (1989) 423.
[2.229] J.A. Magner, *Endocr. Rev.* **11** (1990) 354.
[2.230] J.A. Franklyn, M.C. Sheppard, *J. Endocrinol.* **117** (1988) 161.
[2.231] K. Törnquist, C. Lamberg-Allardt, *Horm. Metabol. Res.* **18** (1986) 69.
[2.232] N.W. Kasting, J.B. Martin, *Neuroendocrinology* **39** (1984) 201.
[2.233] M.D. Lumpkin et al., *Science (Washington, D.C.)* **235** (1987) 1070.
[2.234] V. Rettori et al., *Proc. Natl. Acad. Sci. USA* **86** (1989) 4789.
[2.235] S.W.J Lamberts et al., *J. Endocrinol. Invest.* **7** (1984) 313.
[2.236] F. Sanchez-Franco L. Cacicedo, *Hormone Res.* **24** (1986) 55.
[2.237] R.J. Comi et al., *N. Engl. J. Med.* **317** (1987) 12.
[2.238] M. Parmentier et al., *Science (Washington, D.C.)* **246** (1989) 1620.
[2.239] Y. Nagayama et al., *Biochem. Biophys. Res. Commun.* **165** (1989) 1184.
[2.240] F. Libert et al., *Biochem. Biophys. Res. Commun.* **165** (1989) 1250.
[2.241] M. Misrahi et al., *Biochem. Biophys. Res. Commun.* **166** (1990) 394.
[2.242] N. Bagchi, T.R. Brown, *J. Endocrinol.* **106** (1985) 153.
[2.243] E. Bone et al., *Biochem. Biophys. Res. Commun.* **141** (1986) 1261.
[2.244] P.P. Roger, J.E. Dumont, *Biochem. Biophys. Res. Commun.* **149** (1987) 707.
[2.245] G. Zerek Melen et al., *Res. Exp. Med.* **187** (1987) 415.

[2.246] A. Nilsson, R. Axelsson, *Curr. Ther. Res.* **46** (1989) 85.
[2.247] J.C. Morris III et al., *Endocrinology* **123** (1988) 456.
[2.248] J.C. Morris et al., *J. Biol. Chem.* **265** (1990) 1881.
[2.249] K.L. Hsi et al., *FEBS Lett.* **147** (1982) 261.
[2.250] G.J.M. Martens, *FEBS Lett.* **234** (1988) 160.
[2.251] G.J.M. Martens, *Eur. J. Biochem.* **181** (1989) 75.
[2.252] J.-Y. Deng et al., *Neuroendocrinolgy* **44** (1986) 373-377.
[2.253] J.S.D. Chan et al., *Can. J. Physiol. Pharmacol.* **64** (1986) 1259.
[2.254] U. Leonhardt et al., *Acta Endocrinol. (Copenhagen)* **120** (1989) 289.
[2.255] R.W. Padgett et al., *Nature (London)* **325** (1987) 81.
[2.256] D.L. Weeks, D.A. Melton, *Cell* **51** (1987) 861.
[2.257] K. Lyons et al., *Proc. Natl. Acad. Sci. USA* **86** (1989) 4554.
[2.258] N. Ling et al., *Vitam. Horm. (N.Y.)* **44** (1988) 1.
[2.259] S.-Y. Ying, *Endocr. Rev.* **9** (1988) 267.
[2.260] R.I. McLachlan et al., *Biochem. Biophys. Res. Commun.* **140** (1986) 485.
[2.261] P.E. Sawchenko et al., *Nature (London)* **334** (1988) 615, 2406-2410.
[2.262] T. Suzuki et al., *Mol. Cell. Endocrinol.* **54** (1987) 185-195.
[2.263] C.G. Tsonis et al., *J. Endocrinol.* **112** (1987) R11.
[2.264] F. Petraglia et al., *Science (Washington, D.C.)* **237** (1987) 187.
[2.265] Z. Zhiwen et al., *Endocrinology (Baltimore)* **120** (1987) 1633.
[2.266] C. Rivier, W. Vale, *Endocrinology (Baltimore)* **124** (1989) 195.
[2.267] R.I. McLachlan et al., *J. Clin. Endocrinol. Metab.* **65** (1987) 954-961.
[2.268] C.P. Channing et al., *Proc. Soc. Exp. Biol. Med.* **178** (1985) 339.
[2.269] C.B. Sheckter et al., *J. Clin. Endocrinol. Metab.* **67** (1988) 1221.
[2.270] R.I. McLachlan et al., *J. Clin. Endocrinol. Metab.* **67** (1988) 1305.
[2.271] R.E. Lappöhn et al., *N. Engl. J. Med.* **321** (1989) 790-793.
[2.272] S.A. Shanbhag et al., *J. Endocrinol.* **103** (1984) 389-393.
[2.273] J.E. Mercer et al., *Mol. Cell. Endocrinol.* **53** (1987) 252-254.
[2.274] Q.F. Wang et al., *Endocrinology (Baltimore)* **124** (1989) 363.
[2.275] B.V. Bapat et al., *Int. J. Fertil.* **31** (1986) 71-76.
[2.276] S.-Y. Ying et al., *Biochem. Biophys. Res. Commun.* **136** (1986) 969.
[2.277] W. Vale et al., *Nature (London)* **321** (1986) 776.
[2.278] N. Ling et al., *Biochem. Biophys. Res. Commun.* **138** (1986) 1129.
[2.279] N. Ling et al., *Nature (London)* **321** (1986) 779-782.
[2.280] T. Nakamura et al., *Science (Washington, D.C.)* **247** (1990) 836.
[2.281] R.H. Schwall et al., *Biochem. Biophys. Res. Commun.* **151** (1988) 1099.
[2.282] L.A. Hutchinson et al., *Biochem. Biophys. Res. Commun.* **146** (1987) 1405.
[2.283] J. Massague, *Cell* **49** (1987) 437.
[2.284] J. Yu et al., *Nature (London)* **330** (1987) 765.
[2.285] M.B. Sporn et al., *Science (Washington, D.C.)* **233** (1986) 532.
[2.286] R. Derynck et al., *Nature (London)* **316** (1985) 701.
[2.287] R. Pircher et al., *Biochem. Biophys. Res. Commun.* **136** (1986) 30.
[2.288] G. Ranganathan et al., *Biochem. Biophys. Res. Commun.* **148** (1987) 1503.
[2.289] T. Ikeda et al., *Biochemistry* **26** (1987) 2406.
[2.290] R. Derynck et al., *EMBO J.* **7** (1988) 3737.
[2.291] P. Kondaiah et al., *J. Biol. Chem.* **265** (1990) 1089.
[2.292] L.M. Wakefield et al., *J. Biol. Chem.* **263** (1988) 7646.
[2.293] T.A. McCaffrey et al., *J. Cell. Biol.* **109** (1989) 441.
[2.294] J. Pfeilschifter, G.R. Mundy, *Proc. Natl. Acad. Sci. USA* **84** (1987) 2024-2028.
[2.295] R.O.C. Oreffo et al., *Biochem. Biophys. Res. Commun.* **158** (1989) 817.
[2.296] M. Centrella et al., *Proc. Natl. Acad. Sci. USA* **85** (1988) 5889.
[2.297] S.Y. Ying et al., *Biochem. Biophys. Res. Commun.* **135** (1986) 950.
[2.298] E.Y. Adashi, C.E. Resnick, *Endocrinology (Baltimore)* **119** (1986) 1879.
[2.299] M. Knecht et al., *Endocrinology (Baltimore)* **120** (1987) 1243-1249.
[2.300] M. Hotta, A. Baird, *Proc. Natl. Acad. Sci. USA* **83** (1986) 7795.
[2.301] M. Hotta, A. Baird, *Endocrinology (Baltimore)* **121** (1987) 150-159.
[2.302] Y. Totsuka et al., *Biochem. Biophys. Res. Commun.* **158** (1989) 1060.

[2.303] H. Kurihara et al., *Biochem. Biophys. Res. Commun.* **159** (1989) 1435.
[2.304] E.B. Leof et al., *Proc. Natl. Acad. Sci. USA* **83** (1986) 2453.
[2.305] H.L. Moses et al., *Cancer Cells* **3** (1985) 65.
[2.306] T. Nakamura et al., *Biochem. Biophys. Res. Commun.* **133** (1985) 1042.
[2.307] A.B. Roberts et al., *Proc. Natl. Acad. Sci. USA* **82** (1985) 119.
[2.308] M. Frater-Schröder et al., *Biochem. Biophys. Res. Commun.* **137** (1986) 295.
[2.309] A. Baird, T. Durkin, *Biochem. Biophys. Res. Commun.* **138** (1986) 476.
[2.310] J. Massague et al., *Proc. Natl. Acad. Sci. USA* **83** (1986) 8206.
[2.311] B. Like, J. Massague, *J. Biol. Chem.* **261** (1986) 13426.
[2.312] K. Takehara et al., *Cell* **49** (1987) 4151.
[2.313] N.E. Wikner et al., *J. Invest. Dermatol.* **88** (1987) 524.
[2.314] E. Balza et al., *FEBS Lett.* **228** (1988) 42.
[2.315] W. Douglas Appling et al. *FEBS Lett.* **250** (1989) 541.
[2.316] D.R. Edwards et al., *EMBO J.* **6** (1987) 1899.
[2.317] J. Liu, J.M. Davidson, *Biochem. Biophys. Res. Commun.* **154** (1988) 895.
[2.318] A. Bassols, J. Massague, *J. Biol. Chem.* **263** (1988) 3039.
[2.319] S. Chandrasekhar, A.K. Harvey, *Biochem. Biophys. Res. Commun.* **157** (1988) 1352.
[2.320] J.R. Gamble, M.A. Vadas, *Science (Washington, D.C.)* **242** (1988) 97.
[2.321] S. Tsunawaki et al., *Nature (London),* **334** (1988) 260.
[2.322] H.J. Andrews et al., *Biochem. Biophys. Res. Commun.* **162** (1989) 144.
[2.323] A.M. Lefer et al., *Science (Washington, D.C.)* **249** (1990) 61.
[2.324] J.-C. Carel et al., *Proc. Natl. Acad. Sci. USA* **87** (1990) 1591.
[2.325] M. Centrella et al., *Endocrinology (Baltimore)* **119** (1986) 2306.
[2.326] M. Noda et al., *J. Biol. Chem.* **263** (1988) 13916.
[2.327] C. Chenu et al., *Proc. Natl. Acad. Sci. USA* **85** (1988) 5683.
[2.328] W. Craig Hooper et al., *Biochem. Biophys. Res. Commun.* **165** (1989) 145.
[2.329] R. Montesano, L. Orci, *Proc. Natl. Acad. Sci. USA* **85** (1988) 4894.
[2.330] M.R. Urist, *Science (Washington, D.C.)* **150** (1965) 893.
[2.331] R.W. Katz, A.H. Reddi, *Biochem. Biophys. Res. Commun.* **157** (1988) 1253.
[2.332] H. Heumann, K. Schmidt. *Bild Wiss.* **5** (1986) 46.
[2.333] J.M. Wozney et al., *Science (Washington, D.C.)* **242** (1988) 1528.
[2.334] F.P. Luyten et al., *J. Biol. Chem.* **264** (1989) 13377.
[2.335] E. Özkaynak et al., *EMBO J.* **9** (1990) 2085.
[2.336] T.K. Sampath et al., *Biochem. Biophys. Res. Commun.* **124** (1984) 829.
[2.337] A.H. Reddi, *Isr. J. Med. Sci.* **21** (1985) 312.
[2.338] N. Muthukumaran, A.H. Reddi, *Clin. Orthop. Relat. Res.* **200** (1985) 159.
[2.339] E.A. Wang et al., *Proc. Natl. Acad. Sci. USA* **87** (1990) 2220.
[2.340] J.J. Vandersteenhoven et al., *Calcif. Tissue Int.* **42** (1988) 39.
[2.341] H.S. Talwar et al., *Kidney Int.* **29** (1986) 1038.
[2.342] D.P. Desimone, A.H. Reddi, *Calcif. Tissue Int.* **35** (1983) 732.
[2.343] R. Howes et al., *Calcif. Tissue Int.* **42** (1988) 34.
[2.344] F.W. Thielemann et al., *Z. Orthop. Ihre Grenzgeb.* **122** (1984) 843.
[2.345] R.L. Cate et al., *Cell* **45** (1986) 685.
[2.346] J.P. Coughlin et al., *Mol. Cell. Endocrinol.* **49** (1987) 75.
[2.347] R.B. Pepinsky et al., *J. Biol. Chem.* **263** (1988) 18961.
[2.348] S.-Y. Ying, *Endocr. Rev.* **9** (1988) 267.
[2.349] S. Shimasaki et al., *Proc. Natl. Acad. Sci.* **85** (1988) 4218.
[2.350] T. Nakamura et al., *Science (Washington, D.C.)* **247** (1990) 836.

References for Chapter 3

[3.1] W. König in A. Kleemann et al. (eds.): *Arzneimittel, Fortschritte 1972 bis 1985,* VCH Verlagsgesellschaft, Weinheim 1987.
[3.2] M.E. Joyce et al., *J. Cell. Biol.* **110** (1990) 2195.
[3.3] H.M. Harms et al., *J. Clin. Endocrinol. Metab.* **6** (1989) 843.
[3.4] W. König N. Neubauer, personal communication.

[3.5] L.H. Breimer et al., *Biochem. J.* **255** (1988) 377.
[3.6] J.F. Habener et al., *Physiol. Rev.* **64** (1984) 985.
[3.7] T.J. Martin, L.J. Suva, *Clin. Endocrinol.* **31** (1989) 631.
[3.8] J.J. Orloff et al., *Endocr. Rev.* **10** (1989) 476.
[3.9] S. Balabanova et al., *Klin. Wochenschr.* **64** (1986) 173.
[3.10] L.J. Suva et al., *Science* **237** (1987) 893.
[3.11] E. af Ekenstam et al., *Clin. Endocrinol.* **32** (1990) 323.
[3.12] T. Motokura et al., *Biochem. Biophys. Res. Commun.* **154** (1988) 1182.
[3.13] M.A. Thiede, G.A. Rodan, *Science* **242** (1988) 278.
[3.14] M.A. Thiede et al., *Proc. Natl. Acad. Sci. USA* **85** (1988) 4605.
[3.15] J. Russell, L.M. Sherwood, *Mol. Endocrinol.* **3** (1989) 325.
[3.16] K.L. Insogna et al., *N. Engl. J. Med.* **313** (1985) 1126.
[3.17] M.E. Markowitz et al., *J. Clin. Endocrinol. Metab.* **67** (1988) 1068.
[3.18] S. Ljunghall et al., *Exp. Clin. Endocrinol.* **84** (1984) 319.
[3.19] U. Alon, J.C.M. Chan, *J. Clin Endocrinol. Metab.* **58** (1984) 671.
[3.20] M.S. Calvo et al., *J. Clin. Endocrinol. Metab.* **66** (1988) 823.
[3.21] L.A. Fitzpatrick et al., *Biochem. Biophys. Res. Commun.* **138** (1986) 960.
[3.22] A. Goulding, E. Gold, *Horm. Metab. Res.* **20** (1988) 743.
[3.23] M.S. Seshadri et al., *Endocrinology* **117** (1985) 2417.
[3.24] J. Birnbaum et al., *J. Clin. Endocrinol. Metab.* **66** (1988) 1187.
[3.25] E.W. Seely et al., *Acta Endocrinol. (Copenhagen)* **121** (1989) 174.
[3.26] S. Balabanova et al., *Horm. Res.* **24** (1986) 302.
[3.27] N. Kane-Johnson et al., *Metab. Clin. Exp.* **36** (1987) 580.
[3.28] S. Ljunghall et al., *Exp. Clin. Endocrinol.* **84** (1984) 313.
[3.29] C. Greenberg et al., *Metab. Clin. Exp.* **36** (1987) 151.
[3.30] B. Duarte et al., *J. Clin. Endocrinol. Metab.* **66** (1988) 584.
[3.31] T. Sugimoto et al., *Endocrinology* **125** (1989) 638.
[3.32] J. Toffaletti et al., *Endocrinol. Metab.* **60** (1985) 874.
[3.33] H.-M. Harms et al., *J. Clin. Endocrinol. Metab.* **69** (1989) 843.
[3.34] Y.-L. Chan et al., *Calcif. Tissue Int.* **338** (1986) 27.
[3.35] J. Silver et al., *Proc. Natl. Acad. Sci. USA* **82** (1985) 4270.
[3.36] J. Russell et al., *Endocrinology* **119** (1986) 2864.
[3.37] O. Fermet et al., *J. Endocrinol.* **113** (1987) 117.
[3.38] K.B. Franz, *N. Engl. J. Med.* **320** (1989) 1697.
[3.39] G.A. Williams et al., *Metab. Clin. Exp.* **34** (1985) 612.
[3.40] S.A. Rabbani et al., *J. Biol. Chem.* **259** (1984) 2949.
[3.41] J. Lefebvre, *Horm. Res.* **32** (1989) 104.
[3.42] L.E. Mallette, *Am. J. Med. Sci.* **293** (1987) 239.
[3.43] D.N Kalu et al., *Endocrinology* **122** (1988) 1847.
[3.44] S. Hirschel-Scholz et al., *Calcif. Tissue Int.* **40** (1987) 103.
[3.45] M.J. Atkinson et al., *Clin. Endocrinol.* **19** (1983) 21.
[3.46] E.F. Rittinghaus et al., *Acta Endocrinol.* **111** (1986) 62.
[3.47] V. Fonseca et al., *J. Clin. Endocrinol. Metab.* **64** (1987) 873.
[3.48] T. Ledendecker et al., *J. Clin. Pharmacol.* **27** (1987) 710.
[3.49] V. Fonseca et al., *J. Clin. Endocrinol. Metab.* **64** (1987) 873.
[3.50] H. Yoshida et al., *J. Clin. Endocrinol. Metab.* **67** (1988) 488.
[3.51] T. Andersen et al., *Metab. Clin. Exp.* **35** (1986) 147.
[3.52] G. Herrmann et al., *J. Endocrinol. Invest.* **9** (1986) 256.
[3.53] S. Ljunghall et al., *Exp. Clin. Endocrinol.* **88** (1986) 95.
[3.54] S.J. Silverberg et al., *N. Engl. J. Med.* **320** (1989) 277.
[3.55] K. Ikeda et al., *J. Biol. Chem.* **264** (1989) 15743.
[3.56] D. Wynick et al., *J. Endocrinol.* **123** (1989) Suppl., 36.
[3.57] I. Yamamoto et al., *Endocrinology* **122** (1988) 1208.
[3.58] A.P. Teitelbaum et al., *Endocrinology* **118** (1986) 595.
[3.59] S.B. Rodan et al., *Endocrinology* **115** (1984) 951.
[3.60] B.D. Catherwood, *J. Biol. Chem.* **260** (1985) 736.
[3.61] J.O. Gordeladze et al., *Scand. J. Clin. Lab. Invest.* **46**, Suppl. 185 (1986) 125.

[3.62] S. Fukayama, A.H. Tashjian, Jr., *Endocrinology* **124** (1989) 397.
[3.63] D. ST. J. O'Reilly et al., *J. Endocrinol.* **108** (1986) Suppl. Mar., 256.
[3.64] S. Shapiro et al., *Calcif. Tissue Int.* **46** (1990) 60.
[3.65] A.K. Hall, I.R. Dickson, *Acta Endocrinol.* **108** (1985) 217.
[3.66] T. Okazaki et al., *Metab. Clin. Exp.* **33** (1984) 710.
[3.67] A. van der Plas et al., *Biochem. Biophys. Res. Commun.* **129** (1985) 918.
[3.68] M. De Luise, M. Harker, *J. Endocrinol.* **111** (1986) 61.
[3.69] J.A. Yee, *J. Bone Mineral Res.* **3** (1988) 211.
[3.70] C.W.G.M. Löwik et al., *Biochem. Biophys. Res. Commun.* **162** (1989) 1546.
[3.71] P.M.J. McSheehy, T.J. Chambers, *Endocrinology* **118** (1986) 824.
[3.72] T.J. Chambers et al., *J. Endocrinol.* **102** (1984) 281.
[3.73] P.M.J. McSheehy, T.J. Chambers, *Endocrinology* **119** (1986) 1654.
[3.74] C.W.G.M. Löwik et al., *Calcif. Tissue Int.* **43** (1988) 7.
[3.75] M. Noda, K. Yoon, G.A. Rodan, *J. Biol. Chem.* **263** (1988) 18574.
[3.76] E.M. Spencer et al., *Acta Endocrinol. (Copenhagen)* **121** (1989) 435.
[3.77] M. Pines, S. Hurwitz, *Endocrinology* **123** (1988) 360.
[3.78] M. Noda, G.A. Rodan, *J. Cell Biology* **108** (1989) 713.
[3.79] S.F. Silverton et al., *Am. J. Physiol.* **253** (1987) E670.
[3.80] I.P. Braidman et al., *J. Endocrinol.* **111** (1986) 17.
[3.81] G.E. Hall, A.D. Kenny, *J. Pharmacol. Exp. Therap.* **238** (1986) 778.
[3.82] E.M. Spencer et al., *Acta Endocrinol.* **121** (1989) 435.
[3.83] M. Centrella et al., *Proc. Natl. Acad. Sci. USA* **85** (1988) 5889.
[3.84] Ö. Ljunggren, U.H. Lerner, *Acta Endocrinol. (Copenhagen)* **120** (1989) 357.
[3.85] K. Hove et al., *Endocrinology* **114** (1984) 897.
[3.86] J. Wortsman et al., *J. Clin. Endocrinol. Metab.* **62** (1986) 1305.
[3.87] T. Shigematsu et al., *Endocrinology* **118** (1986) 1583.
[3.88] H.A.P. Pols et al., *Biochem. Biophys. Res. Commun.* **156** (1988) 588.
[3.89] H.N. Hulter, J.C. Peterson, *Metab. Clin. Exp.* **33** (1984) 662.
[3.90] R.D. Wright et al., *J. Endocrinol.* **102** (1984) 375.
[3.91] L.L.S. Mok et al., *Endocr. Rev.* **10** (1989) 420.
[3.92] J.P. Dowe, I.G. Joshua, *Peptides* **8** (1987) 443.
[3.93] P.K.T. Pang et al., *Can. J. Physiol. Pharmacol.* **64** (1986) 1543.
[3.94] M.C.M. Yang et al., *J. Pharmacol. Exp. Therap.* **252** (1990) 840.
[3.95] S. Anderson et al., *Hypertension* **5** (1983) Suppl. I, 59.
[3.96] Y. Saglikes et al., *Am. J. Physiol,* **248** (1985) F674.
[3.97] K. Iseki et al., *Am. J. Physiol.* **250** (1986) F924.
[3.98] T.D. Wingert et al., *Endocrinology* **116** (1985) 2469.
[3.99] G.Z. Fadda et al., *Am. J. Physiol.* **258** (1990) E975.
[3.100] R.H.K. Mak et al., *J. Clin. Endocrinol. Metab.* **60** (1985) 229.
[3.101] J. Rosenberg et al., *Isr. J. Med. Sci.* **23** (1987) 929.
[3.102] M.A. Gorozhankina et al., *Patol. Fiziol. Eksp. Ter.* **2** (1986) 71.
[3.103] G. Clementi et al., *Eur. J. Pharmacol.* **166** (1989) 549.
[3.104] N. Horiuchi et al., *Science* **238** (1987) 1566.
[3.105] D.D. Thompson et al., *Proc. Natl. Acad. Sci. USA* **85** (1988) 5673.
[3.106] R.A. Nissenson et al., *J. Biol. Chem.* **263** (1988) 12866.
[3.107] P.B.J. Burton et al., *Biochem. Biophys. Res. Commun.* **167** (1990) 1134.
[3.108] M.P. Caulfield et al., *Endocrinology* **123** (1988) 2949.
[3.109] J.J. Helwig et al., *Eur. J. Pharmacol.* **140** (1987) 247.
[3.110] A.L. Frelingerl, III, J.E. Zull, *J. Biol. Chem.* **259** (1984) 5507.
[3.111] R.L. Shew, P.K.T. Pang, *Peptides* **5** (1984) 485.
[3.112] S.H. Doppelt et al., *Proc. Natl. Acad. Sci. USA* **83** (1986) 7557.
[3.113] J.W. Hamilton et al., *Mol. Cell. Endocrinol.* **44** (1986) 179.
[3.114] M.E. Goldman et al., *Endocrinology* **123** (1988) 2597.
[3.115] M. Chorev et al., *Biochemistry* **29** (1990) 1580.
[3.116] J. Reeve et al., *Eur. J. Clin. Invest.* **17** (1987) 421.
[3.117] E.F. Rittinghaus et al., *Acta Endocrinol.* **117**, Suppl. 287 (1988) 167.
[3.118] L.E. Mallette, *Ann. Intern. Med.* **109** (1988) 800.

[3.119] D. Goltzman, G.S. Tannenbaum, *Brain Res.* **416** (1987) 1.
[3.120] E. Bucht et al., *Acta Physiol. Scand.* **126** (1986) 289.
[3.121] S. Arver et al., *Acta Physiol. Scand.* **122** (1984) 461.
[3.122] E. Bucht et al., *Acta Endocrinol.* **113** (1986) 529.
[3.123] J.M. Le Moullec et al., *FEBS Lett.* **167** (1984) 93.
[3.124] F. Lasmoles et al., *FEBS Lett.* **180** (1985) 113.
[3.125] J. Glowacki et al., *Endocrinology* **116** (1985) 827.
[3.126] D.M. Burns et al., *Proc. Natl. Acad. Sci. USA* **86** (1989) 9519.
[3.127] P. Laurberg, J.F. Rehfeld. *J. Endocrinol.* **115** (1987) 77.
[3.128] F. Raue et al., *Endocrinology* **115** (1984) 2362.
[3.129] B. Mansson et al., *Acta Endocrinol. (Copenhagen),* **115** (1987) 297.
[3.130] M. Gozariu et al., *Curr. Ther. Res.* **45** (1989) 116.
[3.131] C. Greenberg et al., *Endocrinology* **118** (1986) 2594.
[3.132] H. Scheruebl et al., *Acta Endocrinol.* **117**, Suppl. 287 (1988) 66.
[3.133] O. Torring et al., *Horm. Metab. Res.* **17** (1985) 536.
[3.134] B. Rasmusson, *Acta Endocrinol.* **106** (1984) 112.
[3.135] F. Tagliaro et al., *J. Endocrinol. Invest.* **7** (1984) 331.
[3.136] A. Lundin, K. Öberg, *Acta Med. Scand.* **215** (1984) 281
[3.137] N. Conte et al., *Acta Endocrinol.* **106** (1984) 109.
[3.138] J. D. Zajac et al., *Endocrinology* **116** (1985) 749.
[3.139] J.J. Body et al., *J. Clin Endocrinol. Metab.* **62** (1986) 700.
[3.140] G. Luisetto et al., *J. Endocrinol. Invest.* **9** (1986) 239.
[3.141] J. Zseli et al., *Calcif. Tissue Int.* **39** (1986) 297.
[3.142] J.-Y. Reginster et al., *Gynecol. Endocrinol.* **2** (1988) 195.
[3.143] C. Foresta et al., *Horm. Metab. Res.* **19** (1987) 275.
[3.144] N. Kübler et al., *Exp. Clin. Endocrinol.* **90** (1987) 324.
[3.145] N. Takahashi et al., *Endocrinology* **123** (1988) 1504.
[3.146] H.-G. Schneider et al., *Mol. Cell. Endocrinol.* **58** (1988) 9.
[3.147] C.M. Rotella et al., *Eur. J. Pharmacol.* **107** (1985) 347.
[3.148] J. Barsony, S.J. Marx, *Endocrinology* **122** (1988) 1218.
[3.149] A.N.K. Yusufi et al., *Am. J. Physiol.* **252** (1987) F598.
[3.150] F. Pansini et al., *J. Endocrinol.* **116** (1988) 155.
[3.151] D.R. Roy, *Can. J. Physiol. Pharmacol.* **63** (1985) 89.
[3.152] J. Puig et al., *Acta Pharmacol. Toxicol.* **59** (1986), Suppl. 5, Pt. 2, 126.
[3.153] M. Yamada et al., *Endocrinology* **116** (1985) 693.
[3.154] T.J. Chambers et al., *J. Clin. Endocrinol. Metab.* **63** (1986) 1080.
[3.155] A. Malgaroli et al., *J. Biol. Chem.* **264** (1989) 14342.
[3.156] H.I. Khouja et al., *J. Endocrinol.* **119** (1988) Suppl. 152.
[3.157] J.R. Farley et al., *Calcif. Tissue Int.* **45** (1989) 214.
[3.158] W.M. Burch, *Endocrinology* **114** (1984) 1196.
[3.159] G.P. Ceda et al., *Acta Endocrinol. (Copenhagen)* **120** (1989) 416.
[3.160] C. Beglinger et al., *Gut* **29** (1988) 243.
[3.161] G.H. Greeley, Jr. et al., *Regul. Pept.* **24** (1989) 259.
[3.162] M. Yamaguchi, M. Toyoizumi, *Horm. Metab. Res.* **18** (1986) 378.
[3.163] M. Yamaguchi, K. Momose, *Horm. Metab. Res.* **18** (1986) 22.
[3.164] M. Yamaguchi et al., *Horm. Metab. Res.* **17** (1985) 346.
[3.165] L. Laurian et al., *Horm. Metab. Res.* **18** (1986) 268.
[3.166] E. Rapisarda et al., *Pharmacol. Res. Commun.* **16** (1984) 1151.
[3.167] R. Franceschini et al., *Med. Sci. Res.* **16** (1988) 1279.
[3.168] B.J. Looij, Jr. et al., *Ned. Tijdschr. Geneeskd.* **132** (1988) 1558.
[3.169] G.V. Shah et al., *J. Endocrinol.* **116** (1988) 279.
[3.170] I. Zofkova, J. Nedvidkova, *Exp. Clin. Endocrinol.* **92** (1989) 262.
[3.171] T.S. Mahrous, A.M. Nakhla, *Acta Endocrinol. (Copenhagen)* **119** (1988) 525.
[3.172] I. Zofkova, J. Bednar, *Exp. Clin. Endocrinol.* **92** (1988) 268.
[3.173] I. Zofkova, J. Bednar, *Exp. Clin. Endocrinol.* **92** (1989) 268.
[3.174] G. Clementi et al., *J. Endocrinol. Invest.* **8** (1985) 543.
[3.175] H. Ohno et al., *Jpn. J. Pharmacol.* **37** (1985) 67-75.

[3.176] Y. Goto et al., *Can. J. Physiol. Pharmacol.* (1986) July, Suppl., 110.
[3.177] W. Woloszczuk et al., *Horm. Metab. Res.* **18** (1986) 197.
[3.178] P. Demol et al., *Gastroenterology* **86** (1984) Pt. 2, 1060.
[3.179] D.V. Kleist et al., *Gastroenterology* **86** (1984) Pt. 2, 1290.
[3.180] C.J. Fimmel et al., *Experientia* **39** (1983) 632.
[3.181] A.S. Levine et al., *Psychopharmacol. Bull.* **20** (1984) 459.
[3.182] M.J. Fargeas et al., *Science* **225** (1984) 1050.
[3.183] H.J. Lenz, *Am. J. Physiol.* **254** (1988) G920.
[3.184] Y. Tache et al., *Can. J. Physiol. Pharmacol.* (1986) July, Suppl., 53.
[3.185] P. Bauerfeind et al., *Dig. Dis. Sci.* **30** (1985) 368.
[3.186] M.J. Fargeas et al., *Regul. Pept.* **11** (1985) 95.
[3.187] C. Gatto et al., *Am. Rev. Respir. Dis.* **137** (1988) Pt. 2, 195.
[3.188] V. di Marzo et al., *Neuropeptides* **11** (1988) 169.
[3.189] R.C. Orlowski et al., *Eur. J. Biochem.* **162** (1987) 399.
[3.190] K.E. Schwartz et al., *Endocrinology* **108** (1981) 831.
[3.191] D.M. Findlay et al., *Endocrinology* **117** (1985) 801.
[3.192] C.S. D'Santos et al., *Endocrinology* **123** (1988) 1483.
[3.193] D.M. Findlay et al., *Endocrinology* **112** (1983) 1288.
[3.194] R.M. Epand et al., *J. Med. Chem.* **31** (1988) 1595.
[3.195] R.M. Epand et al., *Eur. J. Biochem.* **159** (1986) 125.
[3.196] G.R. Moe et al., *J. Am. Chem. Soc.* **105** (1983) 4100.
[3.197] G.R. Moe, E.T. Kaiser, *Biochemistry* **24** (1985) 1971.
[3.198] R.M. Epand et al., *Biochemistry* **25** (1986) 1964.
[3.199] R.M. Epand et al., *Biochem. Biophys. Res. Commun.* **152** (1988) 203.
[3.200] M.J. Twery et al., *Eur. J. Pharmacol.* **155** (1988) 285.
[3.201] T. Buclin et al., *Calcif. Tissue Int.* **41** (1987) 252.
[3.202] J.Y. Reginster et al., *Lancet II* (1987) 1481.
[3.203] G. Thamsborg et al., *Lancet I* (1988) 413.
[3.204] M.T. McDermott, G.S. Kidd, *Endocr. Rev.* **8** (1987) 377.
[3.205] L. Nogarin, *Clin. Trials J.* **23** (1986), Suppl. 1, 87.
[3.206] K. Krueger, M. Schattenkirchner, *Z. Rheumatol.* **47** (1988) 287.
[3.207] A. Fremioti et al., *Curr. Therap. Res.* **36** (1984) 627.
[3.208] G. Schiraldi et al., *Eur. J. Resp. Dis.* **69** (1986) Suppl. 146, A150.
[3.209] L.H. Breimer et al., *Biochem. J.* **255** (1988) 377.
[3.210] T. Inui et al., *Life Sci.* **45** (1989) 1199.
[3.211] A. Bjurhom et al., *Peptides* **9** (1988) 165.
[3.212] P.H. Steenbergh et al., *FEBS Lett.* **209** (1986) 97.
[3.213] S. Kimura et al., *Neuropeptides* **9** (1987) 75.
[3.214] K. Morita et al., *Horm. Metab. Res.* **21** (1989) 666.
[3.215] X.-Y. Hua, J.M. Lundberg, *Acta Physiol. Scand.* **128** (1986) 453.
[3.216] A.F. Russo et al., *J. Biol. Chem.* **263** (1988) 5.
[3.217] P. Popper, P.E. Micevych *Neuroendocrinology* **50** (1989) 338.
[3.218] M. Pohl et al., *Neuropeptides* **14** (1989) 151.
[3.219] S. Evangelista et al., *Eur. J. Pharmacol.* **164** (1989) 389.
[3.220] T. Dennis et al., *J. Pharmacol. Exper. Therap.* **251** (1989) 718.
[3.221] H. Kawasaki et al., *Nature* **335** (1988) 164.
[3.222] A. Haegerstrand et al., *Proc. Natl. Acad. Sci. USA* **87** (1990) 3299.
[3.223] A. Yamamoto et al., *Biochem. Biophys. Res. Commun.* **155** (1988) 1452.
[3.224] D. Crossman et al., *Br. J. Pharmacol.* **92** (1987) 695.
[3.225] J.B.D. Palmer et al., *Br. J. Pharmacol.* **91** (1987) 95.
[3.226] J. Wallengren, R. Hakanson, *Eur. J. Pharmacol.* **143** (1987) 267.
[3.227] R. Oku et al., *Brain Res.* **403** (1987) 350.
[3.228] A. Pecile et al., *Regul. Pept.* **18** (1987) 189.
[3.229] C. Netti et al., *Neuroendocrinology* **49** (1989) 242.
[3.230] C. Elie et al., *Neuropeptides* **16** (1990) 109.
[3.231] A. Fahim et al., *Neuroendocrinology* **51** (1990) 688.
[3.232] S.R. Blood et al., *J. Physiol. (London)* **409** (1989) 29.

[3.233] J.P. Hinson, G.P. Vinson, *Neuropeptides* **16** (1990) 129.
[3.234] H.J. Lenz et al., *Regul. Pept.* **12** (1985) 81.
[3.235] H.J. Lenz et al., *Gut* **26** (1985) 550.
[3.236] Y. Tache et al., *Life Sci.* **35** (1984) 871.
[3.237] P. Bauerfeind et al., *Am. J. Physiol.* **256** (1989) G145.
[3.238] C. Beglinger et al., *Gastroenterology* **95** (1988) 958.
[3.239] B.E. Dunning, G.J. Taborsky, Jr., *Endocrinology* **120** (1987) 1774.
[3.240] N.W. Bunnett et al., *Am. J. Physiol.* **258** (1990) G316.
[3.241] W.S. Helton et al., *Am. J. Physiol.* **256** (1989) G715-G720.
[3.242] C.A. Maggi et al., *Pharmacol.* **18** (1987) 33.
[3.243] I.T. Lippe et al., *Regul. Pept.* **26** (1989) 35.
[3.244] L. Bartho et al., *Eu. J. Pharmacol.* **135** (1987) 449.
[3.245] H.E. Raybould et al., *Peptides* **9** (1988) 735.
[3.246] H.J. Lenz, *Am. J. Physiol.* **254** (1988) G920.
[3.247] H.M. Cox et al., *Br. J. Pharmacol.* **97** (1989) 996.
[3.248] P.G. Reasbeck et al., *Gastroenterology* **95** (1988) 966.
[3.249] M. Pettersson et al., *Endocrinology* **119** (1986) 865.
[3.250] J. Ishizuka et al., *Regul. Pept.* **20** (1988) 73.
[3.251] B. Ahren et al., *Diabetologia* **30** (1987) 354.
[3.252] B. Leighton, G.J.S. Cooper, *Nature* **335** (1988) 632.
[3.253] G.J.S. Cooper et al., *Proc. Natl. Acad. Sci. USA* **85** (1988) 7763.
[3.254] B. Leighton et al., *FEBS Lett.* **249** (1989) 357.
[3.255] C. Beglinger et al., *Gut* **29** (1988) 243.
[3.256] W.H. Nealon et al., *Dig. Dis. Sci.* **30** (1985) 984.
[3.257] Y. Umeda et al., *Biochem. Biophys. Res. Commun.* **154** (1988) 227.
[3.258] J.R. Tippins et al., *Biochem. Biophys. Res. Commun.* **134** (1986) 1306.
[3.259] C.A. Maggi et al., *Eur. J. Pharmacol.* **179** (1990) 217.
[3.260] S. Chakder, S. Rattan, *J. Pharmacol. Exp. Ther.* **253** (1990) 200.
[3.261] M. Nishi et al., *J. Biol. Chem.* **265** (1990) 4173.
[3.262] P. Westermark et al., *Proc. Natl. Acad. Sci. USA* **84** (1987) 3881.
[3.263] S. Mosselman et al., *FEBS Lett.* **239** (1988) 227.
[3.264] A.N. Roberts et al., *Proc. Natl. Acad. Sci. USA* **86** (1989) 9662.
[3.265] T. Sanke et al., *J. Biol. Chem.* **263** (1988) 17243.
[3.266] J. Asai et al., *Biochem. Biophys. Res. Commun.* **164** (1989) 400.
[3.267] A. Clark et al., *Lancet II* (1987) 231.
[3.268] H.C. Fehmann et al., *FEBS Lett.* **262** (1990) 279.
[3.269] A. Ogawa et al., *J. Clin. Invest.* **85** (1990) 973.
[3.270] E. Hartter et al., *Lancet I* (1990) 854.
[3.271] B. Leighton, G.J.S. Cooper, *Nature* **335** (1988) 632.
[3.272] G.J.S. Cooper et al., *Proc. Natl. Acad. Sci. USA* **85** (1988) 7763.
[3.273] H.K. Datta et al., *Biochem. Biophys. Res. Commun.* **163** (1989) 876.

References for Chapter 4

[4.1] W. König in A. Kleemann et al. (eds.): *Arzneimittel, Fortschritte 1972 bis 1985*, VCH Verlagsgesellschaft, Weinheim 1987, 700.
[4.2] J.R. Dave et al., *Endocrinology* **120** (1987) 305.
[4.3] M. Fenger, A.J. Johnsen, *Biochem. J.* **250** (1988) 781.
[4.4] A. Bjartell et al., *Peptides* **11** (1990) 149.
[4.5] F.E. Estivariz et al., *J. Endocrinol.* **116** (1988) 207.
[4.6] F.E. Estivariz et al., *J. Endocrinol.* **116** (1988) 201.
[4.7] H.P.J. Bennett, *Peptides* **7** (1986) 615.
[4.8] C.-L.C. Chen et al., *Endocrinology* **118** (1986) 2382.
[4.9] J.E. Pintar et al., *Science* **225** (1984) 632.
[4.10] E. Gizang-Ginsberg, D.J. Wolgemuth, *Proc. Natl. Acad. Sci. USA* **84** (1987) 1600.
[4.11] C.-L.C. Chen et al., *Proc. Natl. Acad. Sci. USA* **81** (1984) 5672.

[4.12] E.M. Smith et al., *Nature* **321** (1986) 881.
[4.13] F. Petraglia et al., *Endocrinology* **119** (1986) 2445.
[4.14] C. Rivier et al., *Science* **231** (1986) 607.
[4.15] D.C. Cumming et al., *J. Clin. Endocrinol. Metab.* **60** (1985) 810.
[4.16] M.H. Melner et al., *Endocrinology* **119** (1986) 2082.
[4.17] C.-L.C. Chen and M.B. Madigan, *Endocrinology* **121** (1987) 590.
[4.18] S.L. Wardlaw, *Neuroendocrinology* **47** (1988) 164.
[4.19] J.N. Wilcox, J.L. Roberts, *Endocrionology* **117** (1985) 2392-2396.
[4.20] C.L. Coe et al., *Endocrinology* **118** (1986) 435.
[4.21] C.B. Newman et al., *Endocrinology* **120** (1987) 962.
[4.22] J.P. Loeffler et al., *Neuroendocrinology* **47** (1988) 95.
[4.23] A. Poteliakhoff, *Lancet II* (1985) 326.
[4.24] A.C.Y. Chang et al., *Proc. Natl. Acad. Sci. USA* (1980) 4890.
[4.25] J. Drouin et al., *FEBS Lett.* **193** (1985) 54.
[4.26] S. Nakanishi et al., *Nature* **278** (1979) 423.
[4.27] G.J.M. Martens et al., *J. Biol. Chem.* **260** (1985) 13685.
[4.28] W. König in A. Kleemann et al. (eds.): *Arzneimittel, Fortschritte 1972 bis 1985*, VCH Verlagsgesellschaft, Weinheim 1987, 625.
[4.29] J.N. Crawley et al., *Peptides* **6** (1985) 891.
[4.30] S. Pretel, D.T. Piekut, *Peptides* **11** (1990) 621.
[4.31] M.H. Whitnall et al., *Neuroendocrinology* **45** (1987) 420.
[4.32] P. Petrusz et al., *Proc. Natl. Acad. Sci. USA* **80** (1983) 1721.
[4.33] T.O. Bruhn et al., *Endocrinology* **120** (1987) 25.
[4.34] A. Sasaki et al., *J. Clin. Endocrinol. Metab.* **67** (1988) 768.
[4.35] D.J. Yoon et al., *Endocrinology* **122** (1988) 759.
[4.36] J.R. Dave, R.L. Eskay, *Biochem. Biophys. Res. Commun.* **134** (1986) 255.
[4.37] H. Jingami et al., *FEBS Lett.* **191** (1985) 63.
[4.38] N. Ling et al., *Biochem. Biophys. Res. Commun.* **122** (1984) 1218.
[4.39] F. Esch et al., *Biochem. Biophys. Res. Commun.* **122** (1984) 899.
[4.40] M. Patthy et al., *Proc. Natl. Acad. Sci. USA* **82** (1985) 8762.
[4.41] M.A. Ronsberg et al., *IRCS Med. Sci.* **14** (1986) 350.
[4.42] T. Watabe et al., *Life Sci.* **40** (1987) 1651.
[4.43] S. Suemaru et al., *Life Sci.* **39** (1986) 1161.
[4.44] C. Rivier, W. Vale, *Endocrinology* **114** (1984) 2409.
[4.45] J. Dohanics et al., *Peptides* **11** (1990) 51.
[4.46] V. Guillaume et al., *Neuroendocrinology* **49** (1989) 676.
[4.47] P.M. Plotsky, *Endocrinology* **121** (1987) 924.
[4.48] T. Suda et al., *Life Sci.* **40** (1987) 1645.
[4.49] P.M. Plotsky et al., *Endocr. Rev.* **10** (1989) 437.
[4.50] E. Redel et al., *Endocrinology* **123** (1988) 2736.
[4.51] S. Suemaru et al., *Endocrinol. Jpn.* **33** (1986) 441.
[4.52] T. Suda et al., *Life Sci.* **40** (1987) 673.
[4.53] M.A. Smith et al., *Endocrinology* **118** (1986) 1997.
[4.54] A. Bateman et al., *Endocr. Rev.* **10** (1989) 92.
[4.55] R. Buzzetti et al., *J. Endocrinol.* **119** (1988) Suppl., 95.
[4.56] T. Suda et al., *Endocrinology* **114** (1984) 851.
[4.57] P.M. Plotsky et al., *Endocrinology* **119** (1986) 1126.
[4.58] K.J. Kovacs, E. Mezey, *Neuroendocrinology* **46** (1987) 365.
[4.59] G.K. Adler et al., *J. Biol. Chem.* **263** (1988) 5846.
[4.60] T. Suda et al., *Endocrinology* **118** (1986) 459.
[4.61] T. Suda et al., *J. Clin. Endocrinol. Metab.* **64** (1987) 909.
[4.62] F. Yajima et al., *Life Sci.* **39** (1986) 181.
[4.63] J.C. Buckingham, T.A. Cooper, *Neuroendocrinology* **44** (1986) 36.
[4.64] J.C. Buckingham, T.A. Cooper, *Br. J. Pharmacol.* **90** (1987) Suppl. 167P.
[4.65] F. Petraglia et al., *Nature* **328** (1987) 717.
[4.66] T. Laatikainen et al., *Neuropeptides* **10** (1987) 343.
[4.67] A. Sasaki et al., *J. Clin. Endocrinol. Metabol.* **64** (1987) 224.

[4.68] M. Grino et al., *Biochem. Biophys. Res. Commun.* **148** (1987) 1208.
[4.69] E.B. De Souza et al., *Nature* 319 (1986) 593.
[4.70] M.M. Mouradian et al., *Neuropeptides* **8** (1986) 393.
[4.71] M. Hotta et al., *J. Clin. Endocrinol. Metab.* **62** (1986) 319.
[4.72] T. Shibasaki et al., *Life Sci.* **43** (1988) 1103.
[4.73] W.H. Kaye et al., *Endocrinology* **64** (1987) 203.
[4.74] A.E. Calogero et al., *Clin. Res.* **35** (1987) 393A.
[4.75] G. Aguilera et al., *Neuroendocrinology* **43** (1986) 79.
[4.76] P.C. Wynn et al., *Endocrinology* **116** (1985) 1653.
[4.77] M.C. Holmes et al., *Endocrinology* **121** (1987) 2093.
[4.78] J. Schwartz et al., *Endocrinology* **119** (1986) 2376.
[4.79] R.L. Hauger et al., *Endocrinology* **120** (1987) 1527.
[4.80] A.N. Margioris et al., *J. Clin. Endocrinol. Metab.* **66** (1988) 922.
[4.81] J. Kraicer et al., *Neuroendocrinology* **41** (1985) 363.
[4.82] J.H. Meador-Woodruff et al., *Neuropeptides* **9** (1987) 269.
[4.83] J.V. Conaglen et al., *J. Clin. Endocrinol. Metab.* **58** (1984) 463.
[4.84] C. Maser-Gluth et al., *Life Sci.* **35** (1984) 879.
[4.85] E. De Souza, G.R. Van Loon, *Experimentia* **40** (1984) 1004.
[4.86] D.O. Sobel, *Peptides* **7** (1986) 443.
[4.87] S. Guild and T. Reisine, *J. Pharmacol. Exp. Therap.* **241** (1987) 125.
[4.88] M.J. Cronin et al., *Peptides* **7** (1986) 935.
[4.89] T.O. Bruhn et al., *Endocrinology* **119** (1986) 1558.
[4.90] H. Raff et al., *Am. J. Physiol.* **251** (1986) R1235.
[4.91] D.M. Gibbs, *Endocrinology* **116** (1985) 723.
[4.92] M. Keller-Wood et al., *Am. J. Physiol.* **250** (1986) R396.
[4.93] A.C. Hale et al., *J. Endocrinol.* **102** (1984) R1.
[4.94] B.I. Baker et al., *J. Endocrinol.* **106** (1985) R5.
[4.95] M.V. Graf et al., *Neuroendocrinology* **41** (1985) 353.
[4.96] R.S. Rittmaster et al., *J. Clin. Endocrinol. Metab.* **60** (1985) 891.
[4.97] C.A. Redekopp et al., *Horm. Metabol. Res.* **17** (1985) 646.
[4.98] K.E. Nikolarakis et al., *Brain Res.* **399** (1986) 152.
[4.99] J.G. Kiang, E.T. Wei, *J. Pharmacol. Exp. Therap.* **243** (1987) 517; E.T. Wei et al., *J. Pharmacol. Exp. Therap.* **247** (1988) 1082.
[4.100] E.T. Wei, J. G. Kiang, Euro. J. Pharmacol. **168** (1989) 81.
[4.101] E.T. Wei et al., *J. Pharmacol. Exp. Therap.* **238** (1986) 783.
[4.102] K.M. Hargreaves et al., *Eur. J. Pharmacol.* **170** (1989) 275.
[4.103] C. Rivier et al., *Science* **231** (1986) 607.
[4.104] O.F.X. Almeida et al., *J. Physiol. (London)* **398** (1988) 77P.
[4.105] S. Ulisse et al., *Biol. Chem.* **265** (1990) 1964.
[4.106] O.F.X. Almeida et al., *Endocrinology* **122** (1988) 1034.
[4.107] C. Rivier, W. Vale, *Endocrinology* **114** (1984) 914.
[4.108] B. Miskowiak et al., *Exp. Clin. Endocrinol.* **88** (1986) 25.
[4.109] F. Petraglia et al., *Endocrinology* **120** (1987) 1083.
[4.110] A. Kooy et al., *Neuroendocrinology* **51** (1990) 261.
[4.111] N. Ono et al., *Proc. Natl. Acad. Sci. USA* **82** (1985) 7787.
[4.112] H. Katakami et al., *Neuroendocrinology* **41** (1985) 390.
[4.113] A.R.M.M. Hermus et al., *Clin. Endocrinology* **21** (1984) 589.
[4.114] M. Vlaskovska et al., *Endocrinology* **115** (1984) 895.
[4.115] A.R.M.M. Hermus et al., *Eur. J. Clin. Pharmacol.* **31** (1987) 531.
[4.116] I. Torres-Aleman et al., *Peptides* **5** (1984) 541.
[4.117] S.J. Konturek et al., *Peptides* **8** (1987) 575.
[4.118] H.J. Lenz. *Proc. Natl. Acad. Sci. USA* **86** (1989) 1417.
[4.119] Y. Tache et al., *Gastroenterology* **84**, Pt. 2 (1983) 1382.
[4.120] Y. Tache et al., *Gastroenterology* **86** (1984) 281.
[4.121] R. Murison et al., *Psychoneuroendocrinology* **14** (1989) 331.
[4.122] T. Pappas et al., *Regul. Pept.* **11** (1985) 193.
[4.123] M. Hagiwara et al. *Gastroenterology* **90** (1986) Pt. 2, 1447.

[4.124] M. Burlage et al., *Klin. Wochenschr.* **66**, Suppl. 13 (1988) 207.
[4.125] M. Hagiwara et al., *Jpn. J. Pharmacol.* **43** (1987) Suppl., 162P.
[4.126] C.L. Williams, T.F. Burks, *Gastroenterology* **92** (1987) Pt. 2, 1808.
[4.127] T. Shibasaki et al., *Life Sci.* **43** (1988) 1103.
[4.128] D. Oppermann et al., *Klin. Wochenschr.* **63** (1985) 231.
[4.129] G. Laux, K.P. Lesch, M. Schwab, *Neuropsychobiology* **19** (1988) 40.
[4.130] G.L. Bernardini et al., *Peptides* **5** (1984) 57.
[4.131] K.T. Britton et al., *Life Sci.* **39** (1986) 1281.
[4.132] G.F. Koob, F.E. Bloom, *Federation Proc.* **44** (1985) 259.
[4.133] D.R. Britton et al., *Life Sci.* **38** (1985) 211.
[4.134] A. Mele et al., *Peptides* **8** (1987) 935.
[4.135] J. Rivier et al., *Science* **224** (1984) 889.
[4.136] C.R. DeBold et al., *J. Clin. Endocrinol. Metab.* **60** (1985) 836.
[4.137] J. Schopohl et al., *Acta Endocrinol.* **112** (1986) 157.
[4.138] P.C. Avgerinos et al., *J. Clin. Endocrinol. Metab.* **62** (1986) 816.
[4.139] D. Desir et al., *J. Clin. Endocrinol. Metab.* **63** (1986) 1292.
[4.140] H. Vierhapper, W. Waldhäusl, *Exp. Clin. Endocrinol.* **88** (1986) 355.
[4.141] O.A. Müller et al., *Hormone Res.* **25** (1987) 185.
[4.142] D.N. Orth et al., *Federation Proc.* **44** (1985) 197.
[4.143] P.W. Gold et al., *N. Engl. J. Med.* **314** (1986) 1329.
[4.144] Y. Hayashizaki et al., *Horm. Metabol. Res.* **18** (1986) 849.
[4.145] W. König in A. Kleemann et al. (eds.): *Arzneimittel, Fortschritte 1972 bis 1985*, VCH Verlagsgesellschaft, Weinheim 1987, 702.
[4.146] L. Proulx et al., *J. Endocrinol. Invest.* **7** (1984) 257.
[4.147] M.M. Valenca, A. Negro-Vilar, *Endocrinology* **118** (1986) 32.
[4.148] J. Knudtzon, *Horm. Metabol. Res.* **18** (1986) 579.
[4.149] A.I. Smith et al., *J. Endocrinol.* **115** (1987) R5.
[4.150] M. Carnes et al., *Peptides* **9** (1988) 325.
[4.151] P.M. Horrocks et al., *Clin. Endocrinol.* **32** (1990) 127.
[4.152] J.C. Buckingham, *Brit. Med. Bull.* **41** (1985) 203.
[4.153] M. Carnes et al., *Neuroendocrinology* **45** (1987) 160.
[4.154] M. Familari, J.W. Funder, *Aust. N.Z.J. Med.* **17**, Suppl. 1 (1987) 164.
[4.155] J. Dohanics et al., *Peptides* **11** (1990) 59.
[4.156] A. Pfeiffer et al., *Endocrinology* **116** (1985) 2688.
[4.157] N. Glorioso et al., *Clin. Exp. Hypertens.* **A 9** (1987) 665.
[4.158] P. Linkowski et al., *J. Clin. Endocrinol. Metab.* **65** (1987) 141.
[4.159] J.F. Mortola et al., *J. Clin. Endocrinol. Metab.* **65** (1987) 962.
[4.160] A. Shimatsu et al., *J. Endocrinol. Invest.* **10** (1987) 127.
[4.161] K.L. Bost, J.E. Blalock, *Mol. Cell. Endocrinol.* **44** (1986) 1.
[4.162] E.M. Smith et al., *N. Engl. J. Med.* **317**, (1987) 1266.
[4.163] A. Penhoat et al., *Biochem. Biophys. Res. Commun.* **165** (1989) 355.
[4.164] W.E. Rainey et al., *J. Biol. Chem.* **264** (1989) 21474.
[4.165] R. Cheitlin et al., *J. Biol. Chem.* **260** (1985) 5323.
[4.166] H. Sekihara et al., *J. Clin. Endocrinol. Metab.* **61** (1985) 291.
[4.167] S.W.J. Lamberts et al., *Mol. Cell. Endocrinol.* **52** (1987) 243.
[4.168] M.-T. Pham-Huu-Trung et al., *J. Clin. Endocrinol. Metab.* **61** (1985) 467.
[4.169] G. Ramirez et al., *J. Clin. Endocrinol. Metab.* **66** (1988) 46.
[4.170] L.J. Valenta et al., *Horm. Res.* **23** (1986) 16.
[4.171] R.A. Vergona et al., *Peptides* **6** (1985) 581.
[4.172] A. Singh et al., *Biochem. Biophys. Res. Commun.* **155** (1988) 524.
[4.173] A. Singh et al., *Biochem. Biophys. Res. Commun.* **155** (1988) 524.
[4.174] N. Fujii et al., *11th Amer. Peptide Symp.* San Diego (1989).
[4.175] A. Bertolini et al., *Eur. J. Pharmacol.* **122** (1986) 387.
[4.176] S. Guarini et al., *Neuropeptides* **8** (1986) 19.
[4.177] S. Guarini et al., *Pharmacol. Res. Commun.* **19** (1987) 511.
[4.178] A. Bertolini et al. *Eur. J. Pharmacol.* **130** (1986) 19.
[4.179] Pi-hsueh Shirley Li, *Life Sci.* **41** (1987) 2493.

[4.180] M. Bernier et al., *Can. J. Physiol. Pharmacol.* **62** (1984) 1166.
[4.181] A. Armario et al., *Life Sci.* **39** (1986) 1119.
[4.182] P.E. Juniewicz et al., *Endocrinology* **122** (1988) 891.
[4.183] J. Chariot et al., *Seances Acad. Sci. Ser.* **3,** 307 (1988) 235.
[4.184] M. Kollind et al., *Diabetologia* **27** (1984) 298A.
[4.185] C. Day et al., *Diabetologia* **30** (1987) 512A.
[4.186] M.G.A. Baggen et al.,, *J. Endocrinol.* **114** (1987) 369.
[4.187] C.A. Hendrie, *Br. J. Pharmacol.* **89**, Dec. Suppl. (1986) 795P.
[4.188] R. Poggioli et al., *Arch. Int. Pharmacodyn. Ther.* **293** (1988) 265.
[4.189] A.J. Dunn, G. Vigle, *Neuropharmacology* **24** (1985) 329.
[4.190] A.V. Vergoni et al., *Neuropeptides* **7** (1986) 153.
[4.191] R. Poggioli et al., *Peptides* **7** (1986), 843.
[4.192] W.S. Saunders, J.A. Thornhill, *Peptides* **7** (1986) 597.
[4.193] N. Chastrette et al., *Neuropeptides* **15** (1990) 61.
[4.194] R. Carnuccio et al., *Br. J. Pharmac.* **90** (1987) 443.
[4.195] J. Fauvel et al., *FEBS Lett.* **216** (1987) 45.
[4.196] B. Rothhut et al., *Biochem. J.* **263** (1989) 929.
[4.197] L. Miele et al., *Nature* **335** (1988) 726.
[4.198] B.P. Wallner et al., *Nature* **320** (1986) 77.
[4.199] K.-S. Huang et al., *Cell* **46** (1986) 191.
[4.200] K.-S. Huang et al., *J. Biol. Chem.* **262** (1987) 7639.
[4.201] J.-L. Fauchere et al., *FEBS Lett.* **183** (1985) 283.
[4.202] M.B. Stolz, J.L. Fauchere, *Helv. Chim. Acta* **71** (1988) 1421.
[4.203] S.J. Dunmore et al., *Biochem. J.* **244** (1987) 797.
[4.204] T. Izumi, Y. Fukuyama, *Life Sci.* **34** (1984) 1023.
[4.205] S.L. Wardlaw, C.E. Markowitz, *Clin. Res.* **34** (1986) 435A.
[4.206] J. Hinson et al., *J. Endocrinol.* **115** (1987) 61.
[4.207] N. Colbert et al., *Presse Med.* **12** (1983) 1077.
[4.208] W. König in A. Kleemann et al. (eds.): *Arzneimittel, Fortschritte 1972 bis 1985*, VCH Verlagsgesellschaft, Weinheim, 1987, 706.
[4.209] X. Bertagna et al., *Proc. Natl. Acad. Sci. USA* **83** (1986) 9719.
[4.210] T. Shibasaki et al., *Regul. Pept.* **2** (1981) 43.
[4.211] M. Fenger, A.H. Johnson, *Biochem. J.* **250** (1988) 781.
[4.212] M. Fenger, *Regul. Pept.* **20** (1988) 345-357.
[4.213] O. Khorram et al., *Endocrinology* **117** (1985) 2483.
[4.214] O. Khorram et al., *Peptides* **6** (1985) 503.
[4.215] R.C. Bell, J.M. Lipton, *Am. J. Physiol.* **252** (1987) R1152.
[4.216] J. M. Danger et al., *Life Sci.* **40** (1987) 1875.
[4.217] B.G. Jenks et al., *Peptides* **6** (1985) 913.
[4.218] M. Fenger, L. Hilsted, *Acta Endocrinol. (Copenhagen)* **118** (1988) 119.
[4.219] K.A. Gruber, M.F. Callahan, *Am. J. Physiol.* **257** (1989) R681.
[4.220] J. Wortsman, et al., *J. Clin. Endocrinol. Metab.* **61** (1985) 355.
[4.221] A.M. de Lauro Castrucci et al., *Drugs of the Future* **15** (1990) 41-55.
[4.222] G.G. Nussdorfer et al., *Biochem. Biophys. Res. Commun.* **141** (1986) 1279.
[4.223] O. Khorram, S.M. McCann, *Endocrinology* **119** (1986) 1071.
[4.224] S.L. Wardlaw, C.E. Markowitz, *Clin. Res.* **34** (1986) 435A.
[4.225] C.B. Newman et al., *Life Sci.* **36** (1985) 1661.
[4.226] R.L. Reid et al., *J. Clin. Endocrinol. Metab.* **58** (1984) 773.
[4.227] R.A. Daynes et al., *J. Invest. Dermatol.* **88** (1987) 483.
[4.228] D.B. Richards, J.M. Lipton, *Peptides* **5** (1984) 815.
[4.229] F. Di Giovine et al., *Br. J. Rheumatol.* **26** (1987) Suppl. 2, 62.
[4.230] A.V. Kotov et al., *Byull. Eksp. Biol. Med.* **97** (1984) 265.
[4.231] J. Knudtzon, *Life Sci.* **34** (1984) 547-54.
[4.232] S.Y. Lin et al., *Hypertension* **10** (1987) 619.
[4.233] M.-T. Pham-Huu-Trung et al., *J. Clin. Endocrinol. Metab.* **61** (1985) 467.
[4.234] K.A. Gruber, M.F. Callahan, *Am. J. Physiol.* **257** (1989) R681.
[4.235] V.J. Hruby et al., *J. Med. Chem.* **30** (1987) 2126.

[4.236] M. Lebl et al., *Collect. Cech. Chem. Commun.* **49** (1984) 2680.
[4.237] N. Levine et al., *J. Invest. Dermatol.* **87** (1986) 416.
[4.238] D.G. Klemes et al., *Biochem. Biophys. Res. Commun.* **137** (1986) 722.
[4.239] F. Al-Obeidi et al., *J. Med. Chem.* **32** (1989) 2555.
[4.240] W.L. Cody et al., *J. Med. Chem.* **28** (1985) 583.
[4.241] G.P. Vinson et al., *Life Sci.* **35** (1984) 603.
[4.242] *Drugs of the Future* **7** (1982) 319; **8** (1983) 468; **9** (1984) 381; **10** (1985) 431; **11** (1986) 432-433; **12** (1987) 505-506; **13** (1988) 489-490; **14** (1989) 480.
[4.243] H. Goldman, R.F. Berman, *Peptides* **5** (1984) 1061.
[4.244] P. De Koning, W.H. Gispen, *Peptides* **8** (1987) 415.
[4.245] C.E.E.M. Van der Zee et al., *Eur. J. Pharmacol.* **147** (1988) 351.
[4.246] J. Verhaagen et al., *Peptides* **8** (1987) 581.
[4.247] C.E.E.M. van der Zee et al., *Eur. J. Pharmacol.* **177** (1990) 211.
[4.248] *Drugs of the Future* **14** (1989) 514-516, **15** (1990) 621, **16** (1991) 572.
[4.249] R. Geiger et al., EP 179332 (Hoechst. A.G.)
[4.250] D. de Wied, personal communication (26. 2. 88)
[4.251] F.J. Hock et al., *Neuroscience* (Oxford) **22** (1987), Suppl. S506.
[4.252] G. Wiemer et al., *Arch. Pharmacol.* **335** (1987) Suppl. R94.
[4.253] G. Wolterink, J.M. van Ree, *Brain Res.* **421** (1987) 41.
[4.254] T.K. Sawyer et al., *Peptides* **11** (1990) 351.
[4.255] M.E. Hadley et al., *Life Sci.* **40** (1987) 1889.
[4.256] W.R. Kiessling, *Arch. Neurol.* **44** (1987) 995.
[4.257] P.K. Sorensen et al., *Prog. Neuropsychopharmacol. Biol. Psychiatry* **10** (1986) 479.
[4.258] W. König in A. Kleemann et al. (eds.): *Arzneimittel, Fortschritte 1972 bis 1985*, VCH Verlagsgesellschaft, Weinheim 1987, 707.
[4.259] K.R. Rao et al., *Proc. Natl. Acad. Sci. USA* **82** (1985) 5319.
[4.260] K.R. Rao et al., *J. Biol. Chem.* **262** (1987) 2672.
[4.261] H. Kawauchi, *Life Sci.* **45** (1989) 1133.
[4.262] B.I. Baker et al., *J. Endocrinol.* **106** (1985) R5.
[4.263] W. König in A. Kleemann et al. (eds.): *Arzneimittel, Fortschritte 1972 bis 1985*, VCH Verlagsgesellschaft, Weinheim 1987, 654.
[4.264] R.B. Rothman et al., Eur. J. Pharmacol. **167** (1989) 345.
[4.265] I. Kitchen, *Gen. Pharmac.* **16** (1985) 79.
[4.266] Y. Shimohigashi et al., *FEBS Lett.* **233** (1988) 289.
[4.267] R. Schwyzer, *Biochemistry* **25** (1986) 6335.
[4.268] A.T. McKnight et al., *Neuropharmacology* **24** (1985) 1011.
[4.269] R.S. Rapka, *Life Sci.* **39** (1986) 1825.
[4.270] P.W. Schiller et al., *Biol. Chem. Hoppe Seyler* **366** (1985) 846.
[4.271] P.W. Schiller, T. M.-D. Nguyen, *Neuropeptides* **5** (1984) 165.
[4.272] E.L. Newman et al., *Neuropeptides* **8** (1986) 305.
[4.273] *Drugs of the Future* **9** (1984) 416.
[4.274] S. Benyhe et al., *Neuropeptides* **8** (1987) 225.
[4.275] L.H. Lazarus et al., *J. Biol. Chem.* **264** (1989) 354.
[4.276] M. Marastoni et al., *J. Med. Chem.* **30** (1987) 1538.
[4.277] P.W. Schiller et al., *J. Med. Chem.* **32** (1989) 698.
[4.278] P.W. Schiller et al., *J. Med. Chem.* **28** (1985) 1766.
[4.279] G. Landis et al., *J. Med. Chem.* **32** (1989) 638.
[4.280] G.B. Glavin et al., *Life Sci.* **46** (1990) 1075.
[4.281] H. Mihara et al., *FEBS Lett.* **193** (1985) 35.
[4.282] G. Gacel et al., *J. Med. Chem.* **31** (1988) 1891.
[4.283] J.L. Martin et al., *Neuropeptides* **6** (1985) 293.
[4.284] J.D. Leander et al., *Neuropeptides* **8** (1986) 119.
[4.285] F. Porreca et al., *J. Pharmacol. Exp. Therap.* **241** (1987) 393.
[4.286] A.D. Corbett et al., *Br. J. Pharmacol.* **83** (1984) 271.
[4.287] V.J. Hruby et al., in G. Jung and E. Bayer (Eds.): *Peptides* 1988, Walter de Gruyter, Berlin, New York, 1989, 616.
[4.288] G. Toth et al., *J. Med. Chem.* **33** (1990) 249.

[4.289] T. Costa et al., *Biochem. Pharmacol.* **34** (1985) 25.
[4.290] A. Mor et al., *FEBS Lett.* **255** (1989) 269.
[4.291] K. Richter et al., *Proc. Natl. Acad. Sci. USA* **87** (1990) 4836.
[4.292] A.H. Mulder et al., *Neuropeptides* **14** (1989) 99.
[4.293] J. Gairin et al., *J. Pharmacol. Exp. Therap.* **245** (1988) 995.
[4.294] R. Quirion, A.S. Weiss, *Peptides* **4** (1983) 445.
[4.295] P. Sanchez-Blazquez et al., *Neuropeptides* **5** (1984) 181.
[4.296] G.W. Pasternak, *Biochem. Pharmacol.* **35** (1986) 361.
[4.297] G. Feuerstein et al., *Peptides* **6** (1985) 11.
[4.298] S. Nakayama et al., *J. Pharmacol. Exp. Therap.* **254** (1990) 792.
[4.299] L. Krulich et al., *Neuroendocrinology* **42** (1986) 82.
[4.300] A. Pfeiffer et al., *Endocrinology* **116** (1985) 2688.
[4.301] T.F. Burks et al., *Life Sci.* **43** (1988) 2177.
[4.302] D. Bartolini et al., *Br. J. Pharmacol.* **86**, Suppl. (1985), 640P.
[4.303] I.M. Coupar, *Br. J. Pharmacol.* **80** (1983) 371.
[4.304] J.L. Vaught et al., *Eur. J. Pharmacol.* **109** (1985) 43.
[4.305] L.D. Hirning et al., *Eur. J. Pharmacol.* **109** (1985) 49.
[4.306] A.H. Hassen et al., *Peptides* **4** (1983) 621.
[4.307] N. Yonehara, D.H. Clouet, *J. Pharmacol. Exp. Therap.* **231** (1984) 38.
[4.308] A.H. Mulder et al., *Nature* **308** (1984) 278.
[4.309] S.B. Weinberger et al., *Regul. Pept.* **26** (1989) 323.
[4.310] Z. Khalil, R.D. Helme, *Neuropeptides* **17** (1990) 45.
[4.311] A. Jackson, S.J. Cooper, *Neuropharmacology* **25** (1986) 653.
[4.312] A.I. Faden et al., *Peptides* **6** (1985) 15.
[4.313] M.S. Ahmed, M.A. Horst, *Life Sci.* **39** (1986) 535.
[4.314] Y. Ruckebusch et al., *Life Sci.* **35** (1984) 1731.
[4.315] T.P. Blackburn et al., *Br. J. Pharmacol.* **89** (1986) 593.
[4.316] W. König in A. Kleemann et al. (eds.): *Arzneimittel, Fortschritte 1972 bis 1985*, VCH Verlagsgesellschaft, Wienheim 1987, 657.
[4.317] I. Kitchen, *Gen. Pharmacol.* **16** (1985) 79.
[4.318] F.A. Aleem et al., *Fertil. Steril.* **45** (1986) 507.
[4.319] R. Singer et al., *Experientia* **41** (1985) 64.
[4.320] T. Wahlström et al., *Life Sci.* **36** (1985) 987.
[4.321] I. Shimizu et al., *Regul. Pept.* **16** (1986) 331.
[4.322] M. Dennis et al., *J. Chromatogr.* **266** (1983) 163.
[4.323] O. Vuolteenaho, *Biochem. Biophys. Res. Commun.* **130** (1985) 1226.
[4.324] O. Vuolteenaho, *Acta Physiol. Scand. Suppl.* **531** (1984) 1.
[4.325] J.P.H. Burbach et al., *J. Biol. Chem.* **260** (1985) 6663.
[4.326] H. Akil et al., *Science* **227** (1985) 424.
[4.327] C.W. Wilkinson, D.M. Dorsa, *Neuroendocrinology* **43** (1986) 124.
[4.228] R.G. Lorenz et al., *Peptides* **7** (1986) 119.
[4.229] W. Knepel, G. Meyen, *Neuroendocrinology* **43** (1986) 44.
[4.330] D. Copolov et al., *Clin. Endocrinol.* **19** (1983) 575.
[4.331] C. Gianoulakis et al., *Endocrinology* **122** (1988) 817.
[4.332] L.J. Forman et al., *Peptides* **6** (1985) 835.
[4.333] D. Shoupe et al., *J. Clin. Endocrinol. Metab.* **60** (1985) 178.
[4.334] F.R. Caspari, S. Alapin-Rubillovitz, *J. Clin. Endocrinol. Metab.* **60** (1985) 34.
[4.335] C. Shaha et al., *Endocrinology* **115** (1984) 378.
[4.336] A. Fabbri et al., *Endocrinology* **122** (1988) 749.
[4.337] M.M. Valenca, A. Negro-Vilar, *Endocrinology* **118** (1986) 32.
[4.338] F. Petraglia et al., *Clin. Endocrinol.* **24** (1986) 609.
[4.339] T. Barreca et al., *Life Sci.* **38** (1986) 2263.
[4.340] C. Gianoulakis et al., *Life Sci.* **45** (1989) 1097.
[4.341] F. Facchinetti et al., *J. Endocrinol. Invest.* **9** (1986) 11.
[4.342] A.R. Genazzani et al., *J. Clin. Endocrinol. Metab.* **62** (1986) 36.
[4.343] W.O. Richter et al., *Endocrinology* **120** (1987) 1472.
[4.344] S. Itow et al., *Can. J. Physiol. Pharmacol.* **63** (1985) 81.

[4.345] T. Mitsuma et al., *Horm. Metabol. Res.* **21** (1989) 301.
[4.346] W. Knepel et al., *Neuropeptides* **2** (1981) 67.
[4.347] J.C. Buckingham, *Neuroendocrinology* **42** (1986) 148.
[4.348] W.B. Wehrenberg et al., *Neuroendocrinology* **41** (1985) 13.
[4.349] S.L. Wardlaw et al., *Endocrinology* **119** (1986) 112.
[4.350] D. Giugliano et al., *Metab. Clin. Exp.* **36** (1987) 974.
[4.351] M. Camilleri et al., *Gastroenterology* **86** (1984) Pt. 2, 1039.
[4.352] G. Flemstrom et al., *Acta Physiol. Scand.* **123** (1985) 6A.
[4.353] A.Y.S. Lee et al., *Int. J. Pept. Protein Res.* **24** (1984) 525.
[4.354] L.-F. Tseng, *J. Pharmacol. Exp. Therap.* **239** (1986) 160.
[4.355] J.P. Blanc, E.T. Kaiser, *J. Biol. Chem.* **259** (1984) 9549.
[4.356] D. Yamashiro et al., *Int. J. Pept. Protein Res.* **24 (1984) 520.**
[4.357] P. Nicolas et al. *Proc. Natl. Acad. Sci. USA* **81** (1984) 3074.
[4.358] D. Yamashiro et al., *Int. J. Pept. Protein Res.* **24** (1984) 516.
[4.359] J. Blake et al., *Int. J. Pept. Protein Res.* **25** (1985) 575.
[4.360] D. Collado-Escobar et al., *Eur. J. Pharmacol.* **129** (1986) 385.
[4.361] H.H. Suh et al., *Peptides* **8** (1987) 123.
[4.362] A.E. Panerai et al., *Eur. J. Pharmacol.* **99** (1984) 341.
[4.363] M.D. Hirsch, T.L. O'Donohue, *J. Pharmacol. Exp. Therap.* **237** (1986) 378.
[4.364] J.S. Morley, D.M. Ensor, *Neuropeptides* **8** (1986) 45.
[4.365] W.O. Richter et al., *Endocrinology* **120** (1987) 1472.
[4.366] W.O. Richter, P. Schwandt, *Neuropeptides* **7** (1986) 73.
[4.367] W. König in A. Kleemann et al., (eds.): *Arzneimittel, Fortschritte 1972 bis 1985*, VCH Verlagsgesellschaft, Weinheim 1987, 660.
[4.368] K. Yoshikawa et al., *J. Biol. Chem.* **259** (1984) 14301.
[4.369] G.J.M. Martens, E. Herbert, *Nature* **310** (1984) 251.
[4.370] A. Cupo et al., *Neuropeptides* **9** (1987) 237.
[4.371] R.D. Howells et al., *Proc. Natl. Acad. Sci. USA* **83** (1986) 1960.
[4.372] D.L. Kilpatrick, C.F. Millette, *Proc. Natl. Acad. Sci. USA* **83** (1986) 5015.
[4.373] N. Zamir et al., *Proc. Natl. Acad. Sci. USA* **81** (1984) 6886.
[4.374] E.A. Singer et al., *Endocrinology (Baltimore)* **119** (1986) 1527-1533.
[4.375] I. Kitchen, *Gen. Pharmacol. 16* (1985) 79.
[4.376] B.R. Seizinger et al., *Nature* **313** (1985) 57.
[4.377] D.C. Liebisch et al., *Proc. Natl. Acad. Sci. USA* **83** (1986) 1936.
[4.378] R. Micanovic et al., *Peptides* **5** (1984) 853.
[4.379] P.C. Owens et al., *Endocrinology* **122** (1988) 311.
[4.380] T. Kanamatsu et al., *Proc. Natl. Acad. Sci. USA* **83** (1986) 9245.
[4.381] J.D. Watson et al., *J. Endocrinol.* **111** (1986) 329.
[4.382] J.S. Hong et al., *Neuropeptides* **5** (1985) 557.
[4.383] L.-F. Tseng et al., *Regul. Pept.* **14** (1986) 181.
[4.384] B.C. Yoburn et al., *Life Sci.* **40** (1987) 2495.
[4.385] K. Yoshikawa, S.L. Sabol. *Biochem. Biophys. Res. Commun.* **139** (1986) 1.
[4.386] M.L. De Ceballos et al., *Eur. J. Pharmacol.* **130** (1986) 305.
[4.387] J.A. Angulo et al., *Eur. J. Pharmacol.* **130** (1986) 341.
[4.388] S.L. Lightman, W.S. Young III *Proc. Natl. Acad. Sci. USA* **86** (1989) 4306.
[4.389] D.C.-C. Wan et al., *Neuropeptides* **16** (1990) 141.
[4.390] D. Kew, D.L. Kilpatrick, *Mol. Endocrinol.* **3** (1989) 179.
[4.391] D.K. Sommers et al., *Eur. J. Clin. Pharmacol.* **38** (1990) 391.
[4.392] W. Kraemer et al., *Peptides* **6** (1985) 167.
[4.393] A. Pfeiffer, A. Herz, *Horm. Metabol. Res.* **16** (1984) 386.
[4.394] A. Tsunoo et al., *Proc. Natl. Acad. Sci. USA* **83** (1986) 9832.
[4.395] M. Till et al., *Eur. J. Pharmacol.* **175** (1990) 355.
[4.396] S. Madaus et al., *Can. J. Physiol. Pharmacol.* **64** (1986) Suppl. 77.
[4.397] A. Todisco et al., *Can. J. Physiol. Pharmacol.* **64** (1986) Suppl., 76.
[4.398] G. Flemstrom et al., *Acta Physiol. Scand.* **123** (1985) 6A.
[4.399] A. Lavy et al., *Gastroenterology* **86** (1984) Pt. 2, 1155.
[4.400] S.N. Sullivan et al., *Am. J. Gastroenterol.* **78** (1983) 287.

[4.401] S.J. Konturek et al., *Gut* **24** (1983) 740.
[4.402] M. Bickl, *Dig. Dis. Sci.* **29**, Suppl. (1984) 10S.
[4.403] V. Schusdziarra et al., *Neuropeptides* **4** (1984) 507.
[4.404] J.C. Reynolds et al., *Am. J. Physiol.* **246,** Pt. 1, (1984) G130.
[4.405] T.F. Burks, F. Porreca, *Pharmacologist* **25** (1983) 152.
[4.406] A. Dray et al., *Pain (Amsterdam)* **20** (1984) Suppl. 2, S230.
[4.407] M. Bickel, U. Belz, *IRCS Med. Sci.* **13** (1985) 525.
[4.408] J. Janssens et al., *Dig. Dis. Sci.* **30** (1985) 774.
[4.409] M. Bickel, *Dig. Dis. Sci.* **32** (1987) 903.
[4.410] R.F. Crochelt et al., *Gastroenterology* **84** (1983) Pt. 2, 1403.
[4.411] J. Behar, P. Biancani, *Gastroenterology* **86** (1984) 134.
[4.412] P.A. Crooks et al., *Biochem. Phrmacol.* **33** (1984) 4095.
[4.413] A. Dray, R. Metsch, *Eur. J. Pharmacol.* **98** (1984) 155.
[4.414] M.D. Owen et al., *Peptides* **5** (1984) 737.
[4.415] G.E. Sander et al., *Peptides* **6** (1985) 133.
[4.416] S.-H. Chai et al., *Eur. J. Pharmacol.* **130** (1986) 315.
[4.417] D.M. Wright, G. Clarke, *Neuropeptides* **5** (1984) 273.
[4.418] G. Leng et al., *J. Endocrinol.* **106** (1985) 219.
[4.419] K. Hashimoto et al., *Endocrinol.* **33** (1986) 813.
[4.420] P.G. Andreis et al., *Neuropeptides* **12** (1988) 165.
[4.421] B. Allilio et al., *J. Clin. Endocrinol. Metab.* **63** (1986) 1427.
[4.422] F. Facchinetti et al., *J. Clin. Endocrinol. Metab.* **63** (1986) 1222.
[4.423] I. Gerendai et al., *Regul. Pept.* **27** (1990) 107.
[4.424] T. Mitsuma et al., *Horm. Metabol. Res.* **21** (1989) 301.
[4.425] N.F. Plotnikoff et al., *Fed. Am. Soc. Exp.* **44** (1985) 118.
[4.426] Y. Shavit et al., *Science* **223** (1984) 188.
[4.427] J.W. Lewis et al., *Peptides* **4** (1983) 635.
[4.428] I.S. Zagon, P.J. McLaughlin, *Life Sci.* **35** (1984) 409.
[4.429] M. Maeda et al., *Chem. Pharm. Bull.* **37** (1989) 826.
[4.430] J.L. Torres et. al., *Experientia* **45** (1989) 574.
[4.431] M. Maeda et al., *Chem. Pharm. Bull.* **32** (1984) 4157.
[4.431a] G.W. Hardy et al., *J. Med. Chem.* **31** (1988) 960.
[4.431b] B.A. Boucher et al., *J. Clin. Pharmacol.* **27** (1987) 151.
[4.431c] *Drugs of the Future* **11** (1986) 657.
[4.432] *Drugs of the Future* **7** (1982) 803; **8** (1983) 977.
[4.433] *Drugs of the Future* **3** (1978) 511; **4** (1979) 529; **5** (1980) 368; **6** (1981) 443; **7** (1982) 523; **8** (1983) 648; **9** (1984) 548.
[4.434] N. Zamir et al., *Nature* **307** (1984) 643.
[4.435] S. Lemaire et al., *Can. J. Physiol. Pharmacol.* **62** (1984) 484.
[4.436] E. Weihe et al., *Neuropeptides* **5** (1985) 453.
[4.437] H.J. Wolter, *Peptides* **7** (1986) 389.
[4.438] S. Spampinato et al., *Life Sci.* **38** (1986) 403.
[4.439] H.-K. Jiang et al., *Proc. Natl. Acad. Sci. USA* **86** (1989) 2948.
[4.440] A.I. Faden et al., *Regul. Pept.* **11** (1985) 35.
[4.441] A.Z. Zhang et al., *Neuropeptides* **5** (1985) 553.
[4.442] R. Schick et al., *Dig. Dis. Sci.* **29**, Suppl. (1984) 74S.
[4.443] T. Mitsuma et al., *Horm. Metabol. Res.* **21** (1989) 301.
[4.444] N. Falke, *Neuropeptides* **11** (1988) 163.
[4.445] J.A.T. Haaf et al., *Acta Endocrinol. (Copenhagen)* **114** (1987) 96.
[4.446] J.D. Leander, *J. Pharmacol. Exp. Therap.* **227** (1983) 35.
[4.447] A. Rezvani, E.L. Way, *Eur. J. Pharmacol.* **102** (1984) 475.
[4.448] E.E. Codd et al., *Neuropeptides* **15** (1990) 133.
[4.449] A. Sydbom, L. Terenius, *Agents Actions* **16** (1985) 269.
[4.450] L.A. Chahl, J.S. Chahl, *Eur. J. Pharmacol.* **124** (1986) 343.
[4.451] M.A. Della-Fera et al., *Am. J. Physiol.* **258** (1990) R946.
[4.452] H. Shimizu et al., *Life Sci.* **45** (1989) 25.
[4.453] B.H. Herman, A. Goldstein, *Neuropeptides* **2** (1981) 13.

[4.454] H. Tilson et al., *Neuropeptides* **8** (1986) 193.
[4.455] A.I. Faden et al., *Peptides* **4** (1983) 631.
[4.456] D.S. Baskin et al., *Nature* **312** (1984) 551.
[4.457] P. Sanchez-Blazquez et al., *Neuropeptides* **4** (1984) 369.
[4.458] *Drugs of the Future* **6** (1981) 469; **7** (1982) 594; **8** (1983) 728; **10** (1985) 706.
[4.459] A. Turcotte et al., *Int. J. Pept. Protein Res.* **23** (1984) 361.
[4.460] H. Yoshino et al., *J. Med. Chem.* **33** (1990) 206.
[4.461] J.E. Gairin et al., *J. Med. Chem.* **29** (1986) 1913.
[4.462] J.E. Gairin et al., *Br. J. Pharmacol.* **95** (1988) 1023.
[4.463] Y. Kiso et al., *J. Pharmacobio Dyn.* **7** (1984) S-91.
[4.464] H.L. Wen et al., *Peptides* **8** (1987) 191.
[4.464] W. König in A. Kleemann et al. (eds.): *Arzneimittel, Fortschritte 1972 bis 1985*, VCH Verlagsgesellschaft, Weinheim 1987, 670.
[4.465] A. Mor et al., *FEBS Lett.* **255** (1989) 2690.
[4.466] K. Richter et al., *Proc. Natl. Acad. Sci. USA* **87** (1990) 4836.
[4.467] K. Richter et al., *Science* **238** (1987) 200.
[4.468] A. Mor et al., *Neuropeptides* **13** (1989) 51.
[4.469] S.J. Konturek et al., *Gastroenterology* **84**, Pt. 2 (1983) 1213.
[4.470] D. Giudici et al., *Neuroendocrinology* **39** (1984) 236.
[4.471] E.C. degli Uberti et al., *Horm. Res.* **24** (1986) 251.
[4.472] E. Roti et al., *J. Endocrinol. Invest.* **7** (1984) 211.
[4.473] H.G. Gullner et al., *Arch. Int. Pharmacodyn. Ther.* **266** (1983) 155.
[4.474] E.C. degli Uberti et al., *J. Clin. Endocrinol. Metab.* **61** (1985) 1985.
[4.475] F. Petraglia et al., *Peptides* **6** (1985) 869.
[4.476] G. Soldani et al., *Neuropeptides* **10** (1987) 67.
[4.477] F. Pavone et al., *Peptides* **11** (1990) 591.
[4.478] A. Rossi et al., *Peptides* **4** (1983) 577.
[4.479] K.M. Sivanandaiah et al., *Int. J. Pept. Protein Res.* **33** (1989) 463.
[4.480] K. Darlak et al., *Peptides* **5** (1984) 687.
[4.481] Y. Sasaki et al., *Chem. Pharm. Bull.* **33** (1985) 1528.
[4.482] K. Kisara et al., *Br. J. Pharmacol.* **87** (1986) 183.
[4.483] M.A. Cervini et al., *Peptides* **6** (1985) 433.
[4.484] S. Salvadori et al., *Arzneim. Forsch.* **34** (1984) 410.
[4.485] S. Salvadori et al., *Eur. J. Med. Chem.* **18** (1983) 489.
[4.486] S. Salvadori et al., *Hoppe Seyler's Z. Physiol. Chem.* **365** (1984) 1199.
[4.487] K. Suzuki et al., *Chem. Pharm. Bull.* **33** (1985) 4865.
[4.488] K. Chaki et al., *Peptides* **11** (1990) 139.
[4.489] Y. Sasaki et al., *Biochem. Biophys. Res. Commun.* **120** (1984) 214.
[4.490] K. Suzuki et al., *Chem. Pharm. Bull.* **36** (1988) 4834.
[4.491] S. Salvadori et al., *J. Med. Chem.* **29** (1986) 889.
[4.492] M. Marastoni et al., *J. Med. Chem.* **30** (1987) 1538.
[4.493] K. Suzuki et al., in G. Jung und E. Bayer: *Peptides 1988*, Walter de Gruyter, Berlin, New York 1989.
[4.494] G. Balboni et al., *Biochem. Biophys. Res. Commun.* **169** (1990) 617.
[4.495] S. Sagan et al., *J. Biol. Chem.* **264** (1989) 17100.
[4.496] W. König in A. Kleemann et al., (eds.): *Arzneimittel, Fortschritte 1972 bis 1985*, VCH Verlagsgesellschaft, Weinheim 1987, 601.
[4.497] H. Shiomi, H. Ueda, *J. Pharm. Soc. Jpn.* **105** (1985) 531.
[4.498] R. Kase et al., *Biochem. Pharmacol.* **35** (1986) 4499.
[4.499] K. Rolka et al., *Pol. J. Pharmacol. Pharm.* **35** (1984) 473.
[4.500] M.T. Garcia-Lopez et al., *Peptides* **7** (1986) 39.
[4.501] M.T. Garcia-Lopez et al., *J. Med. Chem.* **31** (1988) 295.
[4.502] K. Kitagawa et al., *Chem. Pharm. Bull.* **33** (1985) 377.
[4.503] W. König in A. Kleemann et al. (eds.): *Arzneimittel, Fortschritte 1972 bis 1985*, VCH Verlagsgesellschaft, Weinheim 1987, 671.
[4.504] R. Greenberg et al., *J. Biol. Chem.* **259** (1984) 5132.
[4.505] K.J. Chang et al., *J. Biol. Chem.* **260** (1985) 9706.

[4.506] L. Graf et al., *Neuropeptides* **9** (1987) 113.
[4.507] R. Coomans et al., *Eur. J. Med. Chem.* **12** (1977) 377.
[4.508] *Drugs of the Future* **7** (1982) 847; **9** (1984) 552.
[4.509] V. Schusdziarra et al., *Diabetologia* **24** (1983) 113.
[4.510] D. Tome et al., *Am. J. Physiol.* **253** (1987) G737.
[4.511] V. Schusdziarra et al., *Endocrinology* **112** (1983) 885.
[4.512] J. Nedvidkova et al., *Exp. Clin. Endocrinol.* **85** (1985) 249.
[4.513] T. Mitsuma et al., *Exp. Clin. Endocrinol.* **84** (1984) 324.
[4.514] M. Yoshikawa et al., *Agric. Biol. Chem.* **48** (1984) 3185.
[4.515] K.-J. Chang et al., *J. Pharmacol. Exp. Ther.* **227** (1983) 403.
[4.516] H. Matthies et al., *Peptides* **5** (1984) 463.
[4.517] C. Liebmann et al., *Peptides* **7** (1986) 195.
[4.518] W. König in A. Kleemann et al., (eds.): *Arzneimittel, Fortschritte 1972 bis 1985,* VCH Verlagsgesellschaft, Weinheim 1987, 667.
[4.519] M. Schaefer et al., *Cell* **41** (1985) 457.
[4.520] B.L. Roth et al., *Neuropeptides* **10** (1987) 37.
[4.521] R.H.M. Ebberink et al., *Peptides* **8** (1987) 515.
[4.522] D.A. Price et al., *Peptides* **8** (1987) 533.
[4.523] P.E. Lloyd et al., *Peptides* **8** (1987) 179.
[4.524] B.S. Rothman et al., *Peptides* **6** (1985) 1113.
[4.525] S. Robb et al., *Biochem. Biophys. Res. Commun.* **160** (1989) 850.
[4.526] T.G. Kingan et al., *Peptides* **11** (1990) 849.
[4.527] D. Graff, C.J.P. Grimmelikhuijzen, *FEBS Lett.* **239** (1988) 137.
[4.528] C.J.P. Grimmelikhuijzen, A. Groeger, *FEBS Lett.* **211** (1987) 105.
[4.529] H.Y.T. Yang et al., *Proc. Natl. Acad. Sci. USA* **82** (1985) 7757.
[4.530] R.B. Raffa, *Peptides* **9** (1988) 915.
[4.531] C.S. Barnard, G.J. Dockray, *Regul. Pept.* **8** (1984) 209.
[4.532] T. Austin et al., *Can. J. Physiol. Pharmacol.* **61** (1983) 949.
[4.533] K.E. Doble, M.J. Greenberg, *Neuropeptides* **2** (1982) 157.
[4.534] M. Kavaliers, M. Hirst, *Neuropeptides* **6** (1985) 485.
[4.535] M. Kavaliers, M. Hirst, *Peptides* **6** (1985) 847.
[4.536] X.Z. Zhu, R.B. Raffa, *Neuropeptides* **8** (1986) 55.
[4.537] D.H. Malin et al., *Peptides* **11** (1990) 969.
[4.538] G. Telegdy, I. Bollok, *Neuropeptides* **10** (1987) 157.
[4.539] A. Ottlecz, G. Telegdy, *Neuropeptides* **9** (1987) 161.
[4.540] K. Kouge et al., *Bull. Chem. Soc. Jpn.* **60** (1987) 4343.

References for Chapter 5

[5.1] W. König in A. Kleemann et al., (eds.): *Arzneimittel, Fortschritte 1972 bis 1985,* VCH Verlagsgesellschaft, Weinheim 1987, 626.
[5.2] A. Fukamizu et al., *J. Biol. Chem.* **265** (1990) 7576.
[5.3] E.T. Ben-Ari et al., *J. Biol. Chem.* **264** (1989) 13074.
[5.4] N. Itoh et al., *FEBS Lett.* **244** (1989) 6.
[5.5] E. Coezy et al., *Am. J. Physiol.* **257** (1989) C888.
[5.6] T. Inagami et al., Fed. Proc. *Fed. Am. Soc. Exp. Biol.* **45** (1986) 1414.
[5.7] M. Naruse et al., *J. Clin. Endocrinol. Metab.* **61** (1985) 480.
[5.8] C.F. Deschepper et al., *Endocrinology* **119** (1986) 36.
[5.9] M.C. Chappell et al., *J. Biol. Chem.* **264** (1989) 16518.
[5.10] M. Bumpus et al., *Am. J. Med. Sci.* **295** (1988) 406.
[5.11] M.T. Schiavone et al., Neurobiolog. **85** (1988) 4095.
[5.12] R.T. Fernley et al., *Eur. J. Biochem.* **154** (1986) 597.
[5.13] D.G. Changaris et al., *Biochem. Biophys. Res. Commun.* **138** (1986) 573.
[5.14] H. Kawaguchi, H. Yasuda, *FEBS Lett.* **221** (1987) 305.
[5.15] O. Kurauchi et al., *Horm. Metab. Res.* **21** (1989) 558.
[5.16] D.D. Smyth, H.Y.M. Fung, *Eur. J. Pharmacol.* **102** (1984) 55.
[5.17] J.C. Lee et al., *Am. J. Physiol.* **251** (1986) R258.

[5.18] J. Mann, *Acta Endocrinol.* **102** (1983) Suppl. 253. 10.
[5.19] N.R. Levens, *Am. J. Physiol.* **246** (1984) Pt. 1, G634.
[5.20] H.F. Jansen et al., *Circ. Shock* **23** (1987) 197.
[5.21] A.T. Chiu et al., *Biochem. Biophys. Res. Commun.* **165** (1989) 196.
[5.22] R.S.L. Chang et al., *Biochem. Biophys. Res. Commun.* **171** (1990) 813.
[5.23] D.R. Gehlert et al., *Eur. J. Pharmacol.* **187** (1990) 123.
[5.24] R. Tabrizchi et al., *Life Sci.* **43** (1988) 537.
[5.25] A. Israel et al., *Neuroendocrinology* **42** (1986) 57.
[5.26] E. Castren, J.M. Saavedra, *Endocrinology* **122** (1988) 370.
[5.27] I. Luoveau et al., *Biochem. Biophys. Res. Commun.* **163** (1989) 32.
[5.28] L.M. Plunkett, J.M. Saavedra, *Proc. Natl. Acad. Sci. USA* **82** (1985) 7721.
[5.29] B. Bouscarel et al., *J. Biol. Chem.* **23** (1988) 14920.
[5.30] B.J. Ballermann et al., *Am. J. Physiol.* **247** (1984) F110.
[5.31] G. Andoka et al., *Biochem. Biophys. Res. Commun.* **121** (1984) 441.
[5.32] P.D. Carriere et al., *Neuroendocrinology* **43** (1986) 49.
[5.33] J. Bingham Smith et al., *Proc. Natl. Acad. Sci. USA* **81** (1984) 7812.
[5.34] J.A. Gilbert et al., *Biochem. Pharmacol.* **33** (1984) 2527.
[5.35] H. Satoh et al., *Biochem. Biophys. Res. Commun.* **126** (1985) 464.
[5.36] I. Antonipillai et al., *Clin. Res.* **35** (1987) 135A.
[5.37] A. Nakamura et al., *Am. J. Physiol.* **258** (1990) E1.
[5.38] M.E. Olsen et al., *Clin. Sci.* **72** (1987) 429.
[5.39] K. Yamaguchi et al., *Acta Endocrinol. (Copenhagen)* **118** (1988) 82.
[5.40] V.L. Brooks et al., *Circ. Res.* **58** (1986) 829.
[5.41] P. Roubert et al., *Biochem. Biophys. Res. Commun.* **164** (1989) 809.
[5.42] M.C. Khosla et al., *Hypertension* **11** (1988) Suppl I, I-38.
[5.43] J.E. Hall, *Fed. Proc. Fed. Am. Soc. Exp. Biol.* **45** (1986) 1431.
[5.44] V.L. Schuster, Federation Proc. 45 (1986) 1444.
[5.45] N.R. Levens, *Am. J. Physiol.* **245** (1983) G511.
[5.46] G.G. Nussdorfer et al., *Exp. Clin. Endocrinol.* **88** (1986) 158.
[5.47] G. Mazzocchi, G.G. Nussdorfer, *Exp. Clin. Endocrinol.* **87** (1986) 157.
[5.48] A. Bobik et al., *Biochem. Biophys. Res. Commun.* **166** (1990) 580.
[5.49] D. Stirling et al., *J. Biol. Chem.* **265** (1990) 5.
[5.50] A. Pellicer et al., *Science* **240** (1988) 1660.
[5.51] L.L. Espey et al., *Proc. Soc. Exp. Biol. Med.* **193** (1990) 249.
[5.52] M. Volpe et al., *Hypertension* **12** (1988) 346.
[5.53] S. Makino et al., *Brain Res.* **501** (1989) 84.
[5.54] C. Schramme, C. Denef, *Life Sci.* **34** (1984) 1651.
[5.55] E.C. degli Uberti et al., *Metab. Clin. Exp.* **39** (1990) 1063.
[5.56] H. Haller et al., *Acta Endocrinol.* **112** (1986) 150.
[5.57] D.O. Sobel, *Peptides* **5** (1984) 47.
[5.58] M. Matsumura et al., *Endocrinol. Jpn.* **32** (1985) 233.
[5.59] C.R. Franci et al., *Neuroendocrinology* **51** (1990) 683.
[5.60] J.J. Braszko et al., *Neuroscience (Oxford)* **22** (1987), Suppl. S503.
[5.61] S. Kaneko et al., *Neuropharmacology* **24** (1985) 1131.
[5.62] K. Wisniewski et al., *Therapie* **43** (1988) 153.
[5.63] T. Kono et al., *Endocrinol. Jpn.* **34** (1987) 737.
[5.64] K.Y. Hui et al., *Int. J. Pept. Protein Res.* **34** (1989) 177.
[5.65] T. Kono et al., *Life Sci.* **37** (1985) 365.
[5.66] G. Guillemette et al., *J. Med. Chem.* **.27** (1984) 315.
[5.67] A. Husain et al., *J. Pharmacol. Exp. Ther.* **239** (1986) 71.
[5.68] J. Samanen et al., *J. Med. Chem.* **32** (1989) 466.
[5.69] K.-h. Hsieh et al., *J. Med. Chem.* **32** (1989) 898.
[5.70] K.L. Spear et al., *J. Med. Chem.* **33** (1990) 1935.
[5.71] M. Holck et al., *Biochem. Biophys. Res. Commun.* **160** (1989) 1350.
[5.72] J. Samanen et al., *J. Med. Chem.* **31** (1988) 510.
[5.73] O. Nyeki et al., *J. Med. Chem.* **30** (1987) 1719.
[5.74] M. N. Scanlon et al., *Life Sci.* **34** (1984) 317.

[5.75] M.H. Goghari et al., *J. Med. Chem.* **29** (1986) 1121.
[5.76] J. Samanen et al., *J. Med. Chem.* **32** (1989) 1366.
[5.77] J. Samanen et al., *J. Med. Chem.* **32** (1989) 1366.
[5.78] P.R. Bovy et al., *J. Med. Chem.* **32** (1989) 520.
[5.79] P.R. Bovy et al., *J. Med. Chem.* **33** (1990) 1477.
[5.80] P.C. Wong et al., *Hypertension* **12** (1988) 340.
[5.81] J.V. Duncia et al., *J. Med. Chem.* **33** (1990) 1312.
[5.83] J. Samanen et al., *J. Med. Chem.* **31** (1988) 737.
[5.84] J.E. Taylor et al., *Drug Dev. Res.* **16** (1989) 1.
[5.85] N.B. Oza et al., *J. Clin. Invest.* **85** (1990) 597.
[5.86] D.I. Diz, *Peptides* **6** (1985) 57.
[5.87] W.E. Wilson et al., *J. Biol. Chem.* **264** (1989) 17777.
[5.88] H. Kato et al., *FEBS Lett.* **232** (1988) 252.
[5.89] K.-i. Enjyoji, H. Kato, *FEBS Lett.* **238** (1988) 1.
[5.90] H. Maeda et al., *J. Biol. Chem.* **263** (1988) 1605.
[5.91] S. Furuto-Kato et al., *J. Biol. Chem.* **260** (1985) 12054.
[5.92] K.-i. Enjyoji et al., *J. Biol. Chem.* **263** (1988) 973.
[5.93] A. Barlas et al., *FEBS Lett.* **218** (1987) 266.
[5.94] N. Gutman et al., *Eur. J. Biochem.* **171** (1988) 577.
[5.95] J.A. Veenstra, *FEBS Lett.* **250** (1989) 231-234.
[5.96] J.A. Veenstra, F. Camps, *Neuropeptides* **15** (1990) 107.
[5.97] G. Gäde, G. Rosinski, *Peptides* **11** (1990) 455.
[5.98] B.E. Noyes, M.H. Schaffer, *J. Biol. Chem.* **265** (1990) 483.
[5.99] G. Gäde, R. Kellner, *Peptides* **10** (1989) 1287.
[5.100] Y. Kuroki et al., *Biochem. Biophys. Res. Commun.* **167** (1990) 273.
[5.101] L. Schoofs et al., *Peptides* **11** (1990) 427.
[5.102] H. Wilkens et al. in A. Bertelli, N. Back: *Shock, biochemical, pharmacological and clinical aspects.* Plenum Press, New York 1970, 201-214.
[5.103] D. Proud, A.P. Kaplan, *Ann. Rev. Immunol.* **6** (1988) 49.
[5.104] M. Wicklmayr et al., *Horm. Metab. Res.* **20** (1988) 535.
[5.105] L.L. Espey et al., *Am. J. Physiol.* **251** (1986) E362.
[5.106] G.M. Patrassi et al., *Thromb. Haemostasis* **50** (1983) 228.
[5.107] C.P. Quilley et al., *Hypertension* **10** (1987) 294.
[5.108] H. Okunishi et al., *Hypertension* **8** (1986) Suppl. 1, 114.
[5.109] J. Spragg et al., *Fed. Proc. Fed. Am. Soc. Exp. Biol.* **46** (1987) 1126.
[5.110] J. Spragg et al., *Peptides* 9 **(1988) 203.**
[5.111] G. Caspritz et al., *Arzneim. Forsch.* **36** (1986) 1605.
[5.112] S. Kumakura et al., *J. Pharmacol. Exp. Ther.* **243** (1987) 1067.
[5.113] R.E. Ferner et al., *Br. J. Clin. Pharmaol.* **25** (1988) 105P.
[5.114] E.G. Erdös, *J. Cardiovasc. Pharmacol.* **15** (1990) Suppl. 6, S20.
[5.115] J.M. Stewart, R.J. Vavrek, *J. Cardiovasc. Pharmacol.* **15** (1990) S69.
[5.116] I.A. Sheikh, A.P. Kaplan, *Biochem. Pharmacol.* **38** (1989) 993.
[5.117] K. Shimamoto et al., *J. Cardiovasc. Pharmacol.* **15** (1990) S83.
[5.118] P. Madeddu et al., *J. Hypertens.* **5** (1987) 645.
[5.119] I. Miyamori et al., *Exp. Clin. Endocrinol.* **87** (1986) 169.
[5.120] E.I. Tkachenko et al., *Klin. Med. (Moscow)* **62** (1984) 60.
[5.121] D. Proud et al., Paper presented at the Congress *Kinin 84*, Savannah, USA 1984.
[5.122] R. Kogeyama et al., *J. Biol. Chem.* **260** (1985) 12060.
[5.123] W. Sakamoto et al., *Biochem. Pharmacol.* **35** (1986) 4283.
[5.124] A. Barlas et al., Paper presented at the Congress *Kinin 84,* Savannah, USA 1984.
[5.125] J. Bouhnik et al., *Life Sci.* **44** (1989) 1859.
[5.126] J. Bouhnik et al., *Am. J. Physiol.* **255** (1988) E411.
[5.127] F. Sierra et al., *Mol. Cell. Biol.* **9** (1989) 5610.
[5.128] S. Oh-ishi et al., *Biochem. Biophys. Res. Commun.* **150** (1988) 1069.
[5.129] K.M. Braas et al., *Br. J. Phramacol.* **94** (1988) 3.
[5.130] J.K. Saha et al., *J. Pharmacol. Exp. Ther.* **252** (1990) 1012.
[5.131] A.A. Roscher et al., *J. Cardiovasc. Pharmacol.* **15** (1990) Suppl. 6, S39.

[5.132] T.S. Gaginella, J.F. Kachur, *Am. J. Physiol.* **256** (1989) G1.
[5.133] O.A. Carretero, A.G. Scicli in J.H. Laragh et al., (Eds.): *Endocrine Mechanisms in Hypertension,* Raven Press, New York, 1989, 219.
[5.134] D. Regoli et al., *J. Cardiovasc. Pharmacol.* **15** (1990) Suppl. 6, S30.
[5.135] J.A. Guimaaraes et al., *Eur. J. Pharmacol.* **130** (1986) 177.
[5.136] J.M. Ritter et al., *Br. J. Pharmacol.* **96** (1989) 23.
[5.137] L. Churchill, P.E. Ward, *Hypertension* **9** (1987), Suppl. III, III-1.
[5.138] E. Müller-Schweinitzer, *Naunyn-Schmiedeberg's Arch. Pharmacol. (Suppl.)* **338** (1988) R51.
[5.139] C.R. Mantione, R. Rodriguez, *Br. J. Pharmacol.* **99** (1990) 516.
[5.140] B.S. Dixon et al., *Am. J. Physiol.* **258** (1990) C299.
[5.141] P. Francel, G. Dawson, *Biochem. Biophys. Res. Commun.* **152** (1988) 742.
[5.142] B. Waeber et al., *J. Cardiovasc. Pharmacol.* **15** (1990) Suppl. 6, S78.
[5.143] K. Inokuchi, K.U. Malik, *Am. J. Physiol.* **246** (1984) F387.
[5.144] M. Schachter et al., *Br. J. Pharmacol.* **92** (1987) 851.
[5.145] E.T. Whalley et al., *Naunyn-Schmiedeberg's Arch. Pharmacol.* **336** (1987) 652.
[5.146] B.A. Schölkens et al., *J. Hypertens.* **5** (1987) Suppl. 5, S7.
[5.147] J.J. Smith et al., *Am. J. Physiol.* **258** (1990) L369.
[5.148] A.A. Jaffa et al., *Clin. Res.* **37** (1989) 492A.
[5.149] J.P. Granger, J.E. Hall, *Am. J. Physiol.* **248** (1985) F87.
[5.150] V.L. Schuster, *Am. J. Physiol.* **249** (1985) F645.
[5.151] A. Dray et al., *Br. J. Pharmacol.* **95** (1988) 1008.
[5.152] J.E. Haley et al., *Neurosci. Lett.* **97** (1989) 198.
[5.153] A.H. Costello, K.M. Hargreaves, *Eur. J. Pharmacol.* **171** (1989) 259.
[5.154] Y. Fujiwara et al., *Life Sci.* **44** (1989) 1645.
[5.155] C.J. Lindsey et al., *Br. J. Pharmacol.* **97** (1989) 763.
[5.156] Y. Yoshimura et al., *Endocrinology* **122** (1988) 2540.
[5.157] L.F. Carbonell et al., *Hypertension* **11** (1988) 239.
[5.158] M. Matsumura et al., *Neuroendocrinology* **41** (1985) 101.
[5.159] P. Geppetti et al., *Br. J. Pharmacol.* **94** (1988) 288.
[5.160] ö. Ljunggren, U.H. Lerner, *Br. J. Pharmacol.* **101** (1990) 382.
[5.161] C. Wahlestedt et al., *Eur. J. Pharmacol.* **106** (1984) 577.
[5.162] G. Dietze et al., Paper presented at the Congress *Kinin 84,* Savannah, USA 1984.
[5.163] A. Unterberg et al., *J. Neurosurg.* **64** (1986) 269.
[5.164] O. Laneuville et al., *Eur. J. Pharmacol.* **159** (1989) 273.
[5.165] N.-E. Rhaleb et al., *Br. J. Pharmacol.* **99** (1990) 445.
[5.166] J. Barabe et al., *Can. J. Physiol. Pharmacol.* **62** (1984) 627.
[5.167] G. Drapeau et al., *Eur. J. Pharmacol.* **155** (1988) 193.
[5.168] R.J. Vavrek et al., in E. Giralt D. Andreu (eds.): *Peptides 1990, ESCOM Sci. Publ.,* Leiden 1991, 642.
[5.169] F.J. Hock et al., *Br. J. Pharmacol.* **102** (1991) 769.
[5.170] R.J. Vavrek, J.M. Stewart in C.M. Deber et al. (Eds.): *Peptides. Structure and Function,* Pierce Chem. Comp., Rockford, Ill., 1985, 655.
[5.171] B. Lammek et al., *Peptides* **11** (1990) 1041.
[5.172] J.M. Stewart, R.J. Vavrek, *J. Cardiovasc. Pharmacol.* **15** (1990) Suppl. 6, S69.
[5.173] D. Regoli et al., *Eur. J. Pharmacol.* **123** (1986) 61.
[5.174] D. Regoli et al., *Eur. J. Pharmacol.* **127** (1986) 219.
[5.175] P. Devillier et al., *Eur. J. Pharmacol.* **149** (1988) 137.
[5.176] J.B. Calixto et al., *Br. J. Pharmacol.* **91** (1987) 199.
[5.177] A.S. Dutta, *Drugs of the Future* **12** (1987) 781-792.
[5.178] W. König in A. Kleemann et al., (eds.): *Arzneimittel, Fortschritte 1972 bis 1985,* VCH Verlagsgesellschaft, Weinheim 1987, 620.
[5.179] H. Nawa et al., *Nature* **312** (1984) 729.
[5.180] Y. Kawaguchi et al., *Biochem. Biophys. Res. Commun.* **139** (1986) 1040.
[5.181] J.E. Krause et al., *Proc. Natl. Acad. Sci. USA* **84** (1987) 881.
[5.182] N. Minamino et al., *Neuropeptides* **4** (1984) 157.
[5.183] A.J. Harmar et al., *FEBS Lett.* **208** (1986) 67.
[5.184] E. Theodorsson-Norheim et al., *Biochem. Biophys. Res. Commun.* **131** (1985) 77.

[5.185] C.-R. Martling et al., *Life Sci.* **40** (1987) 1633.
[5.186] J.M. Conlon et al., *Peptides* **9** (1988) 859.
[5.187] E. Theodorsson-Norheim et al., *Eur. J. Biochem.* **166** (1987) 693.
[5.188] R. Kage et al., *J. Neurochem.* **50** (1988) 1412.
[5.189] M.R. MacDonald et al., *J. Biol. Chem.* **264** (1989) 15578.
[5.190] J.M. Conlon et al., *Regul. Pept.* **20** (1988) 171.
[5.191] A. Bucsics et al., *Peptides* **4** (1983) 451.
[5.192] A. Ericsson et al., *Mol. Endocrinol.* **4** (1990) 1211.
[5.193] B. Beding-Barnekow, E. Brodin, *Regul. Pept.* **25** (1989) 199.
[5.194] M. Simmaco et al., *Peptides* **11** (1990) 299.
[5.195] J.M. Conlon et al., *FEBS Lett.* **200** (1986) 111.
[5.196] L. Schoofs et al., *FEBS Lett.* **261** (1990) 397.
[5.197] B.A. Yankner et al., *Science* **250** (1990) 279.
[5.198] J.L. Vaught, *Life Sci.* **43** (1988) 1419.
[5.199] F. Angel et al., *Gastroenterology* **84** (1983) Pt. 2, 1092.
[5.200] J. Wiley, C. Owyang, *Am. J. Physiol.* **252** (1987) G431.
[5.201] A. Mandahl, *Eur. J. Pharmacol.* **114** (1985) 121.
[5.202] P. Tonnesen, O.B. Schaffalitzky de Muckadell, *Allergy* **42** (1987) 146.
[5.203] M.J. Bannon et al., *J. Biol. Chem.* **261** (1986) 6640.
[5.204] R. Coslovsky et al., *Endocrinology* **115** (1984) 2285.
[5.205] I. Norheim et al., *J. Clin. Endocrinol. Metab.* **63** (1986) 605.
[5.206] P. Devillier et al., *N. Engl. J. Med.* **314** (1986) 1323.
[5.207] J.D. Levine et al., *Science* **226** (1984) 547.
[5.208] J.M. Stewart, M.E. Hall in T. Shiba, S. Sakakibara (eds.): *Peptide Chemistry 1987*, Protein Research Foundation, Osaka 1988, 449.
[5.209] J.M. Conlon, L. Sheehan, *Regul. Pept.* **7** (1983) 335.
[5.210] D.J. Dusser et al., *J Clin. Invest.* **84** (1989) 900.
[5.211] W.E. Siems et al., *Pharmazie* **42** (1987) 153.
[5.212] D. Regoli et al., *Trends in Pharmacol. Sci.* **9** (1988) 290.
[5.213] S. Lavielle et al., *Biochem. Pharmacol.* **37** (1988) 41.
[5.214] Y. Yokota et al., *J. Biol. Chem.* **264** (1989) 17649.
[5.215] A.D. Hershey, J.E. Krause, *Science* **247** (1990) 958.
[5.216] Y. Masu et al., *Nature* **329** (1987) 836.
[5.217] J.E. Shook, T.F. Burks, *Life Sci.* **39** (1986) 2533.
[5.218] S.H. Buck, S.A. Shatzer, *Life Sci.* **42** (1988) 2701.
[5.219] S.H. Buck, J.L. Krstenansky, *Eur. J. Pharmacol.* **144** (1987) 109.
[5.220] J.C. Beaujouan et al., *Eur. J. Pharmacol.* **151** (1988) 353.
[5.221] K. Shigematsu et al., *Regul. Pept.* **16** (1986) 147.
[5.222] C.A. Maggi et al., *J. Pharmacol. Exp. Ther.* **246** (1988) 308.
[5.223] U. Holzer-Petsche et al., *Br. J. Pharmacol.* **90** (1987) 273.
[5.224] P.V. Andrews et al., *Br. J. Pharmacol.* **97** (1989) 1232.
[5.225] I. Zachary, E. Rozengurt, *Biochem. Biophys. Res. Commun.* **137** (1986) 135.
[5.226] C.A. Maggi et al., *Br. J. Pharmacol.* **100** (1990) 588.
[5.227] E.E. Daniel et al., *Br. J. Pharmacol.* **97** (1989) 1013.
[5.228] T.S. Gates et al., *Peptides* **9** (1989) 1207.
[5.229] I. Haro et al., *J. Pharm. Sci.* **79** (1990) 74.
[5.230] R.M. Hagan et al., *Br. J. Pharmacol.* **98** (1989) Suppl. 717P.
[5.231] P. Rovero, V. Pestellini, *Neuropeptides* **10** (1987) 355.
[5.232] P. Rovero et al., *Peptides* **10** (1989) 593.
[5.233] S.H. Buck et al., *Life Sci.* **47** (1990) PL-37.
[5.234] M. Chorev et al., *Eur. J. Pharmacol.* **127** (1986) 187.
[5.235] C.J. Mussap, E. Burcher, *Peptides* **11** (1990) 827.
[5.236] J.-M. Qian et al., *J. Biol. Chem.* **28** (1989) 16667.
[5.237] T. Sakurada et al., *Eur. J. Pharmacol.* **174** (1989) 153.
[5.238] P. Ward et al., *J. Med. Chem.* **33** (1990) 1848.
[5.239] J. Mizrahi et al., *Eur. J. Pharmacol.* **118** (1985) 25.
[5.240] D. Regoli et al., *Eur. J. Pharmacol.* **109** (1985) 121.

[5.241] P. Rovero et al., *Eur. J. Pharmacol.* **175** (1990) 113.
[5.242] G. Drapeau et al., *Regul. Pept.* **31** (1990) 125.
[5.243] M. Ziche et al., *Br. J. Pharmacol.* **100** (1990) 11.
[5.244] F. Marceau et al., *Can. J. Physiol. Pharmacol.* **67** (1989) 159.
[5.245] C.A. Maggi et al., *Eur. J. Pharmacol.* **177** (1990) 81.
[5.246] E. Naline et al., *Am. Rev. Respir. Dis.* **140** (1989) 679.
[5.247] S. Brunelleschi et al., *Br. J. Pharmacol.* **100** (1990) 417.
[5.248] S. Evangelista et al., *Peptides* **10** (1989) 79.
[5.249] M. Perfumi et al., *Peptides* **11** (1990) 339.
[5.250] I. Rollandy et al., *Neuropeptides* **13** (1989) 175.
[5.251] M. Arisawa et al., *Proc. Natl. Acad. Sci. USA* **86** (1989) 7290.
[5.252] B. Ahren, *Peptides* **5** (1984) 305.
[5.253] M. Lotz et al., *Science* **235** (1987) 893.
[5.254] J.C. Reynolds et al., *Am. J. Physiol.* **246** (1984) G346.
[5.255] J.E.T. Fox, E.E. Daniel, *Am. J. Physiol.* **250** (1986) G21.
[5.256] L.D. Hirning, T.F. Burks, *Gastroenterology* **84** (1983) Pt. 2, 1188.
[5.257] P. Holzer et al., *Br. J. Pharmacol.* **89** (1986) 453.
[5.258] K. Yokotani, M. Fujiwara, *J. Pharmacol. Exp. Ther.* **232** (1985) 826.
[5.259] K. Iwatsuki et al., *Clin. Exp. Pharmacol. Physiol.* **13** (1986) 663.
[5.260] K. Katoh et al., *Res. Vet. Sci.* **35** (1984) 147.
[5.261] Y. Chiba et al., Endocrinology **117** (1985) 1996.
[5.262] X.W. Fu, A.M. Sun, *Acta Pharmacol. Sin.* **10** (1989) 69.
[5.263] A.F. De Felice, A. Brousseau, *J. Pharmacol. Exp. Ther.* **246** (1988) 183.
[5.264] J.P. McGillis et al., *Fed. Proc. Fed. Am. Soc. Exp.* **46** (1987) 196.
[5.265] I. Paegelow et al., *Pharmazie* **44** (1989) 145.
[5.266] J.R. Naranjo, J.D. Rio, *Neuropharmacology* **23** (1984) 1167.
[5.267] W.L. Dees et al., *Life Sci.* **37** (1985) 1627.
[5.268] G.R. Trimarchi et al., *Life Sci.* **39** (1986) 1579.
[5.269] Z. Chen et al., *Acta Physiol. Scand.* **134** (1988) 153.
[5.270] M.E. Hall, J.M. Stewart, *Peptides* **4** (1983) 763.
[5.271] M.E. Hall, J.M. Stewart, *Peptides* **5** (1984) 85.
[5.272] O. Gaffori et al., *Experientia* **40** (1984) 89.
[5.273] M.E. Hall et al., *Life Sci.* **40** (1987) 1909.
[5.274] T. Sakurada et al., *Jpn. J. Pharmacol.* **43** (1987) Suppl. 272P.
[5.275] K.D. Yench et al., *Patol. Fiziol. Eksp. Ter.* (1984) 3, 75.
[5.276] R. Michelot et al., *Eur. J. Med. Chem. (Chem. Ther.)* **23** (1988) 243.
[5.277] C. Poulos et al., *J. Med. Chem.* **30** (1987) 1512.
[5.278] Y. Higuchi et al., *Eur. J. Pharmacol.* **160** (1989) 413.
[5.279] E. Munekata et al., *Peptides* **8** (1987) 169.
[5.280] A. Ljungqvist et al., *Regul. Pept.* **24** (1989) 283.
[5.281] K. Folkers et al., *Proc. Natl. Acad. Sci. USA* **87** (1990) 4833.
[5.282] R. Hakanson et al., *Regul. Pept.* **31** (1990) 75.
[5.283] Z. Wiesenfeld-Hallin et al., *Regul. Pept.* **29** (1990) 1.
[5.284] J.-M. Qian et al., *J. Biol. Chem.* **264** (1989) 16667.
[5.285] H.P. Erian et al., *Gastroenterology* **90** (1986) Pt. 2, 1455.
[5.286] P.J. Woll, E. Rozengurt, *Proc. Natl. Acad. Sci. USA* **85** (1988) 1859.
[5.287] G.G. Glenner, *Cell* **52** (1988) 307.
[5.288] C.L. Joachim et al., *Nature* **341** (1989) 226.
[5.289] J. Kang et al., *Nature* **325** (1987) 733.
[5.290] R.E. Tanzi et al., *Science* **235** (1987) 880.
[5.291] D. Goldgaber et al., *Science* **235** (1987) 877.
[5.292] M.R. Palmert et al., *Science* **241** (1988) 1080.
[5.293] D. Schubert et al., *Science* **241** (1988) 223.
[5.294] B.D. Shivers et al., *EMBO J.* **7** (1988) 1365.
[5.295] T. Oltersdorf et al., *Nature* **341** (1989) 144.
[5.296] F. Prelli et al., *Biochem. Biophys. Res. Commun.* **151** (1988) 1150.
[5.297] W.M. Pardridge et al., *Biochem. Biophys. Res. Commun.* **145** (1987) 241.

[5.298] K. Zimmermann et al., *EMBO J,* **7** (1988) 367.
[5.299] J. Ghiso et al., *Biochem. Biophys. Res. Commun.* **163** (1989) 430.
[5.300] W.E. van Nostrand et al., *Science* **248** (1990) 745.
[5.301] T. Dyrks et al., *EMBO J.* **7** (1988) 949.
[5.302] D. Allsop et al., *Proc. Natl. Acad. Sci. USA* **85** (1988) 2790.
[5.303] L.M. Refolo et al., *Biochem. Biophys. Res. Commun.* **164** (1989) 664.
[5.304] A.I. Bush et al., *J. Biol. Chem.* **265** (1990) 15977.
[5.305] D. Schubert et al., *Biochem. Biophys. Res. Commun.* **162** (1989) 83.
[5.306] T. Saitoh et al., *Cell* **58** (1989) 615.
[5.307] R.P. Smith et al., *Science* **248** (1990) 1126.
[5.308] B.A. Yankner et al., *Science* **250** (1990) 279-282.
[5.309] W. König in A. Kleemann et al. (eds.): *Arzneimittel, Fortschritte 1972 bis 1985*, VCH Verlagsgesellschaft, Weinheim 1987, 694.
[5.310] V. Geenen et al., *Science* **232** (1986) 508.
[5.311] R. Ivell, D. Richter, *EMBO J.* **3** (1984) 2351.
[5.312] S. Lundin, G. Toresson, Horm. Metab. Res. **19** (1987) 629.
[5.313] M. Rehbein et al., *Biol. Chem. Hoppe-Seyler* **367** (1986) 695.
[5.314] J. Chavet et al., *Neuroendocrinology* **51** (1990) 233.
[5.315] B. Liu et al., *J. Biol. Chem.* **263** (1988) 72.
[5.316] B. Liu, J.P.H. Burbach, *J. Neuroendocrinology* **1** (1989) no. 1.
[5.317] M. Mühlethaler et al., *Experientia* **40** (1984) 777.
[5.318] L.J. Cruz et al., *J. Biol. Chem.* **262** (1987) 15821.
[5.319] J.P. Proux et al., *Biochem. Biophys. Res. Commun.* **149** (1987) 180.
[5.319a] V.K. Batra et al., *Regul. Pept.* **30** (1990) 77.
[5.320] D.P. Brooks et al., *Experientia* **42** (1986) 1012.
[5.321] P. Soelberg Sorensen, M. Hammer, *Am. J. Physiol.* **248** (1985) R78.
[5.322] B.W. Wang, L. Share, K.L. Goetz, *Fed. Proc. Fed. Am. Soc. Exp. Biol.* **44** (1985) 72.
[5.323] B. Waeber et al., *Am. J. Physiol.* **247** (1984) H895.
[5.324] M.R. Murphy et al., *J. Clin. Endocrinol. Metab.* **65** (1987) 738.
[5.325] W.M. Barron et al., *Am. J. Physiol.* **250** (1986) E352.
[5.326] S.W.T. Cheng, W.G. North, *Neuroendocrinology* **42** (1986) 174.
[5.327] C.M. Gregg, *Neuroendocrinology* **40** (1985) 423-429.
[5.328] M.T. Shiavone et al., *Proc. Natl. Acad. Sci. USA* **85** (1988) 4095.
[5.329] R. Cacabelos et al., *Neuroendocrinology* **45** (1987) 368.
[5.330] T. Ivanyi et al., *Exp. Clin. Endocrinol.* **88** (1986) 303.
[5.331] O. Willoughby et al., *Neuroendocrinology* **45** (1987) 219.
[5.332] T. Kimura et al., *Endocrinology* **114** (1984) 1426.
[5.333] L. Share, J.T. Crofton, *Fed. Proc. Fed. Am. Soc. Exp. Biol.* **43** (1984) 103.
[5.334] J. Skjoto et al., *Acta Med. Scand.* **215** (1984) 165.
[5.335] I. Os et al., *Acta Med. Scand.* **220** (1986) 195.
[5.336] G. Geelen et al., *Am. J. Physiol.* **247** (1984) R968.
[5.337] H. Ishizawa et al., *Eur. J. Pharmacol.* **189** (1990) 119.
[5.338] J.R. Sladek, Jr., C.D. Sladek, *Fed. Proc. Fed. Am. Soc. Exp. Biol.* **44** (1985) 66.
[5.339] B. Bardrum et al., *Endocrinology* **123** (1988). 2249
[5.340] T.O. Bruhn et al., *Endocrinology* **119** (1986) 1558.
[5.341] R. Landgraf et al., *Am. J. Physiol.* **259** (1990) R1056.
[5.342] N.W. Kasting, *Can. J. Physiol. Pharmacol.* **64** (1986) 1575.
[5.343] D. Schams et al., *J. Endocrinol.* **102** (1984) 337.
[5.344] S. Thornton et al., *Br. Med. J.* **297** (1988) 167.
[5.345] J.G. Verbalis et al., *Science* **232** (1986) 1417.
[5.346] K. Uvnäs-Moberg et al., *Acta Physiol. Scand.* **124** (1985) 391.
[5.347] M.S. Carmichael et al., *J. Clin. Endocrinol. Metab.* **64** (1987) 27.
[5.348] B. Koch, B. Lutz-Bucher, *Endocrinology* **116** (1985) 671.
[5.349] P. Chang, K. Kimura, *Jpn. J. Pharmacol.* **43** (1987) Suppl. 184P.,
[5.350] A.E. Johnson et al., *Neuroendocrinology* **50** (1989) 199.
[5.351] M. Shumacher et al., *Science* **250** (1990) 691.
[5.352] S. Chin Mah, K.G. Hofbauer, *Drugs of the Future* **12** (1987) 1055.

[5.353] E.A. Woodcock et al., *Endocrinology* **118** (1986) 2432.
[5.354] M.A. Burnatowska Hledin, W.S. Spielman, *Am. J. Physiol.* **253** (1987) Pt. 2, F328.
[5.355] M.J.S. Miller et al., *Clin. Exp. Pharmacol. Physiol.* **13** (1986) 577.
[5.356] M.F. Wilkinson et al., *Am. J. Physiol.* **256** (1989) Pt. 2, R1164.
[5.357] S. Caltabiano et al., *Eur. J. Pharmacol.* **139** (1987) 281.
[5.358] P.M. Plotsky et al., *Endocrinology* **115** (1984) 1639.
[5.359] C.G.J. Sweep et al., *Pharm. Weekbl. Sci. Ed.* **9** (1987) 296.
[5.360] A.M. Naylor et al., *Neuropharmacology* **25** (1986) 787.
[5.361] K.E. Cooper et al., *J. Physiol. (London)* **387** (1987) 163.
[5.362] J. Maslyszko et al., *Thromb. Haemostasis* **61** (1989) 537.
[5.363] P.A. Flordal et al., *Thromb. Haemostasis* **61** (1989) 541.
[5.364] A.H. Tang, P.M. Ho, *Life Sci.* **43** (1988) 399.
[5.365] F.A. Laszlo et al., *Acta Endocrinol.* **106** (1984) 56.
[5.366] J.C. Buckingham, *J. Endocrinol.* **113** (1987) 389.
[5.367] P. Mormede et al., *Regul. Pept.* **12** (1985) 175.
[5.368] L. Hary et al., *Fundam. Clin. Pharmacol.* **3** (1989) 452.
[5.369] J.J. Legros et al., *Prog. Neuropsychopharmacol. Biol. Psychiatry* **12** (1988) Suppl., S71.
[5.370] J. McMurray, A.D. Struthers, *Br. Heart J.* **59** (1988) 627.
[5.371] L.M. Graczak et al., *Am. J. Physiol.* **256** (1989) H925.
[5.372] H. Jin et al., *Am. J. Med. Sci.* **298** (1989) 227.
[5.373] T. Wilson et al., *Prostaglandins* **35** (1988) 771.
[5.374] B.M. Altura, B.T. Altura, *Fed. Proc. Fed. Am. Soc. Exp. Biol.* **43** (1984) 80.
[5.375] S. Stock, K. Uvnäs-Moberg, *Acta Physiol. Scand.* **125** (1985) 205.
[5.376] N. Holst et al., *Acta Physiol. Scand.* **132** (1988) 23.
[5.377] D.P. Brooks et al., *Neuroendocrinology* **38** (1984) 382.
[5.378] A.S. Al Janabi et al., *Indian J. Exp. Biol.* **25** (1987) 293.
[5.379] J.D. Caldwell et al., *Neuropeptides* **7** (1986) 175.
[5.380] M.R. Melis et al., *Brain Res.* **398** (1986) 295.
[5.381] S. Stock et al., *Regul. Pept.* **30** (1990) 1.
[5.382] N. Altszuler et al., *Proc. Soc. Exp. Biol. Med.* **182** (1986) 79.
[5.383] A.N. Margioris et al., *J. Clin. Endocrinol. Metab.* **66** (1988) 922.
[5.384] G.L. Kovacs et al., *Life Sci.* **37** (1985) 17.
[5.385] R. Arletti et al., *Peptides* **10** (1989) 89.
[5.386] J.J. Nieuwenhuis, *Life Sci.* **35** (1984) 1713.
[5.387] R. Brown, M.G. King, *Peptides* **5** (1984) 1135.
[5.388] S. Chin Mah, K.G. Hofbauer, *Drugs of the Future* **12** (1987) 1055.
[5.389] M.R. Melis et al., *Pharmacol. Res. Commun.* **20** (1988) 1117.
[5.390] C. Lin et al., *Peptides* **11** (1990) 633.
[5.390a] A. Buku et. al., *Experientia.* **43** (1987), 1248.
[5.391] P. Rekowski et al., *Acta Chem. Scand. Ser. B* **B39** (1985) 453.
[5.392] Z. Grzonka et al., *J. Med. Chem.* **29** (1986) 96.
[5.393] A. Buku et al., *Int. J. Peptide Res.* **23** (1984) 551.
[5.394] Z. Prochazka et al., *Collect. Czech. Chem. Commun.* **53** (1988) 2604.
[5.395] A. Buku et al., *J. Med. Chem.* **30** (1987) 1509.
[5.396] B. Lammek et al., *Experientia* **43** (1987) 1211.
[5.397] Z. Prochazka et al., *Collect. Czech. Chem. Commun.* **49** (1984) 642.
[5.398] M. Barber et al., *Int. J. Pept. Protein Res.* **14** (1979) 247.
[5.399] M. Lebl et al., *Collect. Czech. Chem. Commun.* **50** (1985) 418.
[5.400] G. Sweeney et al., *Curr. Ther. Res.* **47** (1990) 528.
[5.401] Z. Grzonka et al., *J. Med. Chem.* **26** (1983) 1786.
[5.402] *Drugs of the Future* **9** (1984) 830.
[5.403] M. Lebl et al., *Collect. Czech. Chem. Commun.* **50** (1985) 132.
[5.404] R.A. Johns, *Anesthesiology* **71** (1989) A487.
[5.405] B. Lammek et al., *J. Med. Chem.* **32** (1989) 244.
[5.406] L.B. Kinter et al., *Am. J. Physiol.* **254** (1988) F165.
[5.407] W.Y. Chan et al., *Proc. Soc. Exp. Biol. Med.* **185** (1987) 187.
[5.408] M. Manning et al., *J. Med. Chem.* **32** (1989) 382.

[5.409] D.W. Hahn et al., *Am. J. Obstet. Gynaecol.* **157** (1987) 977.
[5.410] P. Rekowski et al., *Pol. J. Pharmacol. Pharm.* **39** (1987) 303.
[5.411] P. Melin et al., *J. Endocrinol.* **111** (1986) 125.
[5.412] *Drugs of the Future* **13** (1988) 25.
[5.413] B. Lammek et al., *Pol. J. Pharmacol. Pharm.* **40** (1988) 423.
[5.414] B. Lammek et al., *J. Med.·Chem.* **31** (1988) 603.
[5.415] M. Manning et al., *J. Med. Chem.* **27** (1984) 423.
[5.416] M. Manning et al., *J. Med. Chem.* **26** (1983) 1607.
[5.417] W.F. Huffman et al., *J. Med. Chem.* **32** (1989) 880.
[5.418] B. Lammek et al., *Peptides* **10** (1989) 1109.
[5.419] F. El-Fehail Ali et al., *J. Med. Chem.* **29** (1986) 984.
[5.420] M. Manning et al., *Nature* **308** (1984) 652.
[5.421] F. El-Fehail Ali et al., *J. Med. Chem.* **30** (1987) 2291.
[5.422] J.F. Callahan et al., *J. Med. Chem.* **32** (1989) 391.
[5.423] W.F. Huffman et al., *J. Med. Chem.* **28** (1985) 1759.
[5.424] D.P. Brooks et al., *J. Pharmacol. Exp. Ther.* **245** (1988) 211.
[5.425] N. Caldwell et al., *J. Pharmacol. Exp. Ther.* **247** (1988) 897.
[5.426] M.L. Moore et al., *J. Med. Chem.* **31** (1988) 1487.
[5.427] M. Manning et al., *J. Med. Chem.* **30** (1987) 2245.
[5.428] W.H. Sawyer et al., *Peptides* **9** (1988) 157.
[5.429] M. Manning et al., *Nature* **329** (1987) 839.
[5.430] F. Kasprzykowski et al., *Collect. Czech. Chem. Commun.* **53** (1988) 2907.
[5.431] D.J. Pettibone et al., *Endocrinology* **125** (1989) 217.
[5.432] R.M. Freidinger et al., *J. Med. Chem.* **33** (1990) 1845.
[5.433] M.G. Bock et al., *J. Med. Chem.* **33** (1990) 2321.
[5.434] P. Mannucci et al., *Ann. N.Y. Acad. Sci.* **509** (1987) 71.
[5.435] P.R. Martens, *Lancet I* (1989) 1322.
[5.436] B. Waeber et al., *J. Cardiovasc. Pharmacol.* **8** (1986) Suppl. 7, S111.
[5.437] M. Akerlund et al., *Br. J. Obstet. Gynaecol.* **94** (1987) 1040.
[5.438] W. König in A. Kleemann et al. (eds.): *Arzneimittel, Fortschritte 1972 bis 1985,* VCH Verlagsgesellschaft, Weinheim 1987, 699.
[5.439] J.E. Zadina et al., *Brain Res.* **409** (1987) 10.
[5.440] G. Campbell Teskey, M. Kavaliers, *Peptides* **6** (1985) 165.
[5.441] H.N. Bhargava, *Neuropharmacology* **25** (1986) 737.
[5.442] G. Szabo et al., *Acta Physiol. Hung.* **69** (1987) 115.
[5.443] G.L. Kovacs et al., *Pharmacol. Biochem. Behav.* **31** (1988) 833.
[5.444] R.L. Johnson et al., *J. Med. Chem.* **33** (1990) 1828.
[5.445] L. Pulvirenti, A.J. Kastin, *Eur. J. Pharmacol.* **151** (1988) 289.
[5.446] R.L. Johnson et al., *J. Med. Chem.* **29** (1986) 2100.
[5.447] K.-L. Yu et al., *J. Med. Chem.* **21** (1988) 1430.
[5.448] R.K. Mishra et al., *Drugs of the Future* **11** (1986) 303.
[5.449] M. Yanagisawa, T. Masaki, *Biochem. Pharmacol.* **38** (1989) 1877.
[5.450] T. Kosaka et al., *FEBS Lett.* **249** (1989) 42.
[5.451] T. Yoshizawa et al., *Science* **247** (1990) 462.
[5.452] H.C. Lam et al., *FEBS Lett.* **261** (1990) 184.
[5.453] A. Inoue et al., *Proc. Natl. Acad. Sci. USA* **86** (1989) 2863.
[5.454] M. Yanagisawa et al., *Nature* **332** (1988) 411.
[5.455] K. Saida et al., *J. Biol. Chem.* **264** (1989) 14613.
[5.456] A. Bdolah et al., *FEBS Lett.* **256** (1989) 1.
[5.457] Y. Kloog et al., *Science* **242** (1988) 268.
[5.458] A. Inoue et al., *J. Biol. Chem.* **264** (1989) 14954.
[5.459] S. Ohkubo et al., *FEBS Lett.* **274** (1990) 136.
[5.460] K.D. Bloch et al., *J. Biol. Chem.* **264** (1989) 18156.
[5.461] H. Onda et al., *FEBS Lett.* **261** (1990) 327.
[5.462] T. Sawamura et al., *Biochem. Biophys. Res. Comun.* **162** (1989) 1287.
[5.463] M. Yoshizumi et al., *Biochem. Biophys. Res. Commun.* **161** (1989) 859.
[5.464] T. Emori et al., *Biochem. Biophys. Res. Commun.* **160** (1989) 93.

[5.465] K. Yamada et al., *Biochem. Biophys. Res. Commun.* **172** (1990) 178.
[5.466] K. Ohta et al., *Biochem. Biophys. Res. Commun.* **169** (1990) 578.
[5.467] C. Boulanger, T.F. Lüscher, *J. Clin. Invest.* **85** (1990) 587.
[5.468] H. Rakugi et al., *Biochem. Biophys. Res. Commun.* **169** (1990) 973.
[5.469] N. Suzuki et al., *Biochem. Biophys. Res. Commun.* **167** (1990) 941.
[5.470] Y. Saito et al., *N. Engl. J. Med.* **322** (1990) 205.
[5.471] P. Cernacek, D.J. Stewart, *Biochem. Biophys. Res. Commun.* **161** (1989) 562.
[5.472] Y. Masuda et al., *FEBS Lett.* **257** (1989) 208.
[5.473] A. Lecci et al., *Neuropeptides* **16** (1990) 21.
[5.474] Y. Hirata et al., *FEBS Lett.* **239** (1988) 13.
[5.475] P. Roubert et al., *Biochem. Biophys. Res. Commun.* **164** (1989) 809.
[5.476] T.L. Yue et al., *Neuropeptidese* **17** (1990) 7.
[5.477] T.-S. Lee et al., *Biochem. Biophys. Res. Commun.* **162** (1989) 381.
[5.478] T.J. Resink et al., *Biochem. Biophys. Res. Commun.* **157** (1988) 1360.
[5.479] H. Rakugi et al., *Biochem. Biophys. Res. Commun.* **160**(1989) 924.
[5.480] S.-Y. Chou et al., *Am. J. Physiol.* **259** (1990) F645.
[5.481] R.M. Eglen et al., *Br. J. Pharmacol.* **97** (1989) 1297.
[5.482] R. Borges et al., *Eur. J. Pharmacol.* **165** (1989) 223.
[5.483] M.D. Randall et al., *Br. J. Pharmacol.* **98** (1989) 685.
[5.484] N. Fukuda et al., *Biochem. Biophys. Res. Commun.* **167** (1990) 739.
[5.485] R.K. Minkes et al., *Eur. J. Pharmacol.* **164** (1989) 571.
[5.486] P. Vigne et al., *FEBS Lett.* **249** (1989) 143.
[5.487] B.J.R. Whittle, J.V. Esplugues, *Br. J. Pharmacol.* **95** (1988) 1011.
[5.488] T. Katoh et al., *Am. J. Physiol.* **258** (1990) F397.
[5.489] Y. Matsumura et al., *Life Sci.* **44** (1989) 149.
[5.490] G. Mazzocchi et al., *Peptides* **11** (1990) 763.
[5.491] J.R. Hu et al., *Eur. J. Pharmacol.* **158** (1988) 177.
[5.492] Y. Fukuda et al., *Biochem. Biophys. Res. Commun.* **155** (1988) 167.
[5.493] S.S. Stojilkovic et al., *Science* **248** (1990) 1633.
[5.494] W.K. Samson et al., *Biochem. Biophys. Res. Commun.* **169** (1990) 737.
[5.495] C. Thiemermann et al., *Eur. J. Pharmacol.* **158** (1988) 181.
[5.496] P.S. Lidbury et al., *Eur. J. Pharmacol.* **166** (1989) 335.
[5.497] T. Nakaki et al., *Biochem. Biophys. Res. Commun.* **158** (1989) 880.
[5.498] Y. Shibouta et al., *Life Sci.* **46** (1990) 1611.
[5.499] Y. Hirata et al., *Eur. J. Pharmacol.* **176** (1990) 225.
[5.500] R.K. Minkes et al., *Am. J. Physiol.* **259** (1990) H1152.
[5.501] Z. Wollberg et al., *Biochem. Biophys. Res. Commun.* **162** (1989) 371.
[5.502] T. Kashiwabara et al., *FEBS Lett.* **247** (1989) 73.
[5.503] K. Nakajima et al., *Biochem. Biophys. Res. Commun.* **163** (1989) 424.
[5.504] P. Rovero et al., *Br. J. Pharmacol.* **101** (1990) 232.
[5.505] W. König in A. Kleemann et al. (eds.): *Arzneimittel, Fortschritte 1972 bis 1985,* VCH Verlagsgesellschaft, Weinheim 1987, 638.
[5.506] J. Gutowska, M. Nemer, *Endocr. Rev.* **10** (1989) 519.
[5.507] W.G. Forssmann, *Eur. J. Clin. Invest.* **16** (1986) 4391.
[5.508] G. Thibault et al., *Biochem. J.* **241** (1987) 265.
[5.509] P. Needleman, *Fed. Proc. Fed. Am. Soc. Exp. Biol.* **45** (1986) 2096.
[5.510] A. Miyata et al., *Biochem. Biophys. Res. Commun.* **142** (1987) 461.
[5.511] H. Itoh et al., *Biochem. Biophys. Res. Commun.* **143** (1987) 560.
[5.512] R. Eskay et al., *Science* **232** (1986) 636.
[5.513] G. Thibault et al., *Peptides* **9** (1988) 47.
[5.514] L. Meleagros et al., *Peptides* **10** (1989) 545.
[5.515] C.J. Winters et al., *Biochem. Biophys. Res. Commun.* **150** (1988) 231.
[5.516] R. Takayanagi et al., *Biochem. Biophys. Res. Commun.* **142** (1987) 483.
[5.517] N. Zamir et al., *Brain Res.* **365** (1986) 105.
[5.518] T. Shibasaki et al., *Biochem. Biophys. Res. Commun.* **136** (1986).
[5.519] F. Marumo et al., *Biochem. Biophys. Res. Commun.* **143** (1987) 813.
[5.520] S. Inagaki et al., *Regul. Pept.* **14** (1986) 101.

[5.521] J. Gutkowska et al., *Peptides* **8** (1987) 461.
[5.522] S.H. Kim et al., *Life Sci.* **45** (1989) 1581.
[5.523] A.M. Vollmar et al., *Peptides* **9** (1988) 965.
[5.524] O. Vuolteenaho et al., *FEBS Lett.* **233** (1988) 79.
[5.525] A.M. Vollmar, R. Schulz, *Biochem. Biophys. Res. Commun.* **155** (1988) 700.
[5.526] W. Debinski et al., *Biochem. Biophys. Res. Commun.* **134** (1986) 279.
[5.527] R.A. Stone, Ch.C. Glembotski, *Biochem. Biophys. Res. Commun.* **134** (1986) 1022.
[5.528] M. Sakamoto et al., *Biochem. Biophys. Res. Commun.* **135** (1986) 515.
[5.529] A. Miyata et al., *Biochem. Biophys. Res. Commun.* **155** (1988) 1330.
[5.530] Y. Takei et al., *Biochem. Biophys. Res. Commun.* **164** (1989) 537.
[5.531] J.-i. Sakata et al., *Biochem. Biophys. Res. Commun.* **155** (1988) 1338.
[5.532] T. Sudoh et al., *Nature (London)* **332** (1988) 78.
[5.533] K. Maekawa et al., *Biochem. Biophys. Res. Commun.* **157** (1988) 410.
[5.534] J.J. Seilhamer et al., *Biochem. Biophys. Res. Commun.* **165** (1989) 650.
[5.535] M. Kojima et al., *Biochem. Biophys. Res. Commun.* **159** (1989) 1420.
[5.536] M. Aburaya et al., *Biochem. Biophys. Res. Commun.* **165** (1989) 872.
[5.537] T.G. Flynn et al., *Biochem. Biophys. Res. Commun.* **161** (1989) 830.
[5.538] M. Aburaya et al., *Biochem. Biophys. Res. Commun.* **165** (1989) 880.
[5.539] Y. Takei et al., *Biochem. Biophys. Res. Commun.* **170** (1990) 883.
[5.540] T. Sudoh et al., *Biochem. Biophys. Res. Commun.* **168** (1990) 863.
[5.541] G.P. Vlasuk et al., *Biochem. Biophys. Res. Commun.* **136** (1986) 396.
[5.542] M. Gagelmann et al., *FEBS Lett.* **233** (1988) 249.
[5.543] H. Ong et al., *Life Sci.* **38** (1986) 1309.
[5.544] S. Ueda et al., *Biochem. Biophys. Res. Commun.* **149** (1987) 1055.
[5.545] *Drugs of the Future* **7** (1982) 716; **9** (1984) 777; **11** (1986) 878; **13** (1988) 969; **14** (1989) 992; **15** (1990) 1028.
[5.546] M.C. Haufe et al., *Circulation* **76** (1987) Pt. 2, 270.
[5.547] H. Ruskoaho et al., *Biochem. Biophys. Res. Commun.* **139** (1986) 266.
[5.548] K. Naruse et al., *Peptides* **8** (1987) 285.
[5.549] P.P. Shields, C.C. Glembotski, *J. Biol. Chem.* **264** (1989) 9322.
[5.550] D.G. Gardner, H.D. Schultz, *J. Clin. Invest.* **86** (1990) 52.
[5.551] Y. Shenker et al., *J. Cardiovasc. Pharmacol.* **8** (1986) 1300.
[5.552] D. Lachance et al., *Hypertension* **9** (1987) 524.
[5.553] J. Closas et al., *Clin. Res.* **35** (1987) 440A.
[5.554] J.A. Sanfield et al., *Am. J. Physiol.* **252** (1987) Pt. 1, E740.
[5.555] M.G. Currie, W.H. Newman, *Biochem. Biophys. Res. Commun.* **137** (1986) 94.
[5.556] B. Baranowska et al., *Biochem. Biophys. Res. Commun.* **143** (1987) 159.
[5.557] H. Nakaoka et al., *Hypertension* **10** (1987) 221.
[5.558] G. Deray et al., *Eur. J. Clin. Pharmacol.* **38** (1990) 363.
[5.559] L.M.A.B. van Bortel et al., *Arch. Pharmacol.* **339** (1989) R116.
[5.560] B. Baranowska et al., *Hypertension* **9** (1987) 545.
[5.561] B. Baranowska et al., *Biochem. Biophys. Res. Commun.* **145** (1987) 680.
[5.562] J.R. Hu et al., *Eur. J. Pharmacol.* **158** (1988) 177.
[5.563] O. Sergev et al., *J. Endocrinol. Invest.* **13** (1990) 649.
[5.564] M. Weissel et al., *Klin. Wochenschr.* **64** (1986) Suppl. VI, 93.
[5.565] K. Horky et al., *Biochem. Biophys. Res. Commun.* **129** (1985) 651.
[5.566] H. Saxenhofer et al., *Acta Endocrinol. (Copenhagen)* **118** (1988) 179.
[5.567] H. Matsubara et al., *Biochem. Biophys. Res. Commun.* **145** (1987) 336.
[5.568] G. Wambach et al., *Klin. Wochenschr.* **64** (1986) Suppl. VI, 53.
[5.569] K.M. Verburg et al., *Am. J. Physiol.* **251** (1986) R947.
[5.570] A.L. Gerbes et al., Klin. *Wochenschr.* **64** (1986) 666.
[5.571] Y. Shenker et al., *Clin. Res.* **34** (1986), 486A.
[5.572] W. Rascher et al., *Horm. Res.* **28** (1987) 58.
[5.573] D.R.J. Singer et al., *Lancet II* (1987) 1394.
[5.574] M.R. Arendt et al., *Klin. Wochenschr.* **64** (1986) Suppl. VI, 97.
[5.575] M. Braquet et al., *Lancet II* (1986) 456.
[5.576] T. Yamaji et al., *J. Clin. Endocrinol. Metab.* **67** (1988) 348.

[5.577] T. Yamaji et al., *J. Clin. Endocrinol. Metab.* **63** (1986) 815.
[5.578] A.L. Gerbes et al., *N. Engl. J. Med.* **313** (1985) 1609.
[5.579] E. Cogan et al., *N. Engl. J. Med.* **314** (1986) 1258.
[5.580] M. Kohno et al., *Am. J. Med.* **83** (1987) 648.
[5.581] M. Ohashi et al., *Horm. Metab. Res.* **20** (1988) 705.
[5.582] A.E.G. Raine et al., *N. Engl. J. Med.* **315** (1986) 533.
[5.583] Y. Shenker et al., *Life Sci.* **41** (1987) 1635.
[5.584] D.H. Petzl et al., *Klin. Wochenschr.* **65** (1987) 194.
[5.585] A.J. Rankin, T.M. Scott, *Regul. Pept.* **30** (1990) 65.
[5.586] N. Zamir et al., *Proc. Natl. Acad. Sci. USA* **84** (1987) 541.
[5.587] R.S. Zimmerman et al., *J. Clin. Endocrinol. Metab.* **64** (1987) 353.
[5.588] R. Garcia et al., *Biochem. Biophys. Res. Commun.* **136** (1986) 510.
[5.589] T. Kimura et al., *J. Clin. Endocrinol. Metab.* **62** (1986) 1003.
[5.590] S.L. Stephenson, A.J. Kenny, *Biochem. J.* **243** (1987) 183.
[5.591] J.L. Sonnenberg et al., *Peptides* **9** (1988) 173.
[5.592] A. Jardine et al., *J. Am. Coll. Cardiol.* **13** (1989) Suppl. A, 76A.
[5.593] D.B. Northridge et al., *J. Am. Coll. Cardiol.* **13** (1989) Suppl. A., 76A.
[5.594] M.-s. Chang et al., *Nature* **341** (1989) 68.
[5.595] M.B. Anand-Srivastava et al., *J. Biol. Chem.* **265** (1990) 8566.
[5.596] P.R. Bovy et al., *J. Med. Chem.* **32** (1989) 869.
[5.597] T. Maack et al., *Science* **238** (1987) 675.
[5.598] D.L. Vesely et al., *Peptides* **11** (1990) 193.
[5.599] K.L. Lanier-Smith, M.G. Currie, *Eur. J. Pharmacol.* **178** (1990) 105.
[5.600] J.M. Saavedra et al., *Endocrinology* **120** (1987) 426.
[5.601] E.L. Schiffrin et al., *Circulation* **72** (1985) Pt. 2, III-293.
[5.602] Y. Hirata et al., *Biochem. Biophys. Res. Commun.* **138** (1986) 405.
[5.603] A.G. Jardine et al., *Klin. Wochenschr.* **67** (1989) 902.
[5.604] W. Linz et al., *Klin. Wochenschr,* **64** (1986) Suppl. VI, 27.
[5.605] N.C. Trippodo, R.W. Barbee, *Am. J. Physiol.* **252** (1987) R915.
[5.606] P.J.S. Chiu et al., *Eur. J. Pharmacol.* **124** (1986) 277.
[5.607] B.L. Pegram et al., *Fed. Proc. Am. Soc. Exp. Biol.* **45** (1986) 2382.
[5.608] R.J. Schiebinger, J. Linden. *Am. J. Physiol.* **251** (1986) H1095.
[5.609] R. Di Nicolantonio et al., *Clin. Exp. Pharmacol. Physiol.* **13** (1986) 311.
[5.610] T. Opgenorth, E.I. Novosad, *Eur. J. Pharmacol.* **191** (1990) 351.
[5.611] A. Petrillo et al., *Gastroenterology* **92** (1987) Pt. 2, 1764.
[5.612] J. McMurray et al., *Eur. J. Clin. Pharmacol.* **35** (1988) 409.
[5.613] K. Schafferhans et al., *Klin. Wochenschr.* **64** (1986) Suppl. VI, 73.
[5.614] E. Heidbreder et al., *Klin. Wochenschr.* **64** (1986) Supp. VI, 78.
[5.615] B.R. Cole, J. Needleman, *Clin. Res.* **34** (1986) 592A.
[5.616] D.A. Hildebrandt, R.O. Banks, *Am. J. Physiol.* **252** (1987), Pt. 2, F977.
[5.617] S. Akabane et al., *Can. J. Physiol. Pharmacol.* **66** (1988) 601.
[5.618] K. Schafferhans et al., *Klin. Wochenschr.* **66** (1988) Suppl. 13, 233.
[5.619] E. Kubota et al., *Clin. Res.* **34** (1986) 728A.
[5.620] A.H. Morice et al., *Am. Rev. Respir. Dis.* **135** (1987) Pt. 2, A300.
[5.621] S. Adnot et al., *Am. Rev. Respir. Dis.* **137** (1988) Pt. 2, 108.
[5.622] T. Imamura et al., *Life Sci.* **42** (1988) 403.
[5.623] M.F. Sugrue, M.-P. Viader, *Eur. J. Pharmacol.* **130** (1986) 349.
[5.624] M.E. O'Donnell, N.E. Owen, *J. Biol. Chem.* **261** (1986) 15461.
[5.625] J.C. Hogan et al., *Eur. J. Pharmacol.* **165** (1989) 129.
[5.626] M.B. Anand-Srivastava, M. Cantin, *Biochem. Biophys. Res. Commun.* **138** (1986) 427.
[5.627] K. Iitake et al., *Endocrinology* **119** (1986) 438.
[5.628] J. Bingham Smith, T.M. Lincoln, *Am. J. Physiol.* **253** (1987) Pt. 1, C147.
[5.629] M. Ueno et al., *Am. J. Physiol.* **259** (1990) C427.
[5.630] W. Debinski et al., *Proc. Soc. Exp. Biol. Med.* **181** (1986) 318.
[5.631] T. Takao et al., *Life Sci.* **42** (1988) 1199.
[5.632] M.A. Dillingham, R.J. Anderson, *Science* **231** (1986) 1572.
[5.633] Z. Zukowska-Grojec et al., *J. Pharamcol. Exp. Ther.* **239** (1986) 480.

[5.634] P. Nambi et al., *Proc. Natl. Acad. Sci. USA* **83** (1986) 8492.
[5.635] J.P. Briggs, J. Schnermann, *J. Cardiovasc. Pharmacol.* **8** (1986) 1296.
[5.636] H. Vierhapper et al., *Hypertension* **8** (1986) 1040.
[5.637] M.K. Steele, *Neuroendocrinology* **51** (1990) 345.
[5.638] P.J. Harris et al., *Nature* **326** (1987) 697.
[5.639] L. Chartier, E.L. Schiffrin, *Am. J. Physiol.* **252** (1987) E485.
[5.640] F. Bex, A. Corbin, *Eur. J. Pharmacol.* **115** (1985) 125.
[5.641] B. Ahren, *Life Sci.* **47** (1990) 1973.
[5.642] R.S. Zimmerman et al., *Circulation* **80** (1989) Suppl., 360.
[5.643] H. Itoh et al., *J. Clin. Invest.* **86** (1990) 1690.
[5.644] T.J. Opgenorth et al., *Peptides* **11** (1990) 1003.
[5.645] J.J. Gonvers et al., *Gastroenterology* **92** (1987) Pt. 2, 1408.
[5.646] H. Oguchi et al., *Biochem. Biophys. Res. Commun.* **146** (1987) 757.
[5.647] H. Oguchi et al., *Jpn. J. Pharmacol.* **43** (1987) Suppl., 161P.
[5.648] W.K. Samson et al., *Neuroendocrinology* **147** (1988) 268.
[5.649] Y. Murakami et al., *Endocrinology* **122** (1988) 2103.
[5.650] J. Puurunen, H. Ruskoaho, *Eur. J. Pharmacol.* **141** (1987) 493.
[5.651] P.R. Bovy, *Med. Res. Rev.* **10** (1990) 115.
[5.652] T. Kimura, S. Sakakibara, *Jpn. J. Pharmacol.* **40** (1986) Suppl., 77P.
[5.653] J.G. De Mey et al., *J. Pharmacol. Exp. Ther.* **240** (1987) 937.
[5.654] H. Saxenhofer et al., *Am. J. Physiol.* **259** (1990) F832.
[5.655] Y. Kiso et al., *J. Pharm. Sci.* **76** (1987) S172.
[5.656] T.X. Watanabe et al., *Jpn, J. Pharmacol.* **39** (1985) Suppl., 99P.
[5.657] J. Rittenhouse et al., *J. Biol. Chem.* **261** (1986) 7607.
[5.658] C.F. Hassman et al., *Biochem. Biophys. Res. Commun.* **152** (1988) 1070.
[5.659] P.W. Schiller et al., *Biochem. Biophys. Res. Commun.* **143** (1987) 499.
[5.660] D.L. Song et al., *Eur. J. Pharmacol.* **160** (1989) 141.
[5.661] J.D. Mogannam et al., *J. Cardiovasc. Pharmacol.* **8** (1986) 1317.
[5.662] R. Deghenghi et al., *Drug. Dev. Res.* **15** (1988) 87.
[5.663] J. Knolle et al., lecture held at Hoechst AG, 1990.
[5.664] R.F. Nutt et al., *Abstr. Pap. Am. Chem. Soc.* **195** (1988) BIOL1.
[5.665] J. DiMaio et al., *J. Med. Chem.* **33** (1990) 661.
[5.666] M.S. King, A.J. Baertschi, *Endocrinology* **124** (1989) 286.
[5.667] D.L. Vesely et al., *Biochem. Biophys. Res. Commun.* **148** (1987) 1540.
[5.668] D.R. Martin et al., *Am. J. Physiol.* **258** (1990) F1401.
[5.669] Y. Kitajima et al., *Biochem. Biophys. Res. Commun.* **164** (1989) 1295.
[5.670] Y. Kambayaschi et al., *FEBS Lett.* **248** (1989) 28.
[5.671] W. König in A. Kleemann et al. (eds.): *Arzneimittel, Fortschritte 1972 bis 1985*, VCH Verlagsgesellschaft, Weinheim 1987, 737.
[5.672] R. Corder et al., *Biochem. J.* **219** (1984) 699.
[5.673] K. Tatemoto et al., *Biochem. Biophys. Res. Commun.* **157** (1988) 713.
[5.674] W.B. Rhoten, *Experientia* **43** (1987) 428.
[5.675] A.B. Leiter et al., *J. Biol. Chem.* **259** (1984) 14702.
[5.676] D.M. McKay et al., *Regul. Pept.* **31** (1990) 187.
[5.677] J.B. Jaspan et al., *Am. J. Physiol.* **251** (1986) E215.
[5.678] S.J. Konturek et al., *Gastroenterology* **86** (1984) Pt. 2, 1140.
[5.679] G. Ribes et al., *Am. J. Physiol.* **244** (1983) E380.
[5.680] R.D. Beauchman et al., *Gastroenterology* **90** (1986) Pt. 2, 1341.
[5.681] A.J.L. de Jong et al., *Regul. Pept.* **17** (1987) 285.
[5.682] G.R. Greenberg et al., *Dig. Dis. Sci.* **30** (1985) 946.
[5.683] L.R. Gingerich et al., *Metabolism* **34** (1985) 25.
[5.684] E. Hidvegi et al., *Acta Physiol. Hung.* **63** (1984) 237.
[5.685] S.J. Konturek et al., *Gastroenterology* **86** (1984) Pt. 2, 1140.
[5.686] R.A. Silvestre et al., *Life Sci.* **40** (1987) 1829.
[5.687] C. Beglinger et al., *Am. J. Physiol.* **246** (1984) G286.
[5.688] E.F. Coelle et al., *Dig. Dis. Sci.* **28** (1983) 1083.
[5.689] S. Ito et al., *J. Pharmacol.* **33** (1983) Suppl., 129P.

[5.690] R.L. Conter et al., *Clin. Res.* **32** (1984) 24A.
[5.691] W.H. Coltharp et al., *Gastroenterology* **90** (1986) Pt. 2, 1379.
[5.692] K.E. Hall et al., *Am. J. Physiol.* **245** (1983) G178.
[5.693] S.J. Konturek et al., *Am. J. Physiol.* **248** (1985) G687.
[5.694] J.K. McDonald et al., *Peptides* **6** (1985) 79.
[5.695] S.P. Kalra, W.R. Crowley, *Neuroendocrinology* **38** (1984) 511.
[5.696] T.S. Gray, J.E. Morley, *Life Sci.* **38** (1986) 389.
[5.697] P. Ciofi et al., *Neuroendocrinology* **45** (1987) 425.
[5.698] E. Ekblad et al., *Regul. Pept.* **10** (1984) 47.
[5.699] C.D. Minth et al., *J. Biol. Chem.* **261** (1986) 11974.
[5.700] J. Allen et al., *Proc. Natl. Acad. Sci. USA* **84** (1987) 2532.
[5.701] J.M. Allen et al., *Neuropeptides* **6** (1985) 95.
[5.702] K. Kitamura et al., *Biochem. Biophys. Res. Commun.* **169** (1990) 1164.
[5.703] M.M.T. O'Hare et al., *Regul. Pept.* **20** (1988) 293.
[5.704] R. Sillard et al., *FEBS Lett.* **258** (1989) 263.
[5.705] M. Haass et al., *Am. J. Physiol.* **259** (1990) R925.
[5.706] J.M. Lundberg et al., *Biochem. Biophys. Res. Commun.* **133** (1985) 30.
[5.707] J.-T. Cheng, C.L. Shen, *Eur. J. Pharmacol.* **123** (1986) 303.
[5.708] K. Takahashi et al., *Peptides* **9** (1988) 433.
[5.709] A. Rudehill et al., *Acta Physiol. Scand.* **131** (1987) 517.
[5.710] V. Castagne et al., *Regul. Pept.* **19** (1987) 55.
[5.711] L.S. Brady et al., *Neuroendocrinology* **52** (1990) 441.
[5.712] R. Corder et al., *Life Sci.* **43** (1988) 1879.
[5.713] A. Sahu et al., *Endocrinology* **120** (1987) 1831.
[5.714] A. Sahu et al., *Endocrinology* **124** (1989) 410.
[5.715] T. Ogawa et al., *Biochem. Biophys. Res. Commun.* **165** (1989) 1399.
[5.716] A. Balasubramaniam, S. Sheriff, *J. Biol. Chem.* **265** (1990) 14724.
[5.717] J.M. Lundberg et al., *Regul. Pept.* **13** (1986) 169.
[5.718] J.D. Watson et al., *J. Endocrinol.* **116** (1988) 421.
[5.719] N. Pages et al., *Neuropeptides* **17** (1990) 141.
[5.720] A. Sahu et al., *Endocrinology* **122** (1988) 2199.
[5.721] S.P. Sheikh et al., *FEBS Lett.* **245** (1989) 209.
[5.722] S.P. Sheikh, J.A. Williams, *J. Biol. Chem.* **265** (1990) 8304.
[5.723] N.A. Scott et al., *Am. J. Physiol.* **259** (1990) H174.
[5.724] J. Fuhlendorff et al., *Proc. Natl. Acad. Sci. USA* **87** (1990) 182.
[5.725] S.P. Sheikh et al., *Am. J. Physiol.* **257** (1989) F978-F984.
[5.726] J.Ch. Jorgensen et al., *Eur. J. Pharmacol.* **186** (1990) 105.
[5.727] S.-i. Mihara et al., *FEBS Lett.* **259** (1989) 79.
[5.728] B.B. Fredholm et al., *Acta Physiol. Scand.* **124** (1985) 467.
[5.729] S. Kassis et al., *J. Biol. Chem.* **262** (1987) 3429.
[5.730] T. Ishiguro et al., *Chem. Pharm. Bull.* **36** (1988) 2720.
[5.731] J.L. Morris, *Peptides* **11** (1990) 381.
[5.732] E. Hackenthal et al., *Am. J. Physiol.* **252** (1987) F543.
[5.733] J.F. Aubert et al., *J. Pharmacol. Exp. Ther.* **244** (1988) 1109.
[5.734] M.A. Dillingham, R.J. Anderson, *Am. J. Physiol.* **256** (1989) F408.
[5.735] B. Baranowska et al., *Biochem. Biophys. Res. Commun.* **145** (1987) 680.
[5.736] C. Wahlestedt et al., *Acta Physiol. Scand.* **129** (1987) 459.
[5.737] G. Neri et al., *Neuropeptides* **17** (1990) 121.
[5.738] A. Saria, E. Beubler, *Eur. J. Pharmacol.* **119** (1985) 47.
[5.739] G.H. Greeley, Jr. et al., *Am. J. Physiol.* **254** (1988) E513.
[5.740] Y.-S. Guo et al., *Am. J. Physiol.* **253** (1987) G298.
[5.741] L. Jivegard et al., *Gastroenterology* **90** (1986) Pt. 2, 1477.
[5.742] A. Cadieux et al., *Neuropeptides* **13** (1989) 215.
[5.743] V. di Marzo et al., *Neuropeptides* **11** (1988) 169.
[5.744] C.D. Stretton et al., *Neuropeptides* **17** (1990) 163.
[5.745] P. Valet et al., *Therapie* **45** (1990) 72.
[5.746] H.A. Dewar et al., *Life Sci.* **45** (1989) 367.

[5.747] J.-M. Danger et al., *Life Sci.* **39** (1986) 1183.
[5.748] J.F. Rodriguez-Sierra et al., *Peptides* **8** (1987) 539.
[5.749] W.B. Wehrenberg et al., *Neuroendocrinology* **49** (1989) 680.
[5.750] W.R. Crowley et al., *Endocrinology* **120** (1987) 941.
[5.751] J. Guy et al., *Regul. Pept.* **23** (1988) 209.
[5.752] H.E. Albers et al., *Am. J. Physiol.* **258** (1990) R376.
[5.753] M. Jimenez, L. Bueno, *Life Sci.* **47** (1990) 205.
[5.754] A. Dray et al., *Peptides* **7** (1986) 323.
[5.755] B.G. Stanley et al., *Peptides* **6** (1985) 1205.
[5.756] T.N. Pappas et al., *Gastroenterology* **90** (1986) Pt. 2, 1578.
[5.757] M. Heilig, R. Murison, *Eur. J. Pharmacol.* **137** (1987) 127.
[5.758] J.T. Clark et al., *Endocrinology* **117** (1985) 2435.
[5.759] H. Härfstrand, *Acta Physiol. Scand.* **128** (1986) 121.
[5.760] J.A. Aguirre et al., *Acta Physiol. Scand.* **139** (1990) 609.
[5.761] J. Esteban et al., *Life Sci.* **45** (1989) 2395.
[5.762] J.L. Krstenansky et al., *Proc. Natl. Acad. Sci. USA* **86** (1989) 4377.
[5.763] A. Beck et al., *FEBS Lett.* **244** (1989) 119.
[5.764] A.G. Beck-Sickinger et al., *Eur. J. Biochem.* **194** (1990) 449.
[5.765] J.L. Krstenansky et al., *Neuropeptides* **17** (1990) 117.
[5.766] B. Baranowska et al., *Biochem. Biophys. Res. Commun.* **145** (1987) 680.
[5.767] L. Grundemar, R. Hakanson, *Br. J. Pharmacol.* **100** (1990) 190.
[5.768] T.E. Adrian et al., *J. Endocrinol.* **113** (1987) 11.
[5.769] G.W. Aponte et al., *Am. J. Physiol.* **249** (1985) G745.
[5.770] R. Corder et al., *Regul. Pept.* **21** (1988) 253.
[5.771] V.E. Eysselein et al., *Peptides* **11** (1990) 111.
[5.772] T.E. Adrian et al., *Gastroenterology* **92** (1987) Pt. 2, 1289.
[5.773] T.E. Adrian et al., *Regul. Pept.* **9** (1984) 320.
[5.774] F. Lluis et al., *Gastroenterology* **90** (1986) Pt. 2, 1525.
[5.775] T.N. Pappas et al., *Am. J. Physiol.* **248** (1985) G118.
[5.776] T.N. Pappas et al., *Gastroenterology* **89** (1985) 1387.
[5.777] R.J. Playford et al., *Lancet* **335** (1990) 1555.
[5.778] T.E. Adrian et al., *Dig. Dis. Sci.* **29** (1984) Suppl., 3S.
[5.779] Y.S. Guo et al., *Can. J. Physiol. Pharmacol.* (1986) Suppl., 110.
[5.780] W.W. Pawlik et al., *Gastroenterology* **90** (1986) Pt. 2, 1580.
[5.781] T.E. Adrian et al., *J. Clin. Endocrinol. Metab.* **63** (1986) 803.
[5.782] T. Suzuki et al., *Gastroenterology* **85** (1983) 114.
[5.783] P.A. Grace et al., *Gastroenterology* **90** (1986) Pt. 2, 1435.
[5.784] R.K. Harding, T.J. McDonald, *Peptides* **10** (1989) 21.
[5.785] G.H. Greeley et al., *Am. J. Physiol.* **254** (1988) Pt. 1, E513.
[5.786] B. Baranowska et al., *Biochem. Biophys. Res. Commun.* **145** (1987) 680.
[5.787] N. Minamino et al., *Biochem. Biophys. Res. Commun.* **156** (1988) 355.
[5.788] J. Domin et al., *J. Biol. Chem.* **264** (1989) 20881.
[5.789] R. Murphy et al., *Peptides* **11** (1990) 613.
[5.790] J. Domin et al., *Peptides* **8** (1987) 779.
[5.791] C.A. Maggi et al., *Br. J. Pharmacol.* **99** (1990) 186.
[5.792] M.V. Graf, A.J. Kastin, *Peptides* **7** (1986) 1165.
[5.793] D. Schneider-Helmert, *Schweiz. Rundschau Med. (Praxis)* **74** (1985) 1311.
[5.794] G.A. Schoenenberger, *Eur. Neurol.* **23** (1984) 321.
[5.795] A. Bjartell et al., *Neuroendocrinology* **50** (1989) 564.
[5.796] A. Nakamura et al., *Eur. J. Pharmacol.* **121** (1986) 157.
[5.797] A.M.J. Young, B.J. Key, *Neuropharmacology* **23** (1984) 1347.
[5.798] J.E. Taylor, *IRCS Med. Sci.* **14** (1886) 1122.
[5.799] K.V. Sudakov et al., *J. Biol. Sci.* **18** (1983) 1.
[5.800] M.V. Graf, A.J. Kastin, *Neurosci. Biobehav. Rev.* **8** (1984) 83.
[5.801] T. Okajima, G. Hertting, *Horm. Metab. Res.* **18** (1986) 497.
[5.802] M.V. Graf et al., *Neuroendocrinology* **41** (1985) 353.
[5.803] K.S. Iyer, S.M. McCann, *Neuroendocrinology* **46** (1987) 93.

[5.804] A. Sahu, S.P. Kalra, *Life Sci.* **40** (1987) 1201.
[5.805] W. Larbig et al., *Eur. Neurol.* **23** (1984) 372.

References for Chapter 6

[6.1] K. Iwai et al., *J. Biol. Chem.* **262** (1987) 8956.
[6.2] S.-I. Fukuoka et al., *Biochem. Biophys. Res. Commun.* **145** (1987) 646.
[6.3] B. Agerberth et al., *Proc. Natl. Acad. Sci. USA* **86** (1989) 8590.
[6.4] W. König in A. Kleemann et al., (eds.): *Arzneimittel, Fortschritte 1972 bis 1985,* VCH Verlagsgesellschaft, Weinheim 1987, 686.
[6.5] Y. Takahashi et al., *Proc. Natl. Acad. Sci. USA* **82** (1985) 1931.
[6.6] U. Gubler et al., *Proc. Natl. Acad. Sci. USA* **81** (1984) 4307.
[6.7] R.J. Deschenes et al., *Proc. Natl. Acad. Sci. USA* **81** (1984) 726.
[6.8] K. Tatemoto et al., *FEBS Lett.* **174** (1984) 289.
[6.9] V.E. Eysselein et al., *Am. J. Physiol.* **258** (1990) G253.
[6.10] Z.-Z. Zhou et al., *Peptides* **6** (1985) 337.
[6.11] J. Shively et al., *Am. J. Physiol.* **252** (1987) G272.
[6.12] A. Takashima, S. Itoh, *Can. J. Physiol. Pharmacol.* **67** (1989) 228.
[6.13] T. Wakabayashi et al., *Gene* **31** (1984) 295.
[6.14] K. Beardshall et al., *Lancet II* (1989) 1008.
[6.15] Y.F. Chen et al., *Am. J. Physiol.* **249** (1985) G29.
[6.16] J.B.M.J. Jansen, C.B.H.W. Lamers, *Life Sci.* **33** (1983) 2197.
[6.17] J.J. Benoliel et al., *Fundam. Clin. Pharmacol.* **3** (1989) 141.
[6.18] I. Koop, *Eur. J. Clin. Invest.* **20** (1990) Suppl. 1, S51.
[6.19] I.C. Forgacs et al., *Clin. Sci.* **66** (1984) 61P.
[6.20] R.E. Rodriguez, M.P. Sacristan, *FEBS Lett.* **250** (1989) 215.
[6.21] R.B. Simerly, L.W. Swanson, *Proc. Natl. Acad. Sci. USA* **84** (1987) 2087.
[6.22] R.A. Siegel et al., *Neuroendocrinology* **46** (1987) 75.
[6.23] J.B.M.J. Jansen, C.B.H.W. Lamers, *Gut* **27** (1986) A1260.
[6.24] I. Nakano et al., *Regul. Pept.* **14** (1986) 229.
[6.25] S. Kanayama et al., *Life Sci.* **41** (1987) 1915.
[6.26] A. Schafmayer et al., *Digestion* **30** (1984) 95.
[6.27] D.-H. Yu et al., *Am. J. Physiol.* **258** (1990) G86.
[6.28] M. Praissman, M. Walden, *Gastroenterology* **92** (1987) Pt. 2, 1579.
[6.29] M. Otsuki, J.A. Williams, *Gastroenterology* **84** (1983) Pt. 2, 1266.
[6.30] Y.K. Leung et al., *Dig. Dis. Sci.* **30** (1985) 981.
[6.31] M. Otsuki et al., *Gastroenterology* **87** (1984) 882.
[6.32] B.E. Evans, *Drugs of the Future* **14** (1989) 971.
[6.33] L.L. Iversen et al., *J. Endocrinol.* **124** (1990) Suppl., 18.
[6.34] J. Hughes et al., *Proc. Natl. Acad. Sci. USA* **87** (1990) 6728.
[6.35] B. Charpentier et al., *Peptides* **9** (1988) 835.
[6.36] B. Charpentier et al., *J. Med. Chem.* **32** (1989) 1148.
[6.37] I. Maraseigne et al., *J. Med. Chem.* **31** (1988) 966.
[6.38] R.A. Liddle et al., *J. Clin. Invest.* **84** (1989) 1270.
[6.39] M.F. O'Rourke et al., *Am. J. Physiol.* **258** (1990) G179.
[6.40] S.J. Konturek et al., *Am. J. Physiol.* **248** (1985) G687.
[6.41] J.E. Kellow et al., *Am. J. Physiol.* **253** (1987) G650.
[6.42] A. Brown et al., *Gut* **26** (1985) A580.
[6.43] L. Gullo et al., *Dig. Dis. Sci.* **31** (1986) 1345.
[6.44] L. Marzio et al., *Am. J. Gastroenterol.* **80** (1985) 1.
[6.45] W.P.M. Hopman et al., *Gut* **27** (1986) A627.
[6.46] J. Schwarzendrube et al., *Klin. Wochenshcr.* **68** (1990) Suppl. 19, 134.
[6.47] G. Schmidt et al., *Klin. Wochenschr.* **64** (1986) Suppl. 5, 56.
[6.48] F. Makovec et al., *Regul. Pept.* **16** (1986) 281.
[6.49] K.A. Zucker et al., *Am. J. Physiol.* **257** (1989) Pt. 1, G511.
[6.50] G. Varga et al., *Digestion* **30** (1984) 89.

[6.51] R. Maani et al., *Gastroenterology* **94** (1988) A274.
[6.52] R.A. Liddle et al., *Clin. Res.* **37** (1989) 369A.
[6.53] S. Karlsson, B. Ahren, *Acta Physiol. Scand.* **135** (1989) 271.
[6.54] K.C.K. Lloyd et al., *Gastroenterology* **96** (1989) Pt. 2, A305.
[6.55] A. Bill et al., *Acta Physiol. Scand.* **138** (1990) 479.
[6.56] T.W. Vickroy et al., *Eur. J. Pharmacol.* **152** (1988) 371.
[6.57] V. Dauge et al., *Fundam. Clin. Pharmacol.* **3** (1989) 458.
[6.58] T. Zelles et al., *Eur. J. Pharmacol.* **178** (1990) 101.
[6.59] C.T. Dourish et al., *Eur. J. Pharmacol.* **176** (1990) 35.
[6.60] T.N. Pappas et al., *Gastroenterology* **86** (1984) Pt. 2, 1206.
[6.61] J.E. Kellow et al., *Am. J. Physiol.* **252** (1987) G345-G356.
[6.62] T. Ishikawa et al., *Jpn. J. Pharmacol.* **39** (1985) Suppl., 161P.
[6.63] C. Niederau et al., *Gastroenterology* **88** (1985) 1192.
[6.64] F. Makovec et al., *Peptides* **7** (1986) 1159.
[6.65] J. Park et al., *Gastroenterology* **92** (1987) Pt. 2, 1566.
[6.66] M. Matsumura et al., *Horm. Metab. Res.* **17** (1985) 293.
[6.67] M. Matsumura et al., *Horm. Metab. Res.* **16** (1984) 626.
[6.68] E. Mezey et al., *Proc. Natl. Acad. Sci. USA* **83** (1986) 3510.
[6.69] D.K. Meyer et al., *Neuropeptides* **2** (1982) 371.
[6.70] E. Spaeth Schwalbe, *Acta Endocrinol.* **117** (1988) 59.
[6.71] S. Guarini et al., *Pharmacology* **37** (1988) 286.
[6.72] C.D. Stretton, P.J. Barnes, *Am. Rev. Respir. Dis.* **137** (1988) Pt. 2, 201.
[6.73] S. Freier et al., *Gastroenterology* **93** (1987) 1242.
[6.74] E.K. Hong, A.E. Takemori, *J. Pharmacol. Exp. Ther.* **251** (1989) 594.
[6.75] M. Rattray et al., *Neuropeptides* **14** (1989) 263.
[6.76] D.S.K. Magnuson et al., *Neuropeptides* **16** (1990) 213.
[6.77] H.J. Liu, M.T. Lin, *Pharmacology* **31** (1985) 108.
[6.78] J.F. Flood et al., *Science* **236** (1987) 832.
[6.79] T. Kadar et al., *Neuropeptides* **4** (1984) 127.
[6.80] G. Zetler, *Peptides* **5** (1984) 729.
[6.81] S. Rattan, R.K. Goyal, *Gastroenterology* **90** (1986) 94.
[6.82] D. Pelaprat, M. Reibaud, *J. Med. Chem.* **32** (1989) 445.
[6.83] K. Iuchi et al., *Chem. Pharm. Bull.* **36** (1988) 959.
[6.84] C. Beglinger et al., *Regul. Pept.* **8** (1984) 291.
[6.85] M. Ruiz-Gayo et al., *Peptides* **6** (1985) 415.
[6.86] E.E. Sugg et al., *Life Sci.* **39** (1986) 1623.
[6.87] T.K. Sawyer et al., *Can. J. Physiol. Pharmacol.* **64** (1986) Suppl., 10.
[6.88] M. Knight et al., *Peptides* **6** (1985) 631.
[6.89] C. Mendre et al., *J. Biol. Chem.* **263** (1988) 10641.
[6.90] C. Nagain et al., *Peptides* **8** (1987) 1023-1028.
[6.91] C. Nagain et al., *Peptides* **10** (1989) 1041.
[6.92] B. Penke et al., *J. Med. Chem.* **27** (1984) 845.
[6.93] I. Marseigne et al., *Int. J. Pept. Protein Res.* **33** (1989) 230.
[6.94] M.-C. Galas et al., *Am. J. Physiol.* **254** (1988) G176.
[6.95] B.E. Evans, *Drugs of the Future* **14** (1989) 971.
[6.96] P.N. Maton et al., *Horm. Metab. Res.* **18** (1986) 2.
[6.97] M.A. Silverman et al., *Am. J. Gastroenterol.* **82** (1987) 703.
[6.98] R.M. Freidinger, *Med. Res. Rev.* **9** (1989) 271.
[6.99] J.D. Gardner et al., *Am. J. Physiol.* **248** (1985) G98.
[6.100] *Drugs of the Future* **13** (1988) 34.
[6.101] M.-F. Lignon et al., *J. Bio. Chem.* **262** (1987) 7226.
[6.102] R.T. Jensen et al., *Biochem. Biophys. Acta* **761** (1983) 269.
[6.103] W.P.M. Hopman et al., *Br. Med. J.* **292** (1986) 375.
[6.104] R. Tritapepe, C. di Padova, *Br. Med. J.* **293** (1986) 1102.
[6.105] N. Basso et al., *Gastroenterology* **89** (1985) 605.
[6.106] A. Pap, V. Varro, *Digestion* **30** (1984) 118.
[6.107] B.S. Barbaz et al., *Neuropharmacology* **25** (1986) 823.

[6.108] W. König in A. Kleemann et al. (eds.): *Arzneimittel, Fortschritte 1972 bis 1985*, VCH Verlagsgesellschaft Weinheim 1987, 613.
[6.109] E.R. Spindel et al., *Proc. Natl. Acad. Sci. USA* **83** (1986) 19.
[6.110] D. Barra et al., *FEBS Lett.* **182** (1985) 53.
[6.111] T.K. Hayes et al., *Life Sci.* **44** (1989) 1259.
[6.112] J. Price et al., *J. Clin. Endocrinol. Metab.* **60** (1985) 1097.
[6.113] G.D. Jahnke, L.H. Lazarus, *Proc. Natl. Acad. Sci. USA* **81** (1984) 578.
[6.113a] S. Knuhtsen et al., *Am. J. Physiol.* **248** (1985) G281.
[6.113b] G.M. Short et al., *Am. J. Physiol.* **249** (1985) G197.
[6.114] T. von Schrenck et al., *Am. J. Physiol.* **259** (1990) G468.
[6.115] C.B.H.W. Lamers et al., *Dig. Dis. Sci.* **31** (1986) Suppl. 428S.
[6.116] A.E. Pontiroli et al., *Horm. Res.* **23** (1986) 129.
[6.117] A.J.L. de Jong et al., *Regul. Pept.* **17** (1987) 285.
[6.118] C. Scarpignato, B. Micali, *Gut* **27** (1986) 499.
[6.119] V.N. Hari Prasad, T.W. Moody, *Peptides* **9** (1989) 1345.
[6.120] J.R. Lambert et al., *Dig. Dis. Sci.* **29** (1984) 1036.
[6.121] S. Evangelista et al., *Gen. Pharmacol.* **18** (1987) 647.
[6.122] J.G. McCoy, D.D. Avery, *Peptides* **11** (1990) 595.
[6.123] A.R. Chikh-Issa et al., *Peptides* **10** (1989) 281.
[6.124] E.A. Mayer et al., *Am. J. Physiol.* **250** (1986) Pt. 1, G581.
[6.125] J.C. Reynolds et al., *J. Clin. Invest.* **77** (1986) 436.
[6.126] C.E. Sievert, Jr. et al., *Am. J. Physiol.* **254** (1988) G361.
[6.127] C.E. Sievert, Jr. et al., *Gastroenterology* **84** (1983) Pt. 2, 1310.
[6.128] F. Angel et al., *Gastroenterology* **84** (1983) Pt. 2, 1092.
[6.129] A.C. Hale et al., *J. Endocrinol.* **102** (1984) R1.
[6.130] S. Guarini et al., *Life Sci.* **45** (1989) 107.
[6.131] W.A. Murphy et al., *Endocrinology* **117** (1985) 1179.
[6.132] P. Buydens et al., *Life Sci.* **43** (1988) 1755.
[6.133] T. Mitsuma et al., *Exp. Clin. Endocrinol.* **87** (1986) 162.
[6.134] T. Lehy et al., *Gastroenterology* **90** (1986) 1942.
[6.135] C. Stock Damge et al., *Gut* **28** (1987), Suppl., 1.
[6.136] F. Cuttitta et al., *Nature* **316** (1985) 823.
[6.137] E.A.F. van Tol et al., *Neuropeptides* **18** (1991) 15.
[6.138] J.F. Flood, J.E. Morley, *Brain Res.* **460** (1988) 314.
[6.139] G. de Caro et al., *Peptides* **5** (1984) 607.
[6.140] Y. Kabayama et al., *Endocrinology* **115** (1984) 649.
[6.141] E. Widerlöv et al., *Peptides* **5** (1984) 523.
[6.142] F. Porreca et al., *Eur. J. Pharmacol.* **114** (1985) 167.
[6.143] R.J. Koslo et al., *Gastroenterology* **88** (1985) Pt. 2, 1455.
[6.144] G. Falconieri Erspamer et al., *Regul. Pept.* **21** (1988) 1.
[6.145] F. Girard et al., *Eur. J. Pharmacol.* **102** (1984) 489.
[6.146] P. Heinz-Erian et al., *Am. J. Physiol.* **252** (1987) G439.
[6.147] Z.A. Saeed et al., *Peptides* **10** (1989) 597.
[6.148] R. Camble et al., *Life Sci.* **45** (1989) 1521.
[6.149] L.-H. Wang et al., *Biochemistry* **29** (1990) 616.
[6.150] D.C. Heimbrook et al., *J. Biol. Chem.* **264** (1989) 11258.
[6.151] T. von Schrenck et al., *Am. J. Physiol.* **259** (1990) G468.
[6.152] D.H. Coy et al., *J. Biol. Chem.* **263** (1988) 5056.
[6.153] D.H. Coy et al., *J. Biol. Chem.* **264** (1989) 14691.
[6.154] W.S. Saari et al., *Biochem. Biophys. Res. Commun.* **165** (1989) 114.
[6.155] J.C.M. Hafkenscheid et al., *Clin. Chim. Acta* **136** (1984) 235.
[6.156] G. Bonora et al., *Gastroenterology* **84** (1983) Pt. 2, 1111.
[6.157] C.A. Helman, B.I. Hirschowitz *Gastroenterology* **92** (1987) 1926.
[6.158] W. König in A. Kleemann et al. *(eds.): Arzneimittel, Fortschritte 1972 bis 1985*, VCH Verlagsgesellschaft, Weinheim 1987, 691.
[6.159] B.N. Andersen et al., *Regul. Pept.* **10** (1985) 329.
[6.160] R. Dimaline et al., *FEBS Lett.* **205** (1986) 318.

[6.161] R.J. Nachman et al., *Science* **234** (1986) 71.
[6.162] J.A. Veenstra, *Neuropeptides* **14** (1989) 145.
[6.163] T.K. Hayes et al., *Life Sci.* **44** (1989) 1259.
[6.164] R. Nichols et al., *J. Biol. Chem.* **263** (1988) 12167.
[6.165] C. Bonato et al., *Peptides* **7** (1986) 689.
[6.166] C. Bonato et al., *Life Sci.* **39** (1986) 959.
[6.167] Y. Shinomura et al., *Biochem. Biophys. Res. Commun.* **143** (1987) 7.
[6.168] R. Jiang et al., *Peptides* **9** (1988) 763.
[6.169] C. Bonato et al., *Life Sci.* **37** (1985) 2563.
[6.170] M.M. Wolfe et al., *Clin. Res.* **31** (1983) 477.
[6.171] T. Kurose et al., *Life Sci.* **42** (1988) 1995.
[6.172] J.M. Allen et al., *Gut* **26** (1985) A558.
[6.173] T. Chiba et al., *Horm. Metab. Res.* **15** (1983) 516.
[6.174] S.F. Alino et al., *Acta Physiol. Scand.* **126** (1986) 1.
[6.175] K.C. Christensen, *Scand. J. Gastroenterol.* **19** (1984) 339.
[6.176] J. DelValle, T. Yamada. *J. Clin. Invest.* **85** (1990) 139.
[6.177] E.J. Dial et al., *Gastroenterology* **90** (1986) 1018.
[6.178] A. Rentis et al., *Horm. Metab. Res.* **18** (1986) 423.
[6.179] B.I. Hirschowitz, E. Molina, *Peptides* **5** (1984) 35.
[6.180] R.F. Harty et al., *Gastroenterology* **88** (1985) 675.
[6.181] A. Todisco, *Gastroenterology* **86** (1984) Pt. 2, 1356.
[6.182] C.H. Cho et al., *IRCS Med. Sci.* **13** (1985) 629.
[6.183] H.J. de Aizpurua et al., *N. Engl. J. Med.* **313** (1985) 479.
[6.184] M.K. Müller et al., *Horm. Metab. Res* **18** (1986) 675.
[6.185] J. Puurunen, U. Schwabe, *Br. J. Pharmacol.* **90** (1987) 479.
[6.186] R. Hakanson et al., *Regul. Pept.* **28** (1990) 131.
[6.187] L. Holm Rutili, T. Berglindh, *Am. J. Physiol.* **250** (1986) Pt. 1, G575.
[6.188] A.J. Blair III et al., *J. Clin. Invest.* **78** (1986) 779.
[6.189] C. Jönson et al., *Acta Physiol. Scand.* **124** (1985) 631-633.
[6.190] B.S. Tsai et al., *Gastroenterology* **86** (1984) Pt. 2, 1284.
[6.191] I. Zahavi et al., *Gastroenterology* **90** (1986) Pt. 2, 1701.
[6.192] R. Hakanson et al., *Scand. J. Gastroenterol.* **21** (1986) Suppl. 118, 18.
[6.193] T.E. Solomon et al., *Gastroenterology* **92** (1987) 429.
[6.194] A. Imdahl et al., *J. Cancer Res. Clin. Oncol.* **115** (1989) 388.
[6.195] C.J. Kusyk et al., *Am. J. Physiol.* **251** (1986) G597.
[6.196] D.L. Morris et al., *Gut* **30** (1989) A739.
[6.197] A.G. Saakyan, *Ter. Arkh.* **57** (1985) 15.
[6.198] J.B.J. Fozard et al., *Gut* **25** (1984) A1308.
[6.199] I.V. Molchanova, G.A. Belokrylov, *Byull. Eksp. Biol. Med.* **104** (1987) 73.
[6.200] M. Tatsuta et al., *Int. J. Cancer* **41** (1988) 239.
[6.201] S. Hsiao et al., *Life Sci.* **34** (1984) 2165.
[6.202] C. de Montigny, *Arch. Gen. Psychiatry* **46** (1989) 511.
[6.203] G. Katsuura, S. Itoh, *Peptides* **7** (1986) 809.
[6.204] V.E. Eysselein et al., *J. Clin. Invest.* **73** (1984) 1284.
[6.205] W. Göhring et al., *Hoppe-Seyler's Z. Physiol. Chem.* **365** (1984) 83.
[6.206] J. Martinez et al., *J. Med. Chem.* **28** (1985) 1874.
[6.207] N.V. Sadovnikova et al., *Khim. Farm. Zh.* **21** (1987) 1424.
[6.208] D.G. Patel et al., *Diabetes* **38** (1989) Suppl. 2, 206A.
[6.209] J.J. Knittel et al., *Peptide Res.* **3** (1990) 224.
[6.210] J. Martinez et al., *J. Med. Chem.* **27** (1984) 1597.
[6.211] J. Martinez et al., *J. Med. Chem.* **28** (1985) 273.
[6.212] M. Rodriguez et al, *J. Med. Chem.* **32** (1989) 522.
[6.213] B. Petersen et al., *Regul. Peptides* **7** (1983) 323.
[6.214] M. Rodriguez et al., *J. Med. Chem.* **30** (1987) 758.
[6.215] W.F. Schmidt et al., *FEBS Lett.* **191** (1985) 264.
[6.216] J.S. Gill et al., *Life Sci.* **44** (1989) 483-491.
[6.217] W.E. Schmidt et al., *Dig. Dis. Sci.* **29** (1984), Suppl., 75S.

[6.218] W.E. Schmidt et al., *Acta Endocrinol.* **108** (1985) Suppl. 267, 62.

References for Chapter 7

[7.1] W. König in A. Kleemann et al., (eds.): *Arzneimittel, Fortschritte 1972 bis 1985,* VCH Verlagsgesellschaft, Weinheim 1987, 592.
[7.2] T.M. Chang et al., *Peptides* **6** (1985) 193.
[7.3] G. Gafvelin et al., *Proc. Natl. Acad. Sci. USA* **87** (1990) 6781.
[7.4] A.S. Kopin et al., *Proc. Natl. Acad. Sci. USA* **87** (1990) 2299.
[7.5] D. Gossen et al, *Biochem. Biophys. Res. Commun.* **160** (1989) 862.
[7.6] Y. Shinomura et al., *Life Sci.* **41** (1987) 1243.
[7.7] D. Gossen et al., *Peptides* **11** (1990) 123.
[7.8] T. Yamamoto et al., *Chem. Pharm. Bull.* **34** (1986) 3803.
[7.9] J. Leonora et al., *Am. J. Physiol.* **252** (1987) E477.
[7.10] G.E. Pratt et al., *Biochem. Biophys. Res. Commun.* **163** (1989) 1243.
[7.11] A. Argiolas, J.J. Pisano. *J. Biol. Chem.* **260** (1985) 1437.
[7.12] D. Andreu et al., *Eur. J. Biochem.* **149** (1985) 531.
[7.13] W. Hoffmann et al., *EMBO J.* **2** (1983) 711.
[7.14] P. Li et al., *J. Clin. Invest.* **86** (1990) 1474.
[7.15] D.L. Kaminski, Y.G. Deshpande, *Gastroenterology* **85** (1983) 1239.
[7.16] S.N.S. Murthy, A. Lavy, *Biochem. Pharmacol.* **37** (1988) 1027.
[7.17] H.T. Debas et al., *Gastroenterolgy* **90** (1986) Pt. 2, 1389.
[7.18] S. Watanabe et al., *Gastroenterology* **90** (1986) Pt. 2, 1685.
[7.19] S. Watanabe et al., *Gastroenterology* **94** (1988) Pt. 2, A488.
[7.20] S. Watanabe et al., *Dig. Dis. Sci.* **29** (1984) 971.
[7.21] S. Iwakawa et al., *J. Pharmacobio. Dyn.* **8** (1985) S136.
[7.22] M.V. Singer et al., *Am. J. Physiol.* **248** (1985) G532.
[7.23] D.R. Fletcher et al., *Regul. Pept.* **11** (1985) 217.
[7.24] L. Bolondi et al., *Dig. Dis. Sci.* **29** (1984) 802.
[7.25] G.E. Feurle et al., *Pancreas* **2** (1987) 422.
[7.26] H. Haarstad et al., *Scand. J. Gastroenterol.* **21** (1986) 589.
[7.27] E.F. Stange et al., *Horm. Metab. Res.* **18** (1986) 303.
[7.28] I.S. Chung et al., *Clin. Res.* **36** (1988) 394A.
[7.29] M. Murakami et al., *Dig. Dis. Sci.* **30** (1985) 346.
[7.30] F. Yamagishi et al., *Eur. J. Pharmacol.* **118** (1985) 203.
[7.31] T. Hashimoto et al., *Can. J. Physiol. Pharmacol.* (1986) Suppl., 40.
[7.32] J.H. Kleibeuker et al., *Gastroenterology* **94** (1988) 122.
[7.33] B. Schölkens, *Acta Pharmacol. Toxicol.* **29** (1971) 52.
[7.34] P. Gunnes et al., *J. Cardiovasc. Pharmacol.* **7** (1985) 1183.
[7.35] P. Gunnes, O. Reikeras, *J. Pharmacol. Exp. Ther.* **244** (1988) 1057.
[7.36] V. Di Marzo et al., *Neuropeptides (Edinburgh)* **9** (1987) 51.
[7.37] N. Holst et al., *Acta Endocrinol. (Copenhagen)* **122** (1990) 313.
[7.38] H. Olson et al., *Peptides* **9** (1988) 301.
[7.39] H. Kofod et al., *Am. J. Physiol.* **250** (1986) E107.
[7.40] S. Kiyama et al., *Chem. Pharm. Bull.* **33** (1985) 3205.
[7.41] T. Tsuda et al., *J. Pharm. Sci.* **79** (1990) 53.
[7.42] J. Glaser et al., *Scand. J. Gastroenterol.* **24** (1989) 179.
[7.43] B. Krakamp et al., *MMW Münch. Med. Wochenschr.* **130** (1988) 47.
[7.44] Y.S. Malov. *Ter. Arkh.* **57** (1985) 6.
[7.45] B. Glaser et al., *J. Clin. Endocrinol. Metab.* **66** (1988) 1138.
[7.46] W. König in A. Kleemann et al. (eds.): *Arzneimittel, Fortschritte 1972 bis 1985,* VCH Verlagsgesellschaft, Weinheim 1987, 595.
[7.47] S.I. Said, *J. Endocrinol. Invest.* **9** (1986) 191.
[7.48] G. Gafvelin et al., *Peptides* **9** (1988) 469.
[7.49] Y. Yiangou et al, *J. Biol. Chem.* **262** (1987) 14010.
[7.50] G. Gafvelin, *Peptides* **11** (1990) 703.

[7.51] B.-H. Du et al., *Biochem. Biophys. Res. Commun.* **128** (1985) 1093.
[7.52] K. Tatemoto et al., *FEBS Lett.* **174** (1984) 258.
[7.53] M. Nishizawa et al., *FEBS Lett.* **183** (1985) 55.
[7.54] D. Gossen et al., *Peptides* **11** (1990) 123.
[7.55] S. Björck et al., *Acta Physiol. Scand.* **128** (1986) 639.
[7.56] M.A. Arnaout et al., *Endocrinology (Baltimore)* **119** (1986) 2052.
[7.57] D.E. Burleigh, J.B. Furness, *Neuropeptides (Edinburgh)* **16** (1990) 77.
[7.58] M. Eriksson et al., *Peptides* **8** (1987) 411.
[7.59] P.K. Opstad, *Peptides* **8** (1987) 175.
[7.60] T.A. Crozier et al., *Horm. Metab. Res.* **20** (1988) 352.
[7.61] E. Giladi et al., *Isr. J. Med. Sci.* **23** (1987) 924.
[7.62] G. Burns et al., *Life Sci.* **40** (1987) 951.
[7.63] T.V. Nowak et al., *Gastroenterology* **96** (1989) Pt. 2, A368.
[7.64] K.S.L. Lam et al., *Neuroendocrinology* **52** (1990) 417.
[7.65] I. Gozes, Y. Shani, *Endocrinology (Baltimore)* **119** (1986) 2497.
[7.66] P. Heinz-Erian et al., *Science (Washington D.C.)* **229** (1985) 1407.
[7.67] F.E. Bauer et al., *MMW Münch. Med. Wochenschr.* **126** (1984) 1097.
[7.68] L. Woie et al., *Gen. Pharmacol.* **18** (1987) 577.
[7.69] M.S. O'Dorisio, *Fed. Proc. Fed. Am. Soc. Exp. Biol.* **46** (1987) 192.
[7.70] K. McArthur et al., *Gastroenterology* **92** (1987) Pt. 2, 1524.
[7.71] M. Murakami et al., *Gastroenterology* **92** (1987) Pt. 2, 1544.
[7.72] P. Wiik et al., *Peptides* **9** (1988) 181.
[7.73] H. Houchi et al., *Biochem. Pharmacol.* **36** (1987) 1551.
[7.74] M. Raiteri et al., *Eur. J. Pharmacol.* **133** (1987) 127.
[7.75] M.S. Barnette, B. Weiss, *J. Neurochem.* **45** (1985) 640.
[7.76] Y. Karasawa et al., *Eur. J. Pharmacol.* **187** (1990) 9.
[7.77] S.I. Said, *N. Engl. J. Med.* **320** (1989) 1271.
[7.78] T. Iwanaga et al., *Am. Rev. Respir. Dis.* **135** (1987) Pt. 2, A262.
[7.79] H. Berisha et al., *Am. J. Physiol.* **259** (1990) L151.
[7.80] D.A. Dartt et al., *Am. J. Physiol.* **254** (1988) G300.
[7.81] T.W. Mittag et al., *J. Pharmacol. Exp. Ther.* **241** (1987) 230.
[7.82] P.S. Olsen et al., *Regul. Pept.* **15** (1986) 37.
[7.83] J. Ekström, L. Olgart, *Acta Physiol. Scand.* **126** (1986) 25.
[7.84] T.M. Feeley et al., *Can. J. Physiol. Pharmacol.* **62** (1984) 356.
[7.85] A. Bennett et al., *J. Pharm. Pharmacol.* **36** (1984) 787.
[7.86] J. Fontaine et al., *Br. J. Pharmacol.* **89** (1986) 599.
[7.87] B. Ottesen et al., *Regul. Pept.* **11** (1985) 83.
[7.88] K.A. Duggan, G.J. MacDonald, *Clin. Sci.* **72** (1987) 195.
[7.89] A. Green et al., *Am. J. Physiol.* **249** (1985) E608.
[7.90] N. Yanaihara, EP 0 225 639, 1985.
[7.91] D.W. Pincus et al., *Nature (London)* **343** (1990) 564.
[7.92] B. Carlsson et al., *Acta Physiol. Scand.* **129** (1987) 437.
[7.93] D.E. Brenneman, L.E. Eiden, *Proc. Natl. Acad. Sci. USA* **83** (1986) 1159.
[7.94] G.J. Poston et al., *Dig. Dis. Sci.* **32** (1987) 1183.
[7.95] V. Schusdziarra et al., *Peptides* **7** (1986) 127.
[7.96] T. Chiba et al., *Gastroenterology* **88** (1985) Pt. 2, 1348.
[7.97] E. Sandberg et al., *Acta Endocrinol. (Copenhagen)* **117** (1988) 545.
[7.98] H. Abe et al., *Endocrinology* **116** (1985) 1383.
[7.99] L.A. Cunningham, M.A. Holzwarth, *Endocrinology* **122** (1988) 2090.
[7.100] B. Ottesen et al., *Regul. Pept.* **16** (1986) 299.
[7.101] S.P. Wilson, *Life Sci.* **40** (1987) 623.
[7.102] F.W. George, S.R. Ojeda, *Proc. Natl. Acad. Sci. USA* **84** (1987) 5803.
[7.103] C.E. Ahmed et al., *Endocrinology* **118** (1986) 1682.
[7.104] B.G. Kasson et al., *Mol. Cell. Endocrinol.* **48** (1986) 21.
[7.105] P. Wiik, *Regul. Pept.* **125** (1989) 187.
[7.106] M. Mirmiran et al., *Brain Res.* **458** (1988) 192.
[7.107] D.E. Hernandez et al., *Brain Res.* **420** (1987) 129.

[7.108] S. Itoh et al., *Peptides* **6** (1985) 53.
[7.109] J.F. Flood et al., *Peptides* **11** (1990) 933.
[7.110] T. Sakamoto et al., *Am. Rev. Respir. Dis.* **139** (1989) Pt. 2, A614.
[7.111] M. Hamada et al., *Am. Rev. Respir. Dis.* **137** (1988) Pt. 2, 35.
[7.112] V.E. Sutliff et al., *Am. J. Physiol.* **251** (1986) G96.
[7.113] J.T. Turner et al., *Peptides* **7** (1986) 849.
[7.114] G.F. Musso et al., *Biochemistry* **27** (1988) 8174.
[7.115] G. Hallden et al., *Regul. Pept.* **16** (1986) 183.
[7.116] P. Robberecht et al., *Eur J. Biochem.* **159** (1986) 45.
[7.117] S.J. Pandol et al., *Am. J. Physiol.* **250** (1986) G553.
[7.118] D.C. Thompson et al., *Peptides* **9** (1988) 443.
[7.119] Y. Yiangou et al., *J. Biol. Chem.* **262** (1987) 14010.
[7.120] P. Robberecht et al., *Eur. J. Biochem.* **165** (1987) 243.
[7.121] A. Miyata et al., *Biochem. Biophys. Res. Commun.* **164** (1989) 567.
[7.122] A. Miyata et al., *Biochem. Biophys. Res. Commun.* **170** (1990) 643.
[7.123] C. Kimura et al., *Biochem. Biophys. Res. Commun.* **166** (1990) 81.
[7.124] K. Ogi et al., *Biochem. Biophys. Res. Commun.* **173** (1990) 1271.
[7.125] H.-C. Lam et al., *Eur. J. Biochem.* **193** (1990) 725.
[7.126] A. Vandermeers et al., *Eur. J. Biochem.* **164** (1987) 321.
[7.127] P. Robberecht et ał., *Biochem. Biophys. Res. Commun.* **130** (1985) 333.
[7.128] P. Robberecht et al., *FEBS Lett.* **190** (1985) 142.
[7.129] A. Bjartell et al., *Regul. Pept.* **26** (1989) 27.
[7.130] T. Grunditz et al., *Proc. Natl. Acad. Sci. USA* **86** (1989) 1357.
[7.131] Y. Tsutsumi et al., *Regul. Pept.* **31** (1990) 11.
[7.132] F. Sundler et al., *Regul. Pept.* **20** (1988) 83.
[7.133] P. Robberecht et al., *FEBS Lett.* **228** (1988) 351.
[7.134] J.P. Dehaye et al., *Peptides* **5** (1984) 333.
[7.135] B. Ahren, *Peptides* **10** (1989) 709.
[7.136] L. Grundemar, E.D. Högestätt, *Br. J. Pharmacol.* **99** (1990) 526.
[7.137] V. di Marzo et al., *Neuropeptides (Endinburgh)* **11** (1988) 169.
[7.138] J. Pfeilschifter et al., *Biochem. Biophys. Res. Comomun.* **170** (1990) 576.
[7.139] W. König in A. Kleemann et al., (eds.): *Arzneimittel, Fortschritte 1972 bis 1985*, VCH Verlagsgesellschaft Weinheim 1987, 604.
[7.140] G. Heinrich et al., *Endocrinology (Baltimore)* **115** (1984) 2176.
[7.141] G.I. Bell et al., *Nature (London)* **304** (1983) 368.
[7.142] L.C. Lopez et al., *Proc. Natl. Acad. Sci. USA* **80** (1983) 5485.
[7.143] T. Buhl et al., *J. Biol. Chem.* **263** (1988) 8621.
[7.144] S. Seino et al., *FEBS Lett.* **203** (1986) 25.
[7.145] M. Nishi, D.F. Steiner, *Endocrinology* **4** (1990) 1192.
[7.146] L. Thim, A.J. Moody, *Regul. Pept.* **2** (1981) 139.
[7.147] D.J. Drucker et al., *J. Biol. Chem.* **261** (1986) 9637.
[7.148] C. Orskov et al., *J. Biol. Chem.* **264** (1989) 12826.
[7.149] H. Itoh et al., *Biochem. Biophys. Res. Commun.* **165** (1989) 1010.
[7.150] P. Blache et al., *J. Biol. Chem.* **265** (1990) 21514.
[7.151] J.-H. Yu et al., *Peptides* **10** (1989) 1195.
[7.152] J.M. Conlon, J.W. Hicks, *Peptides* **11** (1990) 461.
[7.153] J. Eng. et al., *Peptides* **11** (1990) 683.
[7.154] R. Nichols et al., *Endocrinology (Baltimore)* **123** (1988) 2639.
[7.155] P.C. Andrews, P. Ronner, *J. Biol. Chem.* **260** (1985) 3910.
[7.156] E. Plisetskaya et al., *Regul. Pept.* **14** (1986) 57.
[7.157] J.M. Conlon et al., *FEBS Lett.* **214** (1987) 50.
[7.158] D.S. Weigle, C.J. Goodner, *Endocrinology* **118** (1986) 1606.
[7.159] G. Skoglund et al., *Eur. J. Pharmacol.* **143** (1987) 83.
[7.160] G. Paolisso et al., *Acta Endocrinol. (Copenhagen)* **115** (1987) 161.
[7.161] L.J. Klaff, G.J. Taborsky, Jr., *Clin. Res.* **34** (1986) 60A.
[7.162] R.G. Knowles et al., *Biochem. Pharmacol.* **35** (1986) 4043.
[7.163] D.L. Kaminski et al., *Am. J. Physiol.* **254** (1988) G864.

[7.164] P. Rorsman et al., *Nature (London)* **341** (1989) 233.
[7.165] D.G. Pipeleers et al., *Endocrinology* **117** (1985) 817.
[7.166] A. Ohneda et al., *Diabetologia* **29** (1986) 397.
[7.167] C.G. Ostenson et al., *Diabetologia* **29** (1986) 861.
[7.168] A.A.R. Starke et al., *Diabetes* **33** (1984) 277.
[7.169] K.F. Hanssen et al., *Diabetologia* **27** (1984) 285A.
[7.170] A.E. Farah, *Pharmacol. Rev.* **35** (1983) 181.
[7.171] A. Akatsuka et al., *J. Biol. Chem.* **260** (1985) 3239.
[7.172] K.Yamauchi, K. Hashizume, *Endocrinology (Baltimore)* **119** (1986) 218.
[7.173] S. Marubashi et al., *Acta Endocrinol. (Copenhagen)* **108** (1985) 6.
[7.174] R. Holland et al., *Eur. J. Biochem.* **140** (1984) 325.
[7.175] R.J. Howland, A.D. Benning, *FEBS Lett.* **208** (1986) 128.
[7.176] O.G. Bjornsson et al., *Eur. J. Clin. Invest.* **14** (1984) Pt. 2, 39.
[7.177] C. Guettet et al., *Biochim. Biophys. Acta L* **1005** (1989) 233.
[7.178] C. Bailly et al., *Am. J. Physiol.* **246** (1984) 205.
[7.179] K.R. Wong, M.G. Cogan, *Life Sci.* **40** (1987) 1595.
[7.180] E. Mezey et al., *Gastroenterology* **91** (1986) 1271.
[7.181] G. Jonderko et al., *Scand. J. Clin. Lab. Invest.* **48** (1988) 743.
[7.182] B.G. Weick, S. Ritter, *Am. J. Physiol.* **250** (1986) R676.
[7.183] A. Oshima et al., *Gastroenterology* **92** (1987) Pt. 2, 1563.
[7.184] D.L. Carr Locke et al., *Dig. Dis. Sci.* **28** (1983) 312.
[7.185] B.D. Schirmer et al., *Dig. Dis. Sci.* **31** (1986) 297.
[7.186] D.L. Kaminski et al., *Am. J. Physiol.* **254** (1988) G864.
[7.187] M.S. Sherman et al., *J. Allergy Clin. Immunol.* **81** (1988) 908.
[7.188] U.M. Kabadi et al., *Gastroenterology* **88** (1985) 750.
[7.189] T. Chiba et al., *Gastroenterology* **88** (1985) Pt. 2, 1348.
[7.190] T.E. Adrian et al., *J. Physiol. (London)* **346** (1984) 127P.
[7.191] R.H. Rao, G.S. Spathis, *Metab. Clin. Exp.* **36** (1987) 658.
[7.192] R. Perlman et al., *Biochem. Biophys. Res. Commun.* **151** (1988) 1019.
[7.193] D. Malcolm, G. Zaloga, B. Chernow, J. Holaday, *Life Sci.* **39** (1986) 399.
[7.194] K. Corey Flander et al., *J. Biol. Chem.* **259** (1984) 7031.
[7.195] J.F. Rey et al., *Dig. Dis. Sci.* **31** (1986) 355.
[7.196] G. Jacobson et al., *Lancet II* (1984) 1149.
[7.197] A. Mallat et al., *Nature (London)* **325** (1987) 620.
[7.198] P. Robberecht et al., *Regul. Pept.* **21** (1988) 117.
[7.199] J. Sueiras-Diaz et al., *J. Med. Chem.* **27** (1984) 310.
[7.200] J.L. Krstenansky et al., *Biochemistry* **25** (1986) 3833.
[7.201] K.A. Cornely et al., *Arch. Biochem. Biophys.* **240** (1985) 698.
[7.202] C.G. Unson et al., *Proc. Natl. Acad. Sci. USA* **84** (1987) 4083.
[7.203] C.G. Unson et al., *J. Biol. Chem.* **264** (1989) 789.
[7.204] B. Gysin, V.J. Hruby, *Experientia* **42** (1986) 680.
[7.205] S.W. Trenkner et al., *Radiology (Easton, Pa.)* **149** (1983) 401.
[7.206] A.M.N. Gardner, *Br. Med. J.* **290** (1985) 822.
[7.207] L. van Gaal et al., *Acta Clin. Belg.* **40** (1985) 266.
[7.208] J. Buch, A. Buch, *Acta Pharmacol. Toxicol.* **53** (1983) 188.
[7.209] L. Freychet et al., *Lancet I* (1988) 1364.
[7.210] A.E. Pontiroli et al., *Eur. J. Clin. Pharmacol.* **37** (1989) 427.
[7.211] L.M. Chuang et al., *Pharmacologist* **31** (1989) 125.
[7.212] A. Ohneda et al., *Horm. Metab. Res.* **19** (1987) 85.
[7.213] B.T.G. Schjoldager et al., *Eur. J. Clin. Invest.* **18** (1988) 499.
[7.214] C. Jarrousse et al., *FEBS Lett.* **188** (1985) 81.
[7.215] T.M. Biedzinski et al., *Peptides* **8** (1987) 967.
[7.216] K. Shima et al., *Acta Endocrinol. (Copenhagen)* **123** (1990) 464.
[7.217] B. Kreymann et al., *Lancet II* (1987) 1300.
[7.218] C. Orskov, J.H. Nielsen, *FEBS Lett.* **229** (1988) 175.
[7.219] G. Richter et al., *Acta Endocrinol. (Copenhagen)* **120** (1989) Supp. 1, 191.
[7.220] A.B. Hansen et al., *FEBS Lett.* **236** (1988) 119.

[7.221] D.J. Drucker et al., *Proc. Natl. Acad. Sci. USA* **84** (1987) 3434.
[7.222] H.-C. Fehmann et al., *FEBS Lett.* **252** (1989) 109.
[7.223] E. Yamato et al., *Biochem. Biophys. Res. Commun.* **167** (1990) 431.
[7.224] C. Orskov et al., *Endocrinology (Baltimore)* **123** (1988) 2009.
[7.225] G.K. Hendrick et al., *Diabetes* **37** (1988) Suppl. 1, 49A.
[7.226] M. Gutniak et al., *Diabetologia* **33** (1990) Suppl., A73.
[7.227] G. Kegel et al., *FEBS Lett.* **255** (1989) 10.
[7.228] E.S. Chang et al., *Biochem. Biophys. Res. Commun.* **171** (1990) 818.
[7.229] W. König in A. Kleemann et al. (eds.): *Arzneimittel, Fortschritte 1972 bis 1985*, VCH Verlagsgesellschaft, Weinheim 1987, 601.
[7.230] W.A. Rogers et al., *Dig. Dis. Sci.* **28** (1983) 345.
[7.231] P.C.S. Blom et al., *Acta Physiol. Scand.* **123** (1985) 367.
[7.232] L. Groop et al., *Diabetes* **33** (1984) Suppl. 1, 166A.
[7.233] T. Krarup et al., *Diabetoloiga* **25** (1983) 173.
[7.234] C.K. Lardinois et al., *Diabetes* **33** (1984) 110A.
[7.235] F.J. Service et al., *J. Clin. Endocrinol. Metab.* **58** (1984) 1133.
[7.236] P.R. Flatt et al., *J. Endocrinol.* **101** (1984) 249.
[7.237] R.B. Richeson et al., *Clin. Res.* **33** (1985) 66A.
[7.238] A.J. McCullough et al., *Am. J. Physiol.* **248** (1985) E299.
[7.239] G. Hansen Starich et al., *Am. J. Physiol.* **249** (1985) E603.
[7.240] H. Hartmann et al., *Diabetologia* **29** (1986) 112.
[7.241] E. Sandberg et al., *Acta Physiol. Scand.* **127** (1986) 323.
[7.242] W.E. Schmidt et al., *Eur. J. Clin. Invest.* **16** (1986) Pt. 2, A9.
[7.243] J.W. Adelson et al., *J. Biol. Chem.* **261** (1986) 10569.
[7.244] G.C.M. Steffens et al., *Hoppe-Seyler's Z. Physiol. Chem.* **360** (1979) 1641.
[7.245] C.H.S. McIntosh et al., *Can. J. Physiol. Pharmacol.* **66** (1988) 1407.
[7.246] A. Rökaeus, *Trends Neurosci.* **10** (1987) 158.
[7.247] K. Tatemoto et al., *FEBS Lett.* **164** (1983) 124.
[7.248] A. Rökaeus, M. Carlquist, *FEBS Lett.* **234** (1988) 400.
[7.249] L.M. Kaplan et al., *Proc. Natl. Acad. Sci. USA* **85** (1988) 7408.
[7.250] F.E. Bauer et al., *FEBS Lett.* **201** (1986) 327.
[7.251] F.E. Bauer et al., *Peptides* **7** (1986) 5.
[7.252] F.E. Bauer et al., *Lancet II* (1986) 192.
[7.253] B. Ahren et al., *Biochem. Biophys. Res. Commun.* **140** (1986) 1059.
[7.254] F.E. Bauer et al., *J. Clin. Endocrinol. Metab.* **63** (1986) 1372.
[7.255] I. Merchenthaler et al., *Proc. Natl. Acad. Sci. USA* **87** (1990) 6326.
[7.256] S.M. Gabriel et al., *Peptides* **10** (1989) 3698.
[7.257] D.J. O'Halloran et al., *J. Endocrinol.* **119** (1988) Suppl., 142.
[7.258] S.C. Hooi et al., *Neuroendocrinology* **51** (1990) 351.
[7.259] L.S. Brady et al., *Neuroendocrinology* **52** (1990) 441.
[7.260] E. Ekblad et al., *Br. J. Pharmacol.* **86** (1985) 241.
[7.261] S. Rattan, R.K. Goyal, *Life Sci.* **41** (1987) 2783.
[7.262] T. Ohhashi, D.M. Jacobowitz, *Regul. Pept.* **12** (1985) 163.
[7.263] L.W. Haynes, *Trends Pharmacol. Sci.* **7** (1986) 214.
[7.264] A. Kuwahara et al., *Eur. J. Pharmacol.* **164** (1989) 175.
[7.265] J.E.T. Fox et al., *Life Sci.* **39** (1986) 103.
[7.266] K. Tamura et al., *Eur. J. Pharmacol.* **136** (1987) 445.
[7.267] V. di Marzo et al., *Neuropeptides (Edinburgh)* **11** (1988) 169.
[7.268] F.E. Bauer et al., *Gastroenterology* **97** (1989) 260.
[7.269] P. Miralles et al., *Metab. Clin. Exp.* **37** (1988) 766.
[7.270] Y. Tajiri et al., *Horm. Metab. Res.* **22** (1990) 1.
[7.271] S. Ullrich, C.B. Wollheim, *FEBS Lett.* **247** (1989) 401.
[7.272] G.W.G. Sharp et al., *J. Biol. Chem.* **264** (1989) 7302.
[7.273] S. Lindskog, B. Ahren, *Acta Physiol. Scand.* **129** (1987) 305.
[7.274] T. Messell et al., *Regul. Pept.* **28** (1990) 161.
[7.275] Y.N. Kwok et al., *Eur. J. Pharmacol.* **145** (1988) 49.
[7.276] W.J. Rossowski, D.H. Coy, *Life Sci.* **44** (1989) 1807.

[7.277] G. Soldani et al., *Eur. J. Pharmacol.* **154** (1988) 313.
[7.278] S.G. Cella et al., *Endocrinology (Baltimore)* **122** (1988) 855.
[7.279] Y. Murakami et al., *Eur. J. Pharmacol.* **136** (1987) 415.
[7.280] S.M. Gabriel et al., *Life Sci.* **42** (1988) 1981.
[7.281] N. Guerineau et al., *FEBS Lett.* **276** (1990) 111.
[7.282] D.L. Tempel et al., *Peptides* **9** (1988) 309.
[7.283] Y. Ben Ari, M. Lazdunski, *Eur. J. Pharmacol.* **165** (1989) 331.
[7.284] C. Post et al., *Acta Physiol. Scand.* **132** (1988) 583.
[7.285] J.E.T. Fox et al., *Peptides* **9** (1988) 1183.
[7.286] B. Gallwitz et al., *Biochem. Biophys. Res. Commun.* **172** (1990) 268.
[7.287] X.-J. Xu et al., *Eur. J. Pharmacol.* **182** (1990) 137.
[7.288] A. Kuwahara et al., *Regul. Pept.* **29** (1990) 23.
[7.289] W. König in A. Kleemann et al., (eds.): *Arzneimittel, Fortschritte 1972 bis 1985*, VCH Verlagsgesellschaft, Weinheim 1987, 599.
[7.290] L.A. Frohamn, J.-O. Jansson, *Endocr. Rev.* **7** (1986) 223.
[7.291] K. Chihara et al., *J. Clin. Endocrinol. Metab.* **60** (1985) 269.
[7.292] P. Ciofi et al., *Neuroendocrinology* **45** (1987) 425.
[7.293] B. Meister et al., *Acta Physiol. Scand.* **124** (1985) 133.
[7.294] H. Okamura et al., *Neuroendocrinology* **41** (1985) 177.
[7.295] K. Arase et al., *Endocrinology (Baltimore)* **121** (1987) 1960.
[7.296] Y. Murakami et al., *Eur. J. Pharmacol.* **136** (1987) 415.
[7.297] K. Chihara et al., *J. Clin. Endocrinol. Metab.* **62** (1986) 466.
[7.298] Y. Kashio et al., *J. Clin. Endocrinol. Metab.* **64** (1987) 92.
[7.299] S.M. Gabriel et al., *Neuroendocrinology* **50** (1989) 299.
[7.300] J. Argente et al., *J. Clin. Endocrinol. Metab.* **63** (1986) 680.
[7.301] I. Ganzetti et al., *Peptides* **7** (1986) 1011.
[7.302] T. Shibasaki et al., *Regul. Pept.* **15** (1986) 47.
[7.303] G.P. Ceda et al., *Endocrinology (Baltimore)* **120** (1987) 1658.
[7.304] H. Katakami et al., *Endocrinology (Baltimore)* **118** (1986) 1872.
[7.305] M.C. Aguila, S.M. McCann, *Endocrinology (Baltimore)* **120** (1987) 341.
[7.306] I. Ganzetti et al., *J. Endocrinol. Invest.* **10** (1987) 241.
[7.307] V. de Gennaro Colonna et al., *Peptides* **10** (1989) 705.
[7.308] W.B. Wehrenberg et al., *Neuroendocrinology* **43** (1986) 266.
[7.309] H. Seifert et al., *Endocrinology (Baltimore)* **117** (1985) 424.
[7.310] W.B. Wehrenberg, *Endocrinology (Baltimore)* **118** (1986) 489.
[7.311] C.E. Brain et al., *J. Endocrinol.* **117** (1988) Suppl., 82.
[7.312] R. Valcavi et al., *J. Endocrinol.* **112** (1987) Suppl., 154.
[7.313] V. Locatelli et al., *J. Endocrinol.* **111** (1986) 271.
[7.314] N. Mauras et al., *Metab. Clin. Exp.* **36** (1987) 369.
[7.315] S. Akira et al., *Neuroendocrinology* **47** (1988) 116.
[7.316] K. Chihara et al., *Endocrinology (Baltimore)* **114** (1984) 1402.
[7.317] M.C. White et al., *J. Endocrinol.* **105** (1985) 269.
[7.318] M. Press et al., *Diabetes* **33** (1984) Suppl. 1, 13A.
[7.319] C. Pintor et al., *J. Clin. Endocrinol. Metab.* **62** (1986) 263.
[7.320] N.T. Richards et al., *Diabetologia* **27** (1984) 529.
[7.321] R. Valcavi et al., *Clin. Endocrinol.* **24** (1986) 693.
[7.322] M. Hotta et al., *Life Sci.* **42** (1988) 979.
[7.323] K.-P. Lesch et al., *J. Clin. Endocrinol. Metab.* **65** (1987) 1278.
[7.324] T. Imaki et al., *J. Clin. Endocrinol. Metab.* **60** (1985) 290.
[7.325] V. Locatelli et al., *Brain Res.* **341** (1985) 35.
[7.326] R. Cacabelos et al., *Acta Endocrinol. (Copenagen)* **117** (1988) 295.
[7.327] F.F. Casanueva et al., *Neuroendocrinology* **47** (1988) 46.
[7.328] F.L. Culler et al., *Clin. Res.* **35** (1987) 221A.
[7.329] H. Iguchi et al., *Endocrinol. Jpn.* **36** (1989) 787.
[7.330] F. Zeytin, P. Braceau, *Biochem. Biophys. Res. Commun.* **123** (1984) 497.
[7.331] T. Mitsuma et al., *Exp. Clin. Endocrinol.* **90** (1987) 365.
[7.332] H. Hauner et al., *Horm. Metab. Res.* **17** (1985) 620.

[7.333] T. Lehy et al., *Gastroenterology* **100** (1986) 646.
[7.334] J. Hasegawa et al., *Horm. Metab. Res.* **20** (1988) 130.
[7.335] F. Obal, Jr. et al., *Am. J. Physiol.* **255** (1988) R310.
[7.336] R. Cacabelos et al., *Endocrinology (Baltimore)* **122** (1988) 1269.
[7.337] Z. Josefsberg et al., *Isr. J. Med. Sci.* **23** (1987) 837.
[7.338] N. Ling et al., *Biochem. Biophys. Res. Commun.* **123** (1984) 854.
[7.339] N. Ling et al., *Biochem. Biophys. Res. Commun.* **122** (1984) 304.
[7.340] K. Sato et al., *Biochem. Biophys. Res. Commun.* **149** (1987) 531.
[7.341] D.H. Coy et al., *J. Med. Chem.* **30** (1987) 219.
[7.342] T.J. Aitman et al., *Peptides* **10** (1989) 1.
[7.343] M. Kovacs et al., *Life Sci.* **42** (1988) 27.
[7.344] L. Bokser et al., *Life Sci.* **46** (1990) 999.
[7.345] K. Sato et al., *Biochem. Biophys. Res. Commun.* **167** (1990) 360.
[7.346] S.J. Hocart et al., *J. Med. Chem.* **33** (1990) 1954.
[7.347] M. Waelbroeck et al., *Endocrinology (Baltimore)* **116** (1985) 2643.
[7.348] J.S. Tou et al., *Biochem. Biophys. Res. Commun.* **139** (1986) 763.
[7.349] G. Velicelebi et al., *Proc. Natl. Acad. Sci. USA* **83** (1986) 5397.
[7.350] A. Baird et al., *Neuroendocrinology* **42** (1986) 273.
[7.351] R.J.M. Ross et al., *Lancet I* (1987) 5.
[7.352] M.O. Thorner et al., *N. Engl. J. Med.* **312** (1985) 4.
[7.353] O. Butenandt, B. Staudlt, *Eur. J. Pediatr.* **148** (1989) 393.
[7.354] P. Franchimont et al., *Acta Clin. Belg.* **42** (1987) 143.
[7.355] Beckmann Instruments, Inc., EP 0 018 072, 1979.
[7.356] T.M. Badger et al., *Endocrinology* **115** (1984) 1432.
[7.357] F.A. Momany et al., *Endocrinology* **114** (1984) 1531.
[7.358] C.Y. Bowers et al., *Endocrinology* **114** (1984) 1573.
[7.359] O. Sartor et al., *Endocrinology* **116** (1985) 952.
[7.360] E.E. Codd et al., *Neuropharmacology* **28** (1989) 1139.
[7.361] T.O. Yellin et al., *in J.E. Rivier, G.R. Marshall (eds.): Peptides,* ESCOM, Leiden 1990, 214.
[7.362] B.E. Ilson et al., *J. Clin. Endocrinol. Metab.* **69** (1989) 212.
[7.363] R.F. Walker et al., *Life Sci.* **47** (1990) 29.
[7.364] W. König in A. Kleemann et al. (eds.): *Arzneimittel, Fortschritte 1972 bis 1985,* VCH Verlagsgesellschaft, Weinheim 1987, 711.
[7.365] N.E. Cooke, J. Ray, J.G. Emery, S.A. Liebhaber, *J. Biol. Chem.* **263** (1988) 9001.
[7.366] M. Ono et al., *Proc. Natl. Acad. Sci. USA* **87** (1990) 4330.
[7.367] M. Yamakawa et al., *J. Biol. Chem.* **265** (1990) 8915.
[7.368] I.A. MacFarlane et al., *Acta Endocrinol. (Copenhagen)* **112** (1986) 547.
[7.368a] M. Watahiki et al., *J. Biol. Chem.* **264** (1989) 312.
[7.368b] Y. Yamano et al., *FEBS Lett.* **228** (1988) 301.
[7.368c] J.D. Hulmes et al., *Int. J. Pept. Prot. Res.* **33** (1989) 368.
[7.368d] Y. Harada et al., *Biochem. Biophys. Res. Commun.* **173** (1990) 1200.
[7.369] J.-O. Jansson et al., *Endocr. Rev.* **6** (1985) 128.
[7.370] R.G. Clark, I.C.A.F. Robinson, *Endocrinology (Baltimore)* **122** (1988) 2675.
[7.371] H. Soya, M. Suzuki, *Endocrinology (Baltimore)* **122** (1988) 2492.
[7.372] P. Maertens, C. Denef, *Mol. Cell. Endocrinol.* **54** (1987) 203.
[7.373] H. Yagi et al., *Horm. Metab. Res.* **18** (1986) 723.
[7.374] A.D. Struthers et al., *Neuroendocrinology* **44** (1986) 22.
[7.375] F. Vita et al., *Clin. Trials J.* **24** (1987) 387.
[7.376] F.E. Chang et al., *J. Clin. Endocrinol. Metab.* **62** (1986) 551.
[7.377] M. Marastoni et al., *Arzneim. Forsch.* **39** (1989) 639.
[7.378] G. Delitala et al., *J. Clin. Endocrinol. Metab.* **69** (1989) 356.
[7.379] E.C. degli Uberti et al., *Metabolism* **39** (1990) 1063.
[7.380] L. Altomonte et al., *Exp. Clin. Endocrinol.* **88** (1986) 334.
[7.381] P.A. Graham et al., *Clin. Res.* **35** (1987) 396A.
[7.382] U. Knigge et al., *Acta Endocrinol. (Copenhagen)* **122** (1990) 354.
[7.383] J.R. Peters et al., *Clin. Endocrinol.* **25** (1986) 213.

[7.384] K. Iyer, S.M. McCann, *Peptides* **8** (1987) 45.
[7.385] E.W. Bernton et al., *Science (Washington, D.C.)* **238** (1987) 519.
[7.386] F. Lopez et al., *Acta Endocrinol. (Copenhagen)* **113** (1986) 317.
[7.387] M.L. Vance et al., *J. Clin. Endocrinol. Metab.* **64** (1987) 1136.
[7.388] J.K. Schmitt, *Clin. Res.* **37** (1989) 9A.
[7.389] K.Y. Ho et al., *J. Clin. Endocrinol. Metab.* **64** (1987) 51.
[7.390] M.J. Mansfield et al., *J. Clin. Endocrinol. Metab.* **66** (1988) 3.
[7.391] N. Mauras et al., *Metabolism* **38** (1989) 286.
[7.392] A. Giustina et al., *Horm. Metab. Res.* **21** (1989) 693.
[7.393] A. Giustina et al., *Acta Endocrinol. (Copenhagen)* **122** (1990) 206.
[7.394] R.J.M. Ross et al., *Clin. Endocrinol.* **26** (1987) 117.
[7.395] S. Kentroti et al., *Endocrinology (Baltimore)* **122** (1988) 2407.
[7.396] N. Ono et al., *Proc. Natl. Acad. Sci. USA* **82** (1985) 7787.
[7.397] G.P. Ceda et al., *Acta Endocrinol. (Copenhagen)* **120** (1989) 416.
[7.398] C. Netti et al., *Neuroendocrinology* **49** (1989) 242.
[7.399] H. Katakami et al., *Endocrinology (Baltimore)* **117** (1985) 1139.
[7.400] N.C. Schaper, *Acta Endocrinol. (Copenhagen)* **122** (1990) 7.
[7.401] K.-P. Lesch et al., *J. Clin. Endocrinol. Metab.* **65** (1987) 1278.
[7.402] A. Zarate et al., *Horm. Metab. Res.* **18** (1986) 400.
[7.403] C. Marchesi et al., *Pharmacopsychiatry* **20** (1987) 64.
[7.404] P. Pietschmann et al., *Horm. Metab. Res.* **22** (1990) 109.
[7.405] V. Coiro et al., *Psychoneuroendocrinology (Oxford)* **12** (1987) 313.
[7.406] O. Giampietro et al., *Metabolism* **36** (1987) 1149.
[7.407] D.W. Leung et al., *Nature (London)* **330** (1987) 537.
[7.408] L.S. Mathews et al., *J. Biol. Chem.* **264** (1989) 9905.
[7.409] R.L. Hintz, *Horm. Res.* **33** (1990) 105.
[7.410] P.J. Godowski et al., *Proc. Natl. Acad Sci. USA* **86** (1989) 8083.
[7.411] A. Aguirre et al., *Horm. Res.* **34** (1990) 4.
[7.412] G. Baumann et al., *N. Engl. J. Med.* **320** (1989), 1705.
[7.413] M.C. Slootweg et al., *J. Endocrinol* **116** (1988) R11-R13.
[7.414] J.H. Nielsen et al., *Mol. Endocrinol.* **3** (1989) 165.
[7.415] M.E. Markowtiz et al., *J. Clin. Endocrinol. Metab.* **69** (1989) 420.
[7.416] K.Y. Ho, A.J. Weissberger, *Metab. Clin. Exp.* **39** (1990) 133.
[7.417] J.K. Damm, *Acta Endocrinol. (Copenhagen)* **114** (1987) 124.
[7.418] I. Ganzetti et al., *J. Endocrinol. Invest.* **10** (1987) 241.
[7.419] D.M. Maiter et al., *Neuroendocrinology* **51** (1990) 174.
[7.420] J. Isgaard et al., *Endocrinology (Baltimore)* **123** (1988) 2605.
[7.421] C.T. Roberts, Jr., et al., *J. Biol. Chem.* **261** (1986) 10025.
[7.422] J.H. Nielsen et al., *Acta Endocrinol. (Copenhagen)* **110** (1985) *Suppl.* **273**, 74.
[7.423] L.G. Frigert et al., *Horm. Metab. Res.* **19** (1987) 464.
[7.424] F. Salomon et al., *Diabetologia* **33** (1990) Suppl., A218.
[7.425] M.B. Davidson, *Endocr. Rev.* **8** (1987) 115.
[7.426] I.D. Morris et al., *J. Endocrinol.* **123** (1989) Suppl., 68.
[7.427] R. Horikawa et al., *Eur. J. Pharmacol.* **166** (1989) 87.
[7.428] C.K. Edwards III et al., *Science (Washington, D.C.)* **239** (1988) 769.
[7.429] B. Lawier Goff et al., *Clin. Exp. Immunol.* **68** (1987) 580.
[7.430] S.L. Kaplan et al., *Lancet I* (1986) 697.
[7.431] J.L. Kostyo et al., *Proc. Natl. Acad. Sci. USA* **82** (1985) 4250.
[7.432] S.L. Kaplan et al., *Lancet I* (1986) 697.
[7.433] A. Skottner et al., *Acta Endocrinol (Copenhagen)* **118** (1988) 14.
[7.434] F.L. Culler et al., *Horm. Metab. Res.* **20** (1988) 107.
[7.435] N. Emoto et al., *Acta Endocrinol (Copenhagan)* **114** (1987) 283.
[7.436] A. Gertler et al., *Endocrinology (Baltimore)* **118** (1986) 720.
[7.437] F.M. Ng, J.A. Harcourt, *Diabetologia* **29** (1986) 882.
[7.438] V.M.J. Robson et al., *Biol. Chem. Hoppe-Seyler (Berlin)* **371** (1990) 423.
[7.439] R.W. Stevenson et al., *Acta Endocrinol. (Copenhagen)* **117** (1988) 457.
[7.440] R.W. Stevenson et al., *Metabol.* **36** (1987) 400.

[7.441] C.E. Mondon et al., *Endocrinology Baltimore* **123** (1988) 827.
[7.442] A. Ide et al., *Agric. Biol. Chem.* **52** (1988) 765.
[7.443] P. Gellerfors et al., *J. Pharm. Biomed. Anal.* **7** (1989) 173.
[7.444] C. Singer-Granick et al., *Horm. Res.* **24** (1986) 246.
[7.445] H. Waago, *Lancet I* (1987) 1485.
[7.446] E.A. van der Veen, J.C. Netelenbos, *Horm. Res.* **33** (1990) 65.
[7.447] F. Greig et al., *Horm. Res.* **31** (1989) Suppl. 1, 20.
[7.448] G.A. McGauley et al., *Horm. Res.* **33** (1990) Suppl. 4, 52.
[7.449] D. Rudman et al., *N. Engl. J. Med.* **323** (1990) 1.
[7.450] W. König in A. Kleemann et al. (eds.): *Arzneimittel, Fortschritte 1972 bis 1985*, VCH Verlagsgesellschaft Weinheim 1987, 716.
[7.451] W.S. Oetting et al., *J. Biol. Chem.* **261** (1986) 1649.
[7.452] Yu.A. Pankov, V.Yu, Butnev, *J. Pept. Prot. Res.* **28** (1986) 113.
[7.453] R. Einspanier et al., *FEBS Lett.* **204** (1986) 37.
[7.454] M. Yamakawa et al., *J. Biol. Chem.* **265** (1990) B915.
[7.455] M. Ono et al., *Proc. Natl. Acad. Sci. USA* **87** (1990) 4330.
[7.456] M. Watahiki et al., *J. Biol. Chem.* **264** (1989) 5535.
[7.457] S. Song et al., *Eur. J. Biochem.* **172** (1988) 279.
[7.458] S.H. Shin et al., *Can. J. Physiol. Pharmacol.* **65** (1987) 2036.
[7.459] G. Nagy et al., *Biochem. Biophys. Res. Commun.* 151 (1988) 524.
[7.460] J.E. Merritt, B.L. Brown, *Life Sci.* **35** (1984) 707.
[7.461] F. Bigi et al., *Exp. Clin. Endocrinol.* **95** (1990) 224.
[7.462] J.D. Wark, V. Gurtler, *Biochem. J.* **241** (1987) 397.
[7.463] K. Törnquist, *Acta Endocrinol. (Copenhagen)* **116** (1987) 459.
[7.464] H. Koshiyama et al., *Eur. J. Pharmacol.* **141** (1987) 319.
[7.465] I.S. Kampa et al., *Horm. Metab. Res.* **18** (1986) 419.
[7.466] J.S. Ramsdell, A.H. Tashjian, Jr., *Endocrinology (Baltimore)* **117** (1985) 2050.
[7.467] B.L. Spangelo et al., *Life Sci.* **40** (1987) 283.
[7.468] H.E. Carlson et al., *J. Clin. Endocrinol. Metab.* **60** (1985) 673.
[7.469] P.G. Knight et al., *Neuroendocrinology* **44** (1986) 29.
[7.470] C.A. Leadem, S.V. Yagenova, *Neuroendocrinology* **45** (1987) 109.
[7.471] F. Petraglia et al., *Neuroendocrinology* **45** (1987) 338.
[7.472] H. Abe et al., *Endocrinology (Baltimore)* **116** (1985) 1383.
[7.473] F.E. Chang et al., *J. Clin. Endocrinol. Metab.* **62** (1986) 551.
[7.474] E. Rolandi et al., *Horm. Res.* **21** (1985) 209.
[7.475] K.D. Meirleir et al., *Horm. Metab. Res.* **17** (1985) 380.
[7.476] C.Y. Cheung, *Neuroendocrinology* **39** (1984) 489.
[7.477] P. Buydens et al., *Horm. Metab. Res.* **18** (1986) 575.
[7.478] C. Denef, C. Schramme, *Horm. Res.* **22** (1985) 135.
[7.479] P.L. Canonico, R.M. MacLeod, *Endocrinology (Baltimore)* **118** (1986) 233.
[7.480] A.G. Dupont et al., *Horm. Metab. Res.* **19** (1987) 212.
[7.481] R. Drouhault et al., *Neuroendocrinology* **46** (1987) 360.
[7.482] H. Minakami et al., *Endocrinol. Jpn.* **33** (1986) 511.
[7.483] T.H. Jones et al., *J. Endocrinol.* **111** (1986) Suppl., 134.
[7.484] M. Memo et al., *J. Neurochem.* **47** (1986) 1689.
[7.485] E.R. Rey et al., *Life Sci.* **47** (1990) 1635.
[7.486] I.S. Login, *Life Sci.* **47** (1990) 2269.
[7.487] P. Tatar, M. Vigas, *Neuroendocrinology* **39** (1984) 275.
[7.488] U. Knigge et al., *J. Clin Endocrinol. Metab.* **62** (1986) 491.
[7.489] I. Murai, N. Ben-Jonathan, *Neuroendocrinology* **43** (1986) 453.
[7.490] S.H. Shin, R. Stirling, *J. Endocrinol.* **118** (1988) 287.
[7.491] N. Yonehara, D.H. Clouet, *J. Pharm. Exp. Ther.* **231** (1984) 38.
[7.492] S. Röjdmark et al., *J. Endocrinol. Invest.* 7 (1984) 635.
[7.493] L.S. Frawley, C.L. Clark, *Endocrinology (Baltimore)* **119** (1986) 1462.
[7.494] E.L. Lien et al., *Neuroendocrinology* **44** (1986) 184.
[7.495] J.-P. Loeffler et al., *Neuroendocrinology* **43** (1986) 504.
[7.496] K. Fuxe et al., *Acta Physiol. Scand.* **125** (1985) 437.

[7.497] T. Karashima, A.V. Schally, *Proc. Soc. Exp. Biol. Med.* **185** (1987) 69.
[7.498] K.K. Pun et al., *Acta Endocrinol. (Copenhagen)* **115** (1987) 243.
[7.499] I. Zofkova, J. Nedvidkova, *Exp. Clin. Endocrinol.* **92** (1988) 262.
[7.500] E.W. Bernton et al., *Science (Washington, D.C.)* **238** (1987) 519.
[7.501] D.F. Wood et al., *J. Endocrinol.* **115** (1987) 497.
[7.502] H.E. Carlson, *Life Sci.* **35** (1984) 1747.
[7.503] A.I. Esquifino et al., *J. Endocrinol. Invest.* **12** (1989) 171.
[7.504] U.A. Osunkwo et al., *IRCS Med. Sci.* **14** (1986) 998.
[7.505] P. Mormede et al., *Life Sci.* **39** (1986) 1737.
[7.506] L.V. dePaolo et al., *Peptides* **7** (1986) 541.
[7.507] S.L. Wardlaw, C.E. Markowitz, *Clin. Res.* **34** (1986) 435A.
[7.508] A. Gertler et al., *Endocrinology (Baltimore)* **118** (1986) 720.
[7.509] M.A. Sortino, P.M. Wise, *Endocrinology (Baltimore)* **124** (1989) 90.
[7.510] F.R. Boockfor et al., *Am. J. Physiol.* **250** (1986) E103.
[7.511] C.L. Bethea, E. Yuzuriha, *Endocrinology (Baltimore)* **119** (1986) 771.
[7.512] N. Kimura et al., *Endocrinology (Baltimore)* **119** (1986) 1028.
[7.513] L.J.G. Gooren et al., *J. Clin. Endocrinol. Metab.* **66** (1988) 444.
[7.514] S.W.J. Lamberts, *Horm. Res.* **22** (1985) 172.
[7.515] E. Christiansen et al., *Am. J. Obstet. Gynecol.* **157** (1987) 320.
[7.516] B. Couzinet et al., *Acta Endocrinol. (Copenhagen)* **121** (1989) 235.
[7.517] W.D. Hetzel et al., *Horm. Metab. Res.* **22** (1990) 648.
[7.518] V.C. Musey et al., *N. Engl. J. Med.* **316** (1987) 229.
[7.519] A. Kauppila et al., *J. Clin. Endocrinol. Metab.* **64** (1987) 309.
[7.520] B.B. Arnetz et al., *Life Sci.* **39** (1986) 135.
[7.521] L. Dogliotti et al., *Horm. Res.* **21** (1985) 137.
[7.522] G. Muccioli et al., *Neuroendocrinology* **53** (1991) 47.
[7.523] P.C. Hiestand et al., *Proc. Natl. Acad. Sci. USA* **83** (1986) 2599.
[7.524] J.-M. Boutin et al., *Cell* **53** (1988) 69.
[7.525] M. Edery et al., *Proc. Natl. Acad. Sci. USA* **86** (1989) 2112.
[7.526] P. Guillaumot et al., *Biochem. Biophys. Res. Commun.* **135** (186) 1076.
[7.527] H.G. Klemcke et al., *Endocrinology (Baltimore)* **118** (1986) 773.
[7.528] J.A. Rillema et al., *Horm. Metab. Res.* **18** (1986) 672.
[7.529] R.E. Hruska, *J Neurochem.* **47** (1986) 1908.
[7.530] E. Jungmann et al., *Horm. Metab. Res.* **18** (1986) 704.
[7.531] B.D. Murphy, K. Rajkumar, *Can. J. Physiol. Pharmacol.* **63** (1985) 257.
[7.532] J.E. Fortune, S.E. Vincent, *Biol. Reprod.* **35** (1986) 84.
[7.533] T.H. Welsh, Jr. et al., *Biol. Reprod.* **34** (1986) 796.
[7.534] A.K. Munabi et al., *Horm. Metab. Res.* **17** (1985) 47.
[7.535] A. Garcia et al., *Endocrinology (Baltimore)* **117** (1985) 954.
[7.536] I.R. Cohen-Becker et al., *Endocrinology (Baltimore)* **119** (1986) 1718.
[7.537] S.R. Fox et al., *Neuroendocrinology* **46** (1987) 350.
[7.538] F.T. Murray et al., *J. Clin. Endocrinol. Metab.* **59** (1984) 79.
[7.539] A. Canfriez, *Horm. Res.* **22** (1985) 209.
[7.540] D.E. Mills, R.B. Woods, *Am. J. Physiol.* **249** (1985) E614.
[7.541] W.G. North et al., *Horm. Res.* **15** (1981) 55.
[7.542] S. Kaufmann, *J. Physiol. (London)* **310** (1981) 435.
[7.543] C.T. Stier, Jr., et al., *Endocrinology (Baltimore)* **115** (1984) 362.
[7.544] J.R. Sowers et al., *Endocrinology (Baltimore)* **115** (1984) 2085.
[7.545] M.T. Robertson et al., *Endocrinology (Baltimore)* **119** (1986) 2439.
[7.546] R.L. Sorenson et al., *Endocrinology (Baltimore)* **121** (1987) 1447.
[7.547] T. Gerardo-Gettens et al., *Am. J. Physiol.* **256** (1989) R276.
[7.548] G.A. Overbeckl, *Acta Endocrinol (Copenhagen)* **109** (1985) 393.
[7.549] G.J. Pepe, E.D. Albrecht, *Endocrinology (Baltimore)* **117** (1985) 1968.
[7.550] A.R. Buckley et al., *Biochem. Biophys. Res. Commun.* **138** (1986) 1138.
[7.551] A.R. Buckley et al., *Life Sci.* **37** (1985) 2569.
[7.552] T.G. Muldoon, *Endocrinology (Baltimore)* **121** (1987) 141.
[7.553] F. Drago et al., *Life Sci.* **36** (1985) 191.

[7.554] J.W. Simpkins et al., *Neuroendocrinology* **39** (1984) 321.
[7.555] J.P. Barlet, *J. Endocrinol.* **107** (1985) 171.
[7.556] L. Magliola, L.R. Forte, *Am. J. Physiol.* **247** (1984) E675.
[7.557] C.E. Fiore et al., *Horm. Metab. Res.* **20** (1988) 40.
[7.558] M.P. Caraceni et al., *Calcif. Tissue Int.* **37** (1985) 687.
[7.559] A. Klibanski et al., *J. Clin. Endocrinol. Metab.* **67** (1988) 124.
[7.560] D.N. Luck et al., *Mol. Endocrinol.* **4** (1990) 1011.
[7.561] W. König in A. Kleemann et al. (eds.): *Arzneimittel, Fortschritte 1972 bis 1985,* VCH Verlagsgesellschaft, Weinheim, 1987, 726.
[7.562] E.-M. Rutanen, F. Pekonen, *Acta Endocrinol. (Copenhagen)* **123** (1990) 7.
[7.563] R.E. Humbel, *Eur. J. Biochem.* **190** (1990) 445.
[7.564] W.H. Daughaday, P. Rotwein, *Endocr. Rev.* **10** (1989) 68.
[7.565] W.H. Daughaday et al., *Endocrinology (Baltimore)* **121** (1987) 1911.
[7.566] L. Fryklund, H. Sievertsson, *FEBS Lett.* **87** (1978) 55.
[7.567] S. Suzuki et al., *EMBO J.* **4** (1985) 2519.
[7.568] K. Ramasharma et al., *Biochem. Biophys. Res. Commun.* **140** (1986) 536.
[7.569] L.K. Bachrach et al., *Biochem. Biophys. Res. Commun.* **154** (1988) 861.
[7.570] M.R. Hammerman, *Am. J. Physiol.* **257** (1989) F503.
[7.571] D.C. Costigan et al., *J. Clin. Endocrinol. Metab.* **66** (1988) 1014.
[7.572] A.N. Corps et al., *J. Clin. Endocrinol. Metab.* **67** (1988) 25.
[7.573] J. Wroblewski et al., *Acta Endocrinol. (Copenhagen)* **115** (1987) 37.
[7.574] A. Shimatsu, P. Rotwein, *J. Biol. Chem.* **262** (1987) 7894.
[7.575] A. Honegger, R.E. Humbel, *J. Biol. Chem.* **261** (1986) 569.
[7.577] P. Rotwein, *Proc. Natl. Acad. Sci. USA* **83** (1986) 77.
[7.578] V.R. Sara et al., *Proc. Natl. Acad. Sci. USA* **83** (1986) 4904.
[7.579] L.J. Murphy et al., *Endocrinology (Baltimore)* **122** (1988) 2027.
[7.580] E. Jennische et al., *Acta Physiol. Scand.* **129** (1987) 9.
[7.581] P. Bang et al. *Eur. J. Clin. Invest.* **20** (1990) 285.
[7.582] H.-A. Hansson et al., *Acta Physiol. Scand.* **129** (1987) 165.
[7.583] K. Ramasharma, C.H. Li, *Proc. Natl. Acad. Sci. USA* **84** (1987) 2643.
[7.584] L.J. Murphy, A. Ghahary, *Endocr. Rev.* **11** (1990) 443.
[7.585] T.K. Gray et al., *Biochem. Biophys. Res. Commun.* **158** (1989) 407.
[7.586] M.E. Wilson, *Endocrinology (Baltimore)* **119** (1986) 666.
[7.587] S. Guidoux et al., *Horm. Res.* **24** (1986) 256.
[7.588] T.L. McCarthy et al., *J. Biol. Chem.* **265** (1990) 15353.
[7.589] C.G. Scanes et al., *IRCS Med. Sci.* **14** (1986) 515.
[7.590] S. Goldstein, L.S. Phillips, *Metabolism* **38** (1989) 745.
[7.591] M. Maes et al., *Acta Endocrinol. (Copenhagen)* **117** (1988) 320.
[7.592] W.E. Sonntag, R.L. Boyd, *Life Sci.* **43** (1988) 1325.
[7.593] D.A. Harris et al., *J. Clin. Endocrinol. Metab.* **61** (1985) 152.
[7.594] M. Adamo et al., *Endocrinology (Baltimore)* **123** (1988) 2565.
[7.595] A. Tham et al., *Horm. Metab. Res.* **18** (1986) 706.
[7.596] R.G. Rosefeld et al., *Biochem. Biophys. Res. Commun.* **143** (1987) 199.
[7.597] T.A. Gustafson, W.J. Rutter, *J. Biol. Chem.* **265** (1990) 18663.
[7.598] P. Maly, C. Lüthi, *J. Biol. Chem.* **263** (1988) 7068.
[7.599] D.O. Morgan et al., *Nature (London)* **329** (1987) 301.
[7.600] R.G. MacDonald et al., *Science* **239** (1988) 1134.
[7.601] M.H. Dehoff et al., *Endocrinology (Baltimore)* **122** (1988) 2412.
[7.602] G.A. Werther et al., *Horm. Metab. Res.* **21** (1989) 109.
[7.603] M.-H. Perrard-Sapori et al., *Eur. J. Biochem.* **165** (1987) 209.
[7.604] E.S. Geary et al., *Diabetes* **37** (1988) Suppl. 1, 106A.
[7.605] N.J. Bohannon et al., *Endocrinology (Baltimore)* **122** (1988) 1940.
[7.606] P. Lönnroth et al., *Proc. Natl. Acad. Sci. USA* **84** (1987) 3619.
[7.607] R.L. Hintz, *Horm. Res.* **33** (1990) 105.
[7.608] A.L. Albiston, A.C. Herington, *Biochem. Biophys. Res. Commun.* **166** (1990) 892.
[7.609] S. Shimasaki et al., *Mol. Endocrinol.* **4** (1990) 1451.
[7.610] C. Seneviratne et al., *Mol. Endocrinol.* **4** (1990) 1199.

[7.611] K.M. Thrailkill et al., *J. Clin. Invest.* **86** (1990) 878.
[7.612] S.I. Yeoh, R.C. Baxter, *Acta Endocrinol. (Copenhagen)* **119** (1988) 465.
[7.613] A.M. Taylor et al., *Clin. Endocrinol.* **32** (1990) 229.
[7.614] M. Ross et al., *Biochem. J.* **258** (1989) 267.
[7.615] J.-F. Wang et al., *Biochem. Biophys. Res. Commun.* **157** (1988) 718.
[7.616] I. Kojima et al., *J. Biol. Chem.* **265** (1990) 16846.
[7.617] E.R. Froesch et al., *Klin. Wochenschr.* **65** (1987) Suppl. 9, 141.
[7.618] M. Ernst, E.R. Froesch, *Biochem. Biophys. Res. Commun.* **151** (1988) 142.
[7.619] B.P. Halloran, E.M. Spencer, *Endocrinology (Baltimore)* **123** (1988) 1225.
[7.620] H.-P. Guler et al., *N. Engl. J. Med.* **317** (1987) 137.
[7.621] H.-P. Guler et al., *Proc. Natl. Acad. Sci. USA* **86** (1989) 2868.
[7.622] S. Yamashita et al., *J Clin. Endocrinol. Metab.* **43** (1986) 730.
[7.623] C.G. Goodyer et al., *Endocrinology (Baltimore)* **115** (1984) 1568.
[7.624] K.M. Thrailkill et al., *Endocrinology (Baltimore)* **123** (1988) 2930.
[7.625] A.M. Morera et al., *Horm. Res.* **28** (1987) 50.
[7.626] T. Lin et al., *Biochem. Biophys. Res. Commun.* **137** (1986) 950.
[7.627] B.G. Kasson, A.J.W. Hsueh, *Mol. Cell. Endocrinol.* **52** (1987) 27.
[7.628] J.D. Veldhuis, R.W. Furlanetto, *Endocrinology (Baltimore)* **116** (1985) 1235.
[7.629] J.D. Veldhuis et al., *J. Biol. Chem.* **261** (1986) 2499.
[7.630] S.E. Tollefsen et al., *J. Biol. Chem.* **264** (1989) 13810.
[7.631] Y. Nakanishi et al., *J. Clin. Invest.* **82** (1988) 354.
[7.632] D. Yee et al., *J. Clin. Invest.* **86** (1990) 1806.
[7.633] A.J. Stewart et al., *J. Biol. Chem.* **265** (1990) 21172.
[7.634] M.L. Stracke et al., *J. Biol. Chem.* **264** (1989) 21544.
[7.635] K. Binz et al., *Proc. Natl. Acad. Sci. USA* **87** (1990) 3690.
[7.636] P. Hunt, D.D. Eardley, *J. Immunol.* **136** (1986) 3994.
[7.637] L.Y.-H. Tseng et al., *Biochem. Biophys. Res. Commun.* **149** (1987) 672.
[7.638] L.Y.H. Tseng et al., *Biochem. Biophys. Res. Commun.* **149** (1987) 672.
[7.639] M.L. Bayne et al., *J. Biol. Chem.* **263** (1988) 6233.
[7.640] M.L. Bayne et al., *J. Biol. Chem.* **265** (1990) 15648.
[7.641] C.J. Bagley et al., *Biochem. J.* **259** (1989) 665.
[7.642] J.D. Quin et al., *Diabetologia* **32** (1989) 531A.
[7.643] Z. Laron et al., *Lancet II* (1988) 1170.
[7.644] L. Fryklund, H. Sievertsson, *FEBS Lett.* **87** (1978) 55.
[7.645] S. Suzuki et al., *EMBO J.* **4** (1985) 2519.
[7.646] W. König in A. Kleemann et al., (eds.): *Arzneimittel, Fortschritte 1972 bis 1985*, VCH Verlagsgesellschaft, Weinheim 1987, 728.
[7.647] D.J. Gross et al., *J. Biol. Chem.* **264** (1989) 21486.
[7.648] T. Adachi et al., *J Biol. Chem.* **264** (1989) 7681.
[7.649] A.B. Smit et al., *Nature (London)* **331** (1988) 535.
[7.650] M. Lagueux et al., *Eur. J. Biochem.* **187** (1990) 249.
[7.651] E. Helmerhorst, G.B. Stokes, *Diabetes* **36** (1987) 261.
[7.652] V.K. Naithani et al., *Hoppe Seyler's Z. Physiol. Chem.* **365** (1984) 571.
[7.653] J.-H. Yu et al., *Proc. Natl. Acad. Sci. USA* **87** (1990) 9766.
[7.654] J.M. Conlon, J.W. Hicks, *Peptides* **11** (1990) 461.
[7.655] J.-H. Yu et al., *Peptides* **10** (1989) 1195.
[7.656] M. Bajaj et al., *Biochem. J.* **238** (1986) 345.
[7.657] M. Nishi, D.F. Steiner, *Mol. Endocrinol.* **4** (1990) 1192.
[7.658] E. Pilsetskaya et al., *Regul. Pept.* **11** (1985) 105.
[7.659] J.F. Cutfield et al., *Eur. J. Biochem.* **158** (1986) 117.
[7.660] S.J. Chan et al., *Proc. Natl. Acad. Sci. USA* **87** (1990) 9319.
[7.661] J.N. Fain, *Metabolism* **33** (1984) 672.
[7.662] B. Kreymann et al., *Lancet II* (1987) 1300.
[7.663] C. Jarrousse et al., *Endocrinology (Baltimore)* **115** (1984) 102.
[7.664] T. Matsuyama et al., *Diabetes* **36** (1987) Suppl. 1, 164A.
[7.665] W.H. Hsu et al., *Proc. Soc. Exp. Biol. Med.* **184** (1987) 345.
[7.666] Y. Totsuka et al., *Biochem. Biophys. Res. Commun.* **158** (1989) 1060.

[7.667] S.B. Richardson et al., *Diabetes* **37** (1988) 103A.
[7.668] H. Kofod et al., *Am. J. Physiol.* **154** (1988) E454.
[7.669] M.C. d'Emden et al., *Biochem. Biophys. Res. Commun.* **164** (1989) 413.
[7.670] G.H. Greeley, Jr., et al., *Regul. Pept.* **24** (1989) 259.
[7.671] A. Faure et al., *Horm. Metab. Res.* **17** (1985) 378.
[7.672] T. Ikeda et al., *Biochem. Pharmacol.* **40** (1990) 1769.
[7.673] J. Turk et al., *Biochim. Biophys. Acta L* **794** (1984) 125.
[7.674] S.B. Pek, M.F. Walsh, *Proc. Natl. Acad. Sci. USA* **81** (1984) 2199.
[7.675] S.A. Metz, *Life Sci.* **38** (1986) 2069.
[7.676] N.G. Morgan et al., *Biochem. Biophys. Acta C* **845** (1985) 526.
[7.677] G.A. Merrill, J.H. Anderson, Jr., *Diabetes* **36** (1987) 585.
[7.678] G. Holm, *Acta Med. Scand. Suppl.* **672** (1983) 21.
[7.679] D. Giugliano et al., *Diabetologia* **25** (1983) 157.
[7.680] I.C. Green et al., *Diabetologia* **27** (1984) 282A.
[7.681] B. Ahren et al., *Biochem. Biophys. Res. Commun.* **140** (1986) 1054.
[7.682] K. Hermansen, B. Ahren, *Regul. Pept.* **27** (1990) 149.
[7.683] A.F. Perna et al., *Am. J. Physiol.* **259** (1990) F210.
[7.684] S. Raptis et al., *Diabetes* **32** (1983) Suppl. 1, 94A.
[7.685] F. Shimizu et al., *Endocrinology (Baltimore)* **117** (1985) 2081.
[7.686] V.A. Diaz et al., *Diabetes* **35** (1986) Suppl. 1, 100A.
[7.687] C. Southern et al., *FEBS Lett.* **276** (1990) 42.
[7.688] H. Shimizu et al., *Biochem. Biophys. Res. Commun.* **173** (1990) 1280.
[7.689] C. Southern et al., *Biochem. J.* **272** (1990) 243.
[7.690] S. Baekkeskov et al., *Nature (London)* **347** (1990) 151.
[7.691] D.B. Jones et al., *Lancet* **336** (1990) 583.
[7.692] S. Zeuzem, *Aktuel. Endokrinol. Stoffwechsel* **8** (1987) 132.
[7.693] T.A. Gustafson, W.J. Rutter, *J. Biol. Chem.* **265** (1990) 18663.
[7.694] H.E. Tornqvist et al., *J. Biol. Chem.* **263** (1988) 350.
[7.695] C. Correze et al., *Biochem. Biophys. Res. Commun.* **126** (1985) 1061.
[7.696] A. Ullrich et al., *Nature (London)* **313** (1985) 756.
[7.697] M. Taira et al., *Science (Washington, D.C.)* **245** (1989) 63.
[7.698] M. Odawara et al., *Science (Washington, D.C.)* **245** (1989) 66.
[7.699] D.E. Moller et al., *Mol. Endocrinol.* **4** (1990) 1183.
[7.700] W.J. Braund et al., *Lancet I* (1987) 237.
[7.701] E.G. Walters et al., *Lancet I* (1987) 241.
[7.702] D.A. McClain, *J. Biol. Chem.* **265** (1990) 21363.
[7.703] P. Briata et al., *Biochem. Biophys. Res. Commun.* **169** (1990) 397.
[7.704] M.L. Standaert, R.J. Pollet, *J. Biol. Chem.* **259** (1984) 2346.
[7.705] S.A. Rosenzweig et al., *J. Biol. Chem.* **265** (1990) 18030.
[7.706] V. Papa et al., *J. Clin. Invest* **86** (1990) 1503.
[7.707] G.B. Willars, T.W. Atkins, *J. Endocrinol.* **104** (1985) Suppl., 100.
[7.708] J.D. McArmstrong et al., *Acta Endocrinol. (Copenhagen)* **102** (1983) 492.
[7.709] J.C. Jarrett II et al., *Am. J. Obstet. Gynecol.* **149** (1984) 250.
[7.710] K. Kriauciunas, C.R. Kahn, *Clin. Res.* **32** (1984) 401A.
[7.711] G. Kasdorf, R.K. Kalkhoff, *Clin. Res.* **34** (1986) 965A.
[7.712] P. Dent et al., *Nature (London)* **348** (1990) 302.
[7.713] J.S. Marks, L.H. Parker Botelho, *J. Biol. Chem.* **261** (1986) 2781.
[7.714] H. Fukumoto et al., *J. Biol. Chem.* **264** (1989) 7776.
[7.715] M.J. Birnbaum, *Cell* **57** (1989) 305.
[7.716] D.E. James et al., *Nature (London)* **338** (1989) 83.
[7.717] R.V. Farese et al., *Science* **236** (1987) 586.
[7.718] W.G. Lampson et al., *Can. J. Physiol. Pharmacol.* **61** (1983) 457.
[7.719] H.S. Glauber et al., *N. Engl. J. Med.* **316** (1987) 443.
[7.720] C.L. Oppenheimer et al., *J. Biol. Chem.* **258** (1983) 4824.
[7.721] M.P. Cohen et al., *Biochim. Biophys. Acta M* **856** (1986) 182.
[7.722] A.M. Kobbah et al., *Diabetologia* **32** (1989) 729.
[7.723] J. Ben Davoren, A.J.W. Hsueh, *Mol. Cell. Endocrinol.* **35** (1984) 97.

[7.724] T. Lin et al., *Endocrinol. (Baltimore)* **119** (1986) 1641.
[7.725] J. Rodin et al., *Metab. Clin. Exp.* **34** (1985) 826.
[7.726] C.J.H. Ingoldby et al., *Am. J. Gastroenterol.* **79** (1984) 16.
[7.727] K.Y. Lee et al., *Am. J. Physiol.* **258** (1990) G268.
[7.728] E. Mezey et al., *Science (Washington, D.C.)* **226** (1984) 1085.
[7.729] J.H. Pratt et al., *Clin. Res.* **31** (1983) 763A.
[7.730] M.L. Fidelman, C.O. Watlington, *Endocrinology* **115** (1984) 1171.
[7.731] A.J. Spijker et al., *Eur. J. Clin. Invest.* **15** (1985) Pt. 2, A16.
[7.732] V. Schusdziarra et al., *Neuropeptides* **7** (1986) 51.
[7.733] J. Guntupalli et al., *Am. J. Physiol.* **249** (1985) Pt. 2, F610.
[7.734] J.R. Levy et al., *Endocrinology* **119** (1986) 1786.
[7.735] B.E. Kream et al., *Endocrinology* **116** (1985) 296.
[7.736] N. Wongsurawat, H.J. Armbrecht, *Acta Endocrinol. (Copenhagen)* **109** (1985) 243.
[7.737] U.J. Eriksson et al., *Diabetes* **32** (1983) 1141.
[7.738] S. Yagi et al., *Diabetes* **37** (1988) 1064.
[7.739] A. Siani et al., *Eur. J. Clin. Pharmacol.* **38** (1990) 393.
[7.740] R. Sethi et al., *J. Mol. Cell. Cardiol.* **22** (1990) Suppl. 1, S33.
[7.741] T. Mitsuma et al., *Experientia* **40** (1984) 760.
[7.742] A. Dorn et al., *Pain* **29** (1987) Suppl. 4, S26.
[7.743] H.-S. Wang, M.-T. Lin, *Pharmacology* **30** (1985) 96.
[7.744] C. Grunfeld, D.S. Jones, *Diabetes* **32** (1983) 128A.
[7.745] H. Schatz, S. Ammermann, *Horm. Metab. Res.* **20** (1988) Suppl. **18** 1.
[7.746] C. Rosak et al., *Horm. Metab. Res.* **20** (1988) Suppl. **18** 16.
[7.747] G.D. Smith et al., *Proc. Natl. Acad. Sci. USA* **81** (1984) 7093.
[7.748] W.H. Fischer et al., *Biol. Chem. Hoppe-Seyler (Berlin)* **366** (1985) 521.
[7.749] G.P. Schwartz et al., *Proc. Natl. Acad. Sci. USA* **84** (1987) 6408.
[7.750] G.T. Burke et al., *Biochem. Biophys. Res. Commun.* **173** (1990) 982.
[7.751] S.-C. Chu et al., *Sci. Sin. (Engl. Ed.)* **16** (1973) 71.
[7.752] R.G. Mirmira, H.S. Tager, *J. Biol. Chem.* **264** (1989) 6349.
[7.753] J. Brange et al., *Nature (London)* **333** (1988) 679.
[7.754] H.P. Neubauer et al., *Diabetologia* **27** (1984) 129.
[7.755] K. Kitagawa et al., *Biochemistry* **23** (1984) 4444.
[7.756] P. Balschmidt et al., *Diabetes* **38** (1989) Suppl. 2, 157A.
[7.757] M. Hashimoto et al., *Pharm. Res.* **6** (1989) 171.
[7.758] S. Jorgensen et al., *Diabetologia* **32** (1989) 500A.
[7.759] P. Balschmidt et al., *Diabetologia* **33** (1990) Suppl., A117.
[7.760] L.D. Monti et al., *Diabetologia* **33** (1990) Suppl., A60.
[7.761] S. Kang et al., *Diabetologia* **32** (1989) 502A.
[7.762] L. Heinemann et al., *Diabetes* **39** (1990) Suppl. 1, 119A.
[7.763] J.N. MacPherson, J. Feely, *Br. Med. J.* **300** (1990) 731-736.
[7.764] M. Rodier, *Presse Med.* **19** (1990) 959-964.
[7.765] W. Foertsch, *Aktuel. Endokrinol. Stoffwechsel* **7** (1986) 33.
[7.766] G. Paolisso et al., *J. Clin. Endocrinol. Metab.* **66** (1988) 1220.
[7.767] A. Roza et al., *Clin. Res.* **31** (1983) 622A.
[7.768] W.T. Chance et al., *JNCI J. Natl. Cancer Inst.* **77** (1986) 497.
[7.769] S.J. Whang, F.L. Greenway, *Clin. Res.* **33** (1985) 67A.
[7.770] K. Tatemoto et al., *Nature (London)* **324** (1986) 476.
[7.771] A.L. Iacangelo et al., *Endocrinology (Baltimore)* **122** (1988) 2339.
[7.772] W.E. Schmidt et al., *Proc. Natl. Acad. Sci. USA* **85** (1988) 8231.
[7.773] H. Tamamura et al., *Eur. J. Biochem.* **191** (1990) 33.
[7.774] A. Funakoshi et al., *Regul. Pept.* **30** (1990) 159.
[7.775] S. Efendic et al., *Proc. Natl. Acad. Sci. USA* **84** (1987) 7257.
[7.776] J. Ishizuka et al., *Dig. Dis. Sci.* **32** (1987) 1170.
[7.777] A. Funakoshi et al., *Regul. Pept.* **24** (1989) 225.
[7.778] A. Funakoshi et al., *Regul. Pept.* **25** (1989) 157.
[7.779] V. Sanchez et al., *Biosci. Rep.* **10** (1990) 87.
[7.780] J.P. Smith, S.T. Kramer, *Gastroenterology* **96** (1989) Pt. 2, A479.

[7.781] J. DelValle et al., *Gastroenterology* **92** (1987) Pt. 2, 1368.
[7.782] T. Hashimoto et al., *Gastroenterology* **92** (1987) Pt. 2, 1428.
[7.783] T. Zhand et al., *Biochem. Biophys. Res. Commun.* **173** (1990) 1157.
[7.784] K. Miyasaka et al., *FEBS Lett.* **263** (1990) 279.
[7.785] A. Funakoshi et al., *Biochem. Biophys. Res. Commun.* **159** (1989) 913.
[7.786] A. Funakoshi et al., *Biochem. Biophys. Res. Commun.* **158** (1989) 844.
[7.787] W. König in A. Kleemann et al., (eds.): *Arzneimittel, Fortschritte 1972 bis 1985,* VCH Verlagsgesellschaft, Weinheim 1987, 723.
[7.788] B.E. Kemp, H.D. Niall, *Vitam. Horm. (N.Y.)* **41** (1984) 79.
[7.789] G. Mazoujian, G.D. Bryant-Greenwood, *Lancet* **335** (1990) 299.
[7.790] P. Hudson et al., *EMBO J.* **3** (1984) 2333.
[7.791] E.E. Büllesbach, C. Schwabe, *Biochem. Biophys. Res. Commun.* **126** (1985) 130.
[7.792] C. Schwabe et al., *J. Biol. Chem.* **264** (1989) 940.
[7.793] E.E. Büllesbach et al., *Eur. J. Biochem.* **161** (1986) 335.
[7.794] J.S. Ottobre et al., *Biol. Reprod.* **31** (1984) 1000.
[7.795] K. Seki et al., *Endocrinol. Jpn.* **33** (1986) 727.
[7.796] M. Ramachandra Rao, B.M. Sanborn, *Endocrinology* **119** (1986) 435.
[7.797] A.I. Musah et al., *Endocrinology (Baltimore)* **120** (1987) 317.
[7.798] A.L. Bernal et al., *Br. J. Obstet. Gynaecol.* **94** (1987) 1045.
[7.799] C.K.L. Too et al., *Endocrinology (Baltimore)* **115** (1984) 1043.
[7.800] B. Viell, H. Struck, *Horm. Metab. Res.* **19** (1987) 415.
[7.801] J. St-Louis, G. Massicotte, *Life Sci.* **37** (1985) 1351.
[7.802] G. Massicotte et al., *Clin. Exp. Hypertens. Part B* **B6** (1987) 255.
[7.803] E.E. Büllesbach, C. Schwabe, *Biochemistry* **25** (1986) 5998.
[7.804] W. König in A. Kleemann et al., (eds.): *Arzneimittel, Fortschritte 1972 bis 1985,* VCH Verlagsgesellschaft, Weinheim 1987, 607.
[7.805] C.H.S. McIntosh, *Life Sci.* **17** (1985) 2043.
[7.806] M. Bersani et al., *J. Biol. Chem.* **264** (1989) 10633.
[7.807] P.G. Burhol et al., *Acta Physiol. Scand.* **121** (1984) 223.
[7.808] H. Werner et al., *Peptides* **6** (1985) 797.
[7.809] D. LeRoith et al., *Endocrinology (Baltimore)* **117** (1985) 2093.
[7.810] S.M. Cutfield et al., *FEBS Lett.* **214** (1987) 57.
[7.811] J.M. Conlon et al., *Endocrinology (Baltimore)* **122** (1988) 1855.
[7.812] P.C. Andrews et al., *J. Biol. Chem.* **263** (1988) 15809.
[7.813] J.M. Conlon et al., *FEBS Lett.* **266** (1990) 37.
[7.814] S.V. Wu et al., *Metabolism* **39** (1990) 125.
[7.815] R.M. Post, D.R. Rubinow, *Lancet II* (1986) 810.
[7.816] B.M. Lewis et al., *Endocrinology (Baltimore)* **119** (1986) 1712.
[7.817] R. Zorrilla et al., *Neuroendocrinology* **52** (1990) 527.
[7.818] J. Argente et al., *Neuroendocrinology* **52** (1990) 342.
[7.819] T. Chiba et al., *Am. J. Physiol.* **253** (1987) G62.
[7.820] M.C. Aguila et al., *Neuroendocrinology* **52** (1990) 238.
[7.821] M. Sato et al., *Neuroendocrinology* **50** (1989) 139.
[7.822] D.E. Scarborough, *Metabolism* **39** (1990) 108.
[7.823] G.J. Cote et al., *J. Biol. Chem.* **261** (1986) 12930.
[7.824] H. Koop, R. Arnold, *Gastroenterology* **84** (1983) Pt. 2, 1214.
[7.825] B. Amiranoff et al., *Eur. J. Pharmacol.* **191** (1990) 401.
[7.826] K. Uvnäs-Moberg, L. Wetterberg, *Acta Physiol. Scand.* **120** (1984) 517.
[7.827] S.M. Webb et al., *Acta Endocrinol. (Copenhagen)* **110** (1985) 145.
[7.828] S.E. Shoelson et al., *Am. J. Physiol.* **250** (1986) E428.
[7.829] G. Strazzulla et al., *Horm. Metab. Res.* **20** (1988) 126.
[7.830] H. Stepien et al., *Horm. Metab. Res.* **18** (1986) 555.
[7.831] M.F. Beal, *Metabolism* **39** (1990) Suppl. 2, 116.
[7.832] F. Sicuteri et al., *IRCS Med. Sci.* **13** (1985) 308.
[7.833] D. Neugebauer, G. Weber, *Z. Hautkrankh.* **63** (1988) 585.
[7.834] V. Schusdziarra et al., *Diabetes* **34** (1985) 595.
[7.835] K. Uvnas Moberg et al., *Acta Obstet. Gynecol. Scand.* **68** (1989) 165.

[7.836] S. Gomez et al., *Life Sci.* **39** (1986) 623.
[7.837] J.M. Radke et al., *Eur. J. Pharmacol.* **134** (1987) 105.
[7.838] G.W. Roberts et al., *Nature (London)* **314** (1985) 92.
[7.839] J.C. Reubi, *Life Sci.* **36** (1985) 1829.
[7.840] J.C. Reubi et al., *Metabolism* **39** (1990) 78.
[7.841] M. Flint Beal et al., *Science (Washington, D.C.)* **229** (1985) 289.
[7.842] J.P. Moreau, F.V. DeFeudis, *Life Sci.* **40** (1987) 419.
[7.843] S. Reichlin, *N. Engl. J. Med.* **309** (1983) 1556.
[7.844] M.J. Toro et al., *Horm. Res.* **29** (1988) 59.
[7.845] S. Natori et al., *Clin. Endocrinol.* **32** (1990) 49.
[7.846] M.R. Brown et al., *Endocrinology (Baltimore)* **114** (1984) 1546.
[7.847] Y. Litvin et al., *Endocrinology (Baltimore)* **119** (1986) 737.
[7.848] P. Chiodera et al., *Metabolism* **38** (1989) 709.
[7.849] V. Schusdziarra et al., *Diabetes* **34** (1985) 595.
[7.850] A. Skamene, Y.C. Patel, *Clin. Endocrinol.* **20** (1984) 555.
[7.851] C. Villaume et al., *Biomed. Pharmacother.* **40** (1986) 61.
[7.852] R.N. Bergman et al., *Am. J. Physiol.* **247** (1984) E370.
[7.853] L. Pullen et al., *IRCS Med. Sci. Pharmacol.* **13** (1985) 1094.
[7.854] J.P.G. Williams, A. Sabokbar, *J. Endocrinol.* **108** (1986) 300.
[7.855] G.S.G. Spencer et al., *Acta Endocrinol. (Copenhagen)* **114** (1987) 470.
[7.856] G.S. Meneilly et al., *Clin. Res.* **34** (1986) 549A.
[7.857] J.N. Baxter et al., *Gut* **25** (1984) A1173.
[7.858] D.E. Schlarman, D.L. Kaminski, *Clin. Res.* **33** (1985) 935A.
[7.859] S. Mulvihill et al., *N. Engl. J. Med.* **310** (1984) 467.
[7.860] S. Bjorck, J. Svanvik, *Scand. J. Gastroenterol.* **19** (1984) 173.
[7.861] P.P. Gazzaniga et al., *Experientia* **44** (1988) 892.
[7.862] D.H. Teitelbaum et al., *Am. J. Physiol.* **246** (1984) G506.
[7.863] A.J. Olinde et al., *Gastroenterology* **84** (1983) Pt. 2, 1263.
[7.864] F. Kennedy et al., *Clin. Res.* **35** (1987) 152A.
[7.865] K. Bech, D. Andersen, *Scand. J. Gastroenterol.* **19** (1984) 740.
[7.866] E. Schippers et al., *Am. J. Physiol.* **250** (1986) G149.
[7.867] M. Mogard et al., *Gastroenterology* **86** (1984) Pt. 2, 1186.
[7.868] G.J. Krejs, *Diabetes* **33** (1984) 548.
[7.869] R.G. Long et al., *Br. Med. J.* **290** (1985) 886.
[7.870] J. Vora et al., *J. Clin. Endocrinol Metab.* **64** (1987) 975.
[7.871] Z. Tulassay et al., *Horm. Metab. Res.* **22** (1990) 555.
[7.872] C. Roy, *FEBS Lett.* **169** (1984) 133.
[7.873] I. Varga et al., *Acta Physiol. Hung.* **66** (1985) 303.
[7.874] R.D. Hoeldtke et al., *Ann. Intern. Med.* **103** (1985) Pt. 1, 889.
[7.875] L. Uccioli et al., *Clin. Res.* **34** (1986) 633A.
[7.876] J. Morisset, *Regul. Pept.* **10** (1984) 11.
[7.877] E.F. Stange et al., *Horm. Metab. Res.* **16** (1984) 74.
[7.878] N. Altorki et al., *Gastroenterology* **86** (1984) Pt. 2, 1014.
[7.879] C. Liebow et al., *Metabolism,* **39** (1990) 163.
[7.880] J. Myren et al., *Can. J. Physiol. Pharmacol.* (1986) *Suppl.,* 123.
[7.881] T.A. Kassessinoff, F.L. Pearce, *Agents Actions* **23** (1988) 211.
[7.882] M. Pawlikowski et al., *Peptides* **8** (1987) 951.
[7.883] M. Pawlikowski et al., *Neuropeptides* **23** (1989) 75.
[7.884] M. Pawlikowski et al., *Biochem. Biophys. Res. Commun.* **129** (1985) 52.
[7.885] P.J. Marie et al., *Metab. Clin. Exp.* **37** (1988) 429.
[7.886] E. Hanisch et al., *Acta Endocrinol. (Copenhagen)* **120** (1989) Suppl. 1, 246.
[7.887] L.Y. Koda et al., *Eur. J. Pharmacol.* **113** (1985) 425.
[7.888] L. Vecsei et al., *Neuropeptides* **4** (1984) 137.
[7.889] D.M. Gaumann, T.L. Yaksh, *Anesthesiology* **68** (1988) 733.
[7.890] J. Chrubasik, *MMW Muench. Med. Wochenschr.* **131** (1989) 119.
[7.891] J.R. Mancillas et al., *Proc. Natl. Acad. Sci. USA* **83** (1986) 7518.
[7.892] M.P. Primi, L. Bueno, *Peptides* **8** (1987) 619.

[7.893] T. Shibasaki et al., *Life Sci.* **42** (1988) 329.
[7.894] J. Danguir, *Peptides* **9** (1988) 211.
[7.895] M.T. Lin et al., *Eur. J. Physiol.* **413** (1989) 528.
[7.896] J.O. Willoughby, R. Kapoor, *Neuropeptides* **15** (1990) 153.
[7.897] A. Gibson, *Br. J. Pharmacol.* **91** (1987) 205.
[7.898] *Drugs of the Future* **9** (1984) 342; **10** (1985) 435; **11** (1986) 430; **12** (1987) 500; **13** (1988) 482; **14** (1989) 479.
[7.899] R.-Z. Cai et al., *Proc. Natl. Acad. Sci. USA* **83** (1986) 1896.
[7.900] *Drugs of the Future* **14** (1989) 1052; **15** (1990) 1144.
[7.901] R. Z. Cai et al., *Proc. Natl. Acad. Sci. USA* **84** (1987) 2502.
[7.902] W.A. Murphy et al., *Life Sci.* **40** (1987) 2515.
[7.903] G.S.G. Spencer, K.G. Hallett, *Life Sci.* **37** (1985) 27.
[7.904] K. Ziegler et al., *Biochem. Biophys. Acta C* **845** (1985) 86.
[7.905] G. Rohr et al., *Klin. Wochenschr.* **64** (1986) Suppl. 7, 90.
[7.906] S. Szabo et al., *Can. J. Physiol. Pharmacol.* **64** (1986) Suppl., 8.
[7.907] J.T. Pelton et al., *Proc. Natl. Acad. Sci. USA* **82** (1985) 236.
[7.908] J.E. Shook et al., *J. Pharmacol. Exp. Ther.* **240** (1987) 772.
[7.909] J.E. Shook et al., *J. Pharmacol. Exp. Ther.* **242** (1987) 1.
[7.910] J.T. Pelton et al., *J. Med. Chem.* **29** (1986) 2370.
[7.911] W. Kazmierski et al., *J. Med. Chem.* **31** (1988) 2170.
[7.912] J.M. Walker et al., *Peptides* **8** (1987) 869.
[7.913] A.G. Harris, *Metabolism* **39** (1990) Suppl. 2, 180.
[7.914] P.E. Battershill, S.P. Clissold, *Drugs* **38** (1989) 658.
[7.915] J.L. Abelson et al., *J. Clin. Psychopharmacol.* **10** (1990) 128.
[7.916] H.J. Balks et al., *Eur. J. Clin. Pharmacol.* **36** (1989) 133.
[7.917] R.D. Penn et al., *Lancet* **335** (1990) 738.
[7.918] J. Gastiaburu et al., *Bull. Cancer* **76** (1989) 503.
[7.919] D. Guarneri, R. Epis, *Invest. New Drugs* **7** (1989) 408.
[7.920] D. Ardid et al., *Therapie* **45** (1990) 69.

References for Chapter 8

[8.1] M. Mogard et al., *Gastroenterology* **86** (1984) Pt. 2, 1186.
[8.2] W. König in A. Kleemann et al., (eds.): *Arzneimittel, Fortschritte 1972 bis 1985*, VCH Verlagsgesellschaft, Weinheim 1987, 617.
[8.3] R. Markstein, P. Emson, *Eur. J. Pharmacol.* **152** (1988) 147.
[8.4] P. Kitabgi et al., *Eur. J. Pharmacol.* **99** (1984) 357.
[8.5] E. Kislauskis et al., *J. Biol. Chem.* **263** (1988) 4963.
[8.6] R.E. Carraway, S.P. Mitra, *J. Biol. Chem.* **265** (1990) 8627.
[8.7] J.-M. Studler et al., *Neuropeptides* **11** (1988) 95.
[8.8] A. Rökaeus et al., *Acta Physiol. Scand.* **120** (1984) 373.
[8.9] S. Komori et al., *Jpn. J. Pharmacol.* **40** (1986) Suppl., 118P.
[8.10] F. Kronenberg, R.E. Carraway, *J. Clin. Endocrinol. Metab.* **60** (1985) 1081.
[8.11] M.H. Mogard et al., *Biochem. Biophys. Res. Commun.* **136** (1986) 983.
[8.12] R.E. Carraway et al., *J. Biol. Chem.* **262** (1987) 5968.
[8.13] R.E. Carraway et al., *Regul. Pept.* **22** (1988) 303.
[8.14] R.E. Carraway et al., *J. Biol. Chem.* **262** (1987) 15886.
[8.15] I. Sures, M. Crippa, *Proc. Natl. Acad. Sci. USA* **81** (1984) 380.
[8.16] C. Shaw et al., *FEBS Lett.* **202** (1986) 187.
[8.17] L. Rusconi et al., *J. Chromatogr.* **349** (1985) 117.
[8.18] K.M. Merchant et al., *J. Pharmacol. Exp. Ther.* **255** (1990) 775.
[8.19] K.W. Eggerman, D.S. Zahm, *Neuropeptides (Edinburgh)* **11** (1988) 125.
[8.20] G.R. Uhl, M.J. Kuhar, *Nature (London)* **309** (1984) 350.
[8.21] T.F. Lee et al., *Am. J. Physiol.* **252** (1987) E102.
[8.22] D.L. Barber et al., *Gastroenterology* **92** (1987) Pt. 2, 1305.
[8.23] A. Rökaeus, *Acta Physiol. Scand.* **122** (1984) 261.
[8.24] A. Al Saffar et al., *Acta Physiol. Scand.* **122** (1984) 1.

[8.25] A. Rökaeus et al., *Acta Physiol. Scand.* **121** (1984) 181.
[8.26] E. Theodorsson-Norheim et al., *Acta Physiol. Scand.* **123** (1985) 331.
[8.27] F. Zeytin, P. Braceau, *Biochem. Biophys. Res. Commun.* **123** (1984) 497.
[8.28] N.A. Colliere et al., *Lancet I* (1984) 538.
[8.29] T.W. Moody et al., *Life Sci.* **36** (1985) 1727.
[8.30] V.E. Eysselein, *Klin. Wochenschr.* **62** (1984) 523.
[8.31] K. Muraki et al., *Life Sci.* **40** (1987) 1365.
[8.32] J. Staley et al., *Peptides* **10** (1989) 1217.
[8.33] S. Amar et al., *Biochem. Biophys. Res. Commun.* **129** (1985) 117.
[8.34] P.L. Canonico et al., *Mol. Cell. Endocrinol.* **42** (1985) 215.
[8.35] A.D. Drumheller et al., *Neuropeptides (Edinburgh)* **15** (1990) 169.
[8.36] F.B. Jolicoer et al., *Neuropeptides (Edinburgh)* **6** (1985) 143.
[8.37] P.W. Kalivas et al., *J. Pharmacol. Exp. Ther.* **238** (1986) 1126.
[8.38] G. von Euler et al., *Acta Physiol. Scand.* **137** (1989) 309.
[8.39] D. Jiang et al., *Life Sci.* **43** (1988) 27.
[8.40] F. Battaini et al., *Schiedeberg's Arch. Pharmacol.* **332** (1986) 267.
[8.41] B.G. Stanley et al., *Brain Res.* **343** (1985) 297.
[8.42] M.F. Hawkins, *Life Sci.* **38** (1986) 2383.
[8.43] C.F. Ferris et al., *Neuroendocrinology* **38** (1984) 145.
[8.44] T. Akema et al., *Neuroendocrinology* **46** (1987) 345.
[8.45] T. Akema, F. Kimura, *Neuroendocrinololgy* **50** (1989) 688.
[8.46] G.A. Gudelsky et al., *Neuroendocrinology* **49** (1989) 604.
[8.47] S. Furuta et al., *Br. J. Pharmacol.* **83** (1984) 43.
[8.48] L. Zhang et al., *Gastroenterology* **92** (1987) Pt. 2, 1709.
[8.49] D.E. Hermandez et al., *Regul. Pept.* **8** (1984) 41.
[8.50] S. Rosell, *Acta Endocrinol. (Copenhagen)* **103** (1983) Suppl. 256, 44.
[8.51] R.A. Hammer et al., *Life Sci.* **48** (1991) 333.
[8.52] F. Rioux et al., *Neuropeptides (Edinburgh)* **6** (1985) 1.
[8.53] L. Gozzini et al., *Int. J. Pept. Protein Res.* **25** (1985) 323.
[8.54] G. Perseo et al., *Int. J. Pept. Protein Res.* **25** (1985) 316.
[8.55] J.-L. Sadoul et al., *Biochem. Biophys. Res. Commun.* **120** (1984) 812.
[8.56] I.V. Grinshteine et al., *Bioorg. Khim.* **11** (1985) 1167.
[8.57] I.V. Grinshteine et al., *Bioorg. Khim.* **11** (1985) 1589.
[8.58] J.A. Gilbert et al., *Biochem. Pharmacol.* **35** (1986) 391.
[8.59] S. Amar et al., *Biochem. Biophys. Res. Commun.* **129** (1985) 117.
[8.60] S. Furuta et al., *Br. J. Pharmacol.* **83** (1984) 43.
[8.61] D. Hernandez et al., *Brain Res.* **301** (1984) 153.
[8.62] D.E. Hernandez et al., *Neuropeptides (Edinburgh)* **6** (1985) 561.
[8.63] F.B. Jolicoer et al., *Neuropeptides (Edinburgh)* **4** (1984) 467.
[8.64] F.B. Jolicoer et al., *Neuropeptides (Edinburgh)* **6** (1985) 143.
[8.65] F. Rioux et al., *Neuropeptides (Edinburgh)* **3** (1983) 345.
[8.66] P.L. Canonico et al., *Mol. Cell. Endocrinol.* **42** (1985) 215.
[8.67] Z. Bar-Shavit et al., *Neuropeptides (Edinburgh)* **2** (1982) 325.

References for Chapter 9

[9.1] W. König in A. Kleemann et al. (eds.): *Arzneimittel, Fortschritte 1972 bis 1985,* VCH Verlagsgesellschaft, Weinheim 1987, 616.
[9.2] J.E.T. Fox, *Life Sci.* **35** (1984) 695.
[9.3] Y. Seino et al., *FEBS Lett.* **223** (1987) 74.
[9.4] T. Peeters et al., *Am. J. Physiol.* **257** (1989) G470.
[9.5] N. Inatomi et al., *J. Pharmacol. Exp. Ther.* **251** (1989) 707.
[9.6] K.E. Hall et al., *Gastroenterology* **87** (1984) 76.
[9.7] M. Lemoyne et al., *Can. J. Physiol. Pharmacol.* **62** (1984) 1092.
[9.8] N.S. Track, P.L. Carlen, *Peptides* **7** (1986) 1193.
[9.9] P. Poitras et al., *Gastroenterology* **92** (1987) 658.

[9.10] J.E.T. Fox, E.E. Daniel, *Gastroenterology* **86** (1984) Pt. 2, 1080.
[9.11] M. Bickel, U. Belz, *Peptides* **9** (1988) 501.
[9.12] L. Bueno et al., *Dig. Dis. Sci.* **30** (1985) 760.
[9.13] E.L. Muller et al., *Clin. Res.* **32** (1984) 692A.
[9.14] D.F. Magee, S. Naruse, *J. Physiol. (London)* **335** (1984) 411.
[9.15] K.Y. Lee et al., *Dig. Dis. Sci.* **30** (1985) 980.
[9.16] T.L. Garthwaite, *Peptides* **6** (1985) 41.
[9.17] R. Schick, V. Schusdziarra, *Peptides* **6** (1985) 861.
[9.18] J. Eimerl et al., *Neuropeptides (Edinburgh)* **6** (1985) 157.
[9.19] S. Kuno et al., *Chem. Pharm. Bull.* **34** (1986) 4811.
[9.20] K. Kitagawa et al., *Chem. Pharm. Bull.* **33** (1985) 3307.

References for Chapter 10

[10.1] W. König in A. Kleemann et al. (eds.): *Arzneimittel, Fortschritte 1972 bis 1985*, VCH Verlagsgesellschaft, Weinheim 1987, 607.
[10.2] L. Thim et al., *Biochim. Biophys. Acta* **827** (1985) 410.
[10.3] L. Thim, *FEBS Lett.* **250** (1989) 85.
[10.4] S.B. Jakowlew et al., *Nucleic Acids* **12** (1984) 2861.
[10.5] K. Mori et al., *Biochem. Biophys. Res. Commun.* **155** (1988) 366.
[10.6] M. Gmachl et al., *FEBS Lett.* **260** (1990) 145.
[10.7] W. Hoffmann, *J. Biol. Chem.* **263** (1988) 7686.
[10.8] E.K. Frandsen et al., *Regul. Pept.* **16** (1986) 291.
[10.9] J.F. Rey et al., *Gut* **26** (1985) A581.

Epilog

Dear Reader, I am afraid, you are now more confused than you were before reading this book, but I hope, that this confusion is now on a higher level.

Subject Index